New Physics

with

Lorentz Violation
and
Deeper Structure

Third Edition

Frank Q Sun

ISBN: 978-1-4834-8212-5 (sc)
ISBN: 978-1-4834-8213-2 (e)

Lulu Publishing Services rev. date: 04/19/2018

Preface

3rd Edition

In this edition, I have changed the paper size from 6x9 inch to 6.625x10.25 inch. It has a new ISBN. An appendix is attached. It looks like a research paper, giving another systematic description of **DS** (deeper structure) model. Some or many parts in the appendix are similar or the same as what are written in the third part of the Highlights: (III) SA Physics with DS. SA means subatomic. When reading, one may simply skip or quickly go over all the things seen before.

This book is about the new physics with **LV** (Lorentz violation) and DS. It is also a ***historical record*** of how the author made some mistakes in the way to build his broad theory, found these mistakes later, and made corrections so that he eventually put the new developments into the Highlights, Chapter 8, and the Appendix. But the mistakes and the abandoned components and opinions may still be there unmoved. If the reader finds something in any other part of the book, chapter 1 through chapter 7, which is against the new developments seen in the Highlights, Chapter 8, and the Appendix, please follow the new developments, which contain renewed thoughts and formulations of the theory. Surely, major parts in chapter 1 through chapter 7 are not modified by the new developments, showing the right direction to be closer to the truth and to joyfully explain all the physical phenomena and empirical facts. The new developments were put in the 2nd edition 6 years ago. The 2nd edition was never published. Now the 3rd edition contains them. Please read the preface of the 2nd edition to see brief descriptions of the new developments after 2005.

Thought experiment can reveal the contradiction between the existence of graviton and Einstein's principle of equivalence. How about gravitational waves? Two years ago, I found a proof why gravitational waves may not exist.

A ***wave equation*** was deduced long time ago from the field equation under the condition of weak gravitational field (small deviation from the Minkowski metric). Such deduction is proved to be wrong. The 1st evidence is seen in the Schwarzschild metric for the gravitational field on earth as the solution of the field equation for the earth. The deviation from the Minkowski metric is less than 10^{-8}. The condition of weak gravitational field holds. But the Schwarzschild metric has nothing to do with any wave solution or wave-like weak vibration. If we check the details of the deduction from field equation to wave equation made long time ago, we can see the 2nd evidence of the wrong deduction, where all the quantities of the form $\partial_\tau h_{\mu\nu}\,\partial_\lambda h_{\rho\sigma}$ are abandoned in order to get a wave equation as an approximation. The reason is said to be "***neglecting nonlinear terms in the*** $\boldsymbol{h_{\mu\nu}}$". A simple example is found to disprove such neglect:

A 1-dim weak vibration $x=\varepsilon\ sin\omega t$ ($\varepsilon<<1$) is a special solution of the 1-dim wave eqn $d^2x/dt^2 + \omega^2 x = 0$. Here x and ε can be as small as one wants. Once a nonlinear term $(dx/dt)^2$ enters, we have $d^2x/dt^2 + (dx/dt)^2 + \omega^2 x = 0$. Its solution is $t= \pm \int e^x[\omega^2 e^{2x}(0.5-x) + c_1]^{-1/2}dx + c_2$. The weak vibration and wave solution disappear completely without trace. More details are on pages 161, 162.

After SR and GR became dominating theories in the world, Einstein lost his dashing spirit and sharp mind. He went to a wrong direction: To unite the theory of electromagnetic fields and the theory of gravitational fields in the hope that using such classical theory of unified field with probability interpretation undone one might be able to re-deduce all the quantum energy levels for electrons bound in atoms. He failed. All his great achievements were made in the 1st 30 years of his mature life while there was nearly nothing important in his last 30 years of life. He knew he didn't make discoveries in his last 30 years and he did not publish any new theory in his last 30 years. He made too many great discoveries so he disdained to publish any garbage theory and force his country and the world to believe. Einstein was not only a genius but also had good academic morality. Nevertheless, Einstein had gone along that direction for too long. The wrong direction started from his wrong hypothesis: Gravitational waves exist just like electromagnetic waves. This assumption made him feel good when he began his long journey to formulate a theory that can unite gravitational fields and electromagnetic fields. I still remember a story saying that his last words before he passed away are: All my efforts have failed.

A simple and natural reason of the existence of electromagnetic waves can be seen in the solution of Maxwell equations: Lienard-Wiechert potentials yield "acceleration fields" or "radiation fields" showing quantitatively & explicitly that electromagnetic waves exist if and only if the source charge is moving with acceleration or deceleration. But Einstein's 30 year hard work had told him: Field equations in GR cannot yield the existence of gravitational waves. Then why was the 2017 Nobel Prize in physics awarded to the physicists who claimed that they, using their experimental devices, had discovered gravitational waves? You know why. Many Nobel prizes in physics were awarded to wrong theories and wrong experimental findings: W-particle, Gauge theories, theory of asymptotic freedom with divergent poles, theory of universe expansion with acceleration and inflation interpretation (2011), Higgs particles (2013). In this book, the author carefully proves why they are wrong.

Frank Q Sun
Chicago, October 2017

Preface

2nd Edition

Since this book was published through lulu's POD program and appeared on lulu.com web site in early 2005, further development of the new physics with **LV** (Lorentz Violation) reveals major failures of **GR** (general relativity), seen in the cosmological consequences of the **EF** (Einstein-Friedman) universe that, at large scales, ignores the huge contributions of **CHSP**s (cosmic high-speed particles), such as photons and neutrinos, and extends all-time constant G from only few hundreds of years since the times of Newton, Kepler, Tycho to millions, billions, and even more than 10 billions of years without a running factor, a Mach-factor, or spontaneous breaking of G-symmetry. Among them, the age-failure made Einstein give us one of his lifetime conclusions: "*I see no reasonable solutions*." In chapter 8, a most general form of field equations is given. It looks exactly the same as seen in GR and EF cosmology but it allows different settings. The most general solution based on such most general form of field eqs (equations) has been rigorously obtained. The EF settings (0-LV, no CHSP, all time constant G since the beginning more than 10 billion years ago, and no inflation) yield GR and EF cosmology with zero **CT** (cosmological term). CHSPs do not co-move with receding galaxies. Once CHSP 4-momentum tensor enters as part of the 4-momentum tensor in GR, as proved in the CHSP-theorem, field eqs form an inconsistent system as long as the CT is pre-assumed to be a constant (Einstein's nonzero constant and Friedman's zero constant). A consistency condition eqn (equation) is deduced in the CT theorem and a formula of CT is obtained as part of the solution of the field eqs while the formula of the universe expansion rate *dR/dt* is another part of the solution. Dirac proposed a theory with non-constant G. He didn't complete it. A theory of **EFD**-type gravitation and cosmology with non-constant G is proposed. (**EFD** refers to Einstein-Friedman-Dirac). Details are seen in the Highlights and chapter 8. Dark matter and dark energy are fully explained.

Another advancement has been made in **SA** (sub-atomic) physics after 2007. **DS** (deeper structure) of unstable quarks and leptons is found and important roles of **MF**s (magneton forces) are revealed. A basic framework of SA physics is built to explain all facts, including all tough facts either incorrectly explained before or never explained before.

HQMT (heavy quark mass theorem) tells that the mass of u-quark is larger or even may be much larger than one third of proton's mass. The author found that D.H. Perkins ("Introduction to High Energy Physics", 4th edition, World Web, p. 130-132) has set $m_d \approx m_u = 336\ Mev\ / c^2$. He's explained anomalous **MM**s (magnetic moments) of nucleons and other baryons. DS is not considered in his work. As a possibility, net MMs of DS-electrons may be negligible. DS has shown that d-quark is formed by u-quark and d-leptons $e^- \bar{\nu}_e$. *Asymmetric MF field forbids S-state* for a constituent particle as MF is no longer negligible at SA short distances. MF-attractions require certain spin-correlations. In order for both spin and orbital motions of d-electron to cause attraction of MF field produced by u-quark or u-quark core, d-electron's spin and orbital angular momentum must be in opposite directions. Net MM of d-electron may be much smaller than its spin

MM. Thus, for MM issue, d-electron may contribute little to observed MM of a nucleon or any baryon. But it's also possible that mass of u-quark is heavier than what is assumed by Perkins and d-electron does make certain contribution to the MM of a nucleon or other baryon. I am sure that mass of u-quark cannot be determined by MM-facts alone and so far we still don't know for sure the value of u-mass.

Some details of DS and MFs include the following:

1. The list of all elementary particles $e^+e^-u\bar{u}\,\nu_e\bar{\nu}_e\,\gamma\, g\, Z$ and all the new composite particles previously thought to be elementary. Three non-identical u-quarks have 3 different colors. Their 3 antiquarks also have 3 different colors.
2. Extensions of electric charge to **Wk** (weak) and **St** (strong) charges, e.m. (electromagnetic) field to Wk and St magnetic-like fields, Coulomb force to Wk & St Coulomb-type forces, magnetic force to Wk & St magnetic-like forces, e.m. MM to Wk and St MMs, and e.m. MF to Wk and St MFs, with all extensions supported by experimental facts. SA dynamics with resultant forces of Coulombian or Coulomb-type forces plus MFs. There are qqq and $q\bar{q}$. No qq for opposite St currents repel to nearly cancel attractive St force in near-c speed cases, while in qqq 3-body system, there are no opposite motions. Weak charge explains all the phenomena where neutrinos are involved.
3. Formulas of MF showing: MF btw two MMs $\boldsymbol{\mu}_1$ and $\boldsymbol{\mu}_2$ of same type is non-spherosymmetric, θ-depending, and proportional to r^{-4} (at every fixed inclination angle θ) as well as to $\boldsymbol{\mu}_1.\boldsymbol{\mu}_2$ showing spin-correlation dependence. Two magnetons of any same type have a **MD** (magneton distance) to display MF's unusual r-dependence: MF btw them vanishes much faster than corresponding Coulomb or Coulomb-type force at distances longer than the MD and dominates corresponding Coulomb or Coulomb-type force btw them at distances shorter than the MD.
4. Non-existence of S-states for DS-electrons MF-bound in SA spaces, their **OSOD** (opposite spin-orbital directions), and small net MMs remove the oldest obstacle for DS-electron content to agree with nucleons' MMs that are much smaller than electron's spin MM.
5. Proof of **SM**'s (standard model's) spin misconception and **TAM** (total angular momentum) disability. 'Spin Crises' are trivially explained.
6. Insolvability in getting quantitative solutions to explain a series of basic empirical facts about SA <u>composite</u> particles, e.g., mass spectrum, MMs, lifetimes, and branch ratios. Even equal mass 2-body systems like **HMC** (hadronic meson core) $u\bar{u}$ and **LMC** (leptonic meson core) e^+e^-, much heavier than positronium e^+e^- due to positive **TME** (total mechanical energy) of MF-binding at short SA distance, may be unsolvable due to non-spherosymmetric, θ-depending, and spin-correlation dependent MF field.
7. Mass inequality for u-quark in HQMT and small net MMs of d-electrons explain nucleons' MMs qualitatively and half-quantitatively.
8. Huge number of examples showing unusual ability of DS with MFs and **4TDSP** (4 types of DS-processes) to qualitatively and half-quantitatively

explain all SA experimental facts, including mass spectrum, lifetimes, MMs, hundreds of types of **MSPs** (middle stage particles), thousands of decay modes, branch ratios, all kinds of SA reactions, and ironic **FQA** (free quark absence) fact: In all experiments, no single free (unbound) quark has ever been observed among **ISPs** (initial state particles), MSPs, and **FSPs** (final state particles).

9. **DSE** (DS-explanation) of short lifetimes of all mesons **without a single exception**.
10. DSE of short lifetimes of all non-nucleon baryons **without a single exception**.
11. **1BEx** (1 body extraction) proves **MBT** (many body theorem), showing time-varying non-discrete **ME** (mechanical energy) of any particle in time-varying resultant force field produced by others inside the same SA composite particle and in rapid motions. **MBP** (many body postulate) and experimental evidences of **MBT** and **MBP** are shown.
12. DSE of **5BFs** (5 basic facts) of nucleons: stable free proton, unstable & slowly decaying neutron, continuous energy spectrum of beta particles, MMs of nucleons, and preferred spin direction of beta particles bizarrely *parallel* ('*opposite*' was a mistake in writing) to **EMF** (external magnetic field). Note, if e^- & $\bar{\nu}_e$ in beta decay experiment reversed both moving and spin directions, parity violation and SM handedness would remain true. DS with MF explains why it didn't happen that way.
13. DSE of stability island, where many nuclei even contain more neutrons than protons, under the condition that neutrons are unstable due to inner **ENM** (eagle nest mechanism) no matter neutrons are unbound or bound. It is DS-ENM that helps to explain bizarre uniqueness of stable free protons (among all SA composite particles) and bizarre uniqueness of long life of neutron, about 10^8 times longer than slowest decaying SA unstable particles.
14. DSE of fast decaying DS-symmetric mesons with label J=0 **without a single exception**, slowly decaying DS-asymmetric mesons with label J=0 **without a single exception**, and fast decaying mesons with label J=1 **without a single exception**.
15. DSE of Bizarre different lifetimes of K_L^0 and K_S^0 that have the same DS content and the same quark content and may decay into the same set of FSPs.
16. Example of **PCQM** (perfect classical-quantum match) btw classical and quantum mechanics in atomic physics, achieved when **DBR** (Dirac-Bohr radius) is inputted (as initial condition) into circular motion solutions in classical relativistic mechanics, encouraging **CI** (classical imitation) that appears as a powerful approach to penetrate dense 'fog' covering secrets of SA world and to explain mass spectrum of SA composite particles qualitatively/half-quantitatively. A TME theorem is proved that reveals stunning effects of Coulomb-MF resultant force, such as **IES** (inward excited state) (smaller size-higher energy) including large ***positive*** values of ME and TME in many bound states that may trigger **QTC** (quantum tunnel channel) of decay, **NCB** (near-critical bound) states with small differences in energy, naturally low energy ground state, such as π^0, as **OSTP** (orbital-spin-tilting-

precision) suppresses MFs greatly to cause very negative TME.

17. Ground state masses of pions/nucleons/muons, tau's mass with its LMC e^+e^- in IES, naturally low masses of pions due to OSTP state of HMC $u\bar{u}$ in π^0 and $\pi^\pm$ that suppresses MFs btw u and $\bar{u}$ greatly leading to very negative TME of $u\bar{u}$ 2B-system with heavy u-mass. Heavy u-mass is required by HQMT and it simultaneously helps to explain masses and MMs of nucleons and stable free protons.
18. Ironic FQA fact is explained in two possible cases: 1) **QNC** (quark non-confinement) with **AVI** (absolute vacuum inelasticity): **PAPP** (particle-antiparticle pair production) has absolute priority to consume any impact energy larger than the rest energy of a **PAP** (particle-antiparticle pair) due to AVI and a PAP $u\bar{u}$ is produced right before each u-knockout event with $\bar{u}$ in $u\bar{u}$ relocating in and captured by the knocked out u-quark to form HMC $u\bar{u}$ and u in $u\bar{u}$ relocating and filling the hole left by the knocked out u-quark. 2) **QC** (quark confinement) with **NLVUC** (non-LV ultraviolet cutoff) lower than **QKoE** (quark knockout energy): NLVUC is the hypothetic non-LV upper bound of the 'elastic portions' of all collision energies absorbed by free particles to gain **KE**s (kinetic energies) or by bound particles to jump to higher levels including to be knocked out. Any most energetic particle in prime cosmic rays carries super-high **KkE** (knocking energy) making any bags or binding 'machines' alone unable to confine quarks. If QNC-AVI is untrue and quarks have never been knocked out, the only possibility is the existence of NLVUC that is lower than QKoEs. Evidences that prove or disprove NLVUC are right there among all films and photos that recorded collision events in cosmic showers and accelerators. We expect that experimenters check records for NLVUC issue as soon as possible.

The 1st law of physics is: Physicists make mistakes without exception. I have found **3 mistakes** in the 1st edition: 1) FQA fact was interpreted as QC with no other possible explanation. 2) STB (spacetime bag) was thought to be sufficient to explain QC. 3) Radially excited states were thought to be absent. They are corrected in the Highlights of the 2nd edition. But mistakes may still be seen in the text un-removed as historical records of mistakes and struggles in the way to be closer to the truth of the physical world. FQA fact now is explained mainly in the light of DS with 4TDSP and in two possible cases shown above: QNC with AVI or QC with NLVUC. The author is unable to determine which is true till experimenters finish checking all collision records and make NLVUC issue clear and future study reveals sizes of quark-cores and values of QKoEs. Note: elastic sub-events co-exist with PAPP in all DICs (deep inelastic collisions).

Bohr's great model of atomic structure was soon proved by Schrödinger and Dirac theoretically and quantitatively. Dirac equation in a **MT** (magneton) force field, i.e., MF field, may not be solvable. Dirac equation in any time-dependent non-central non-uniform local magnetic or magnetic-like field caused by fast motion of a charged source constituent particle is not solvable, and many-body problem is surely not solvable. DS model is therefore of no way to prove nor disprove by finding out theoretical solution quantitatively and accurately and comparing experimental data with theoretical outcomes drawn from the solution. But we immediately understand why quantitative explanations of atomic light spectrum were done soon after Bohr's model was proposed while mass spectrum and lifetimes of SA composite particles, after 50 year hard works and intensive researches, remain impossible to explain quantitatively. But it is amazing that DS has demonstrated its power to explain so many experimental facts in front of such insurmountable obstacle. Details of this new development are put in the Highlights and Appendix.

Frank Q Sun
Chicago, August 2011

Note: For some reason the 2nd edition has never been published. All the changes in that unpublished 2nd edition are included in the 3rd edition. The reader may treat the above preface of the 2nd edition as part of the preface of the 3rd edition. I apologize for the inconvenience it may cause.

Frank Q Sun
Chicago, October 2017

Preface

1st Edition

General linear transformations, leaving the speed c frame-invariant (c-**postulate**), obey $dS'^2 = f_S(V_{S'S})dS^2$ with $d\tau \equiv \sqrt{-dS^{*2}/c^2} = dt/\lambda\gamma$. Here $\lambda^{-2} \equiv f_S(\boldsymbol{u})$, $\boldsymbol{u} \equiv V_{S^*S}$. Then

$$\boldsymbol{p} = m\lambda\gamma\,\boldsymbol{u},\quad E = m\lambda\gamma c^2,\quad E^2 = c^2\boldsymbol{p}^2 + m^2\lambda^2c^4.$$

Equations of the laws of physics covariant under the new transformations are deduced. Each returns to its counterpart in current physics iff (if and only if)

$$f_S(V_{S'S}) \equiv 1 \text{ for all } S,\ S'.$$

Infinitely many forms of λ yield **finite** limits $N \equiv \lim_{u\to c} \lambda\gamma$. The setting $N^{-2} = -\varepsilon$ gives **Coleman-Glashow** [1] $E - \boldsymbol{p}$ relation. Since N is finite, all particles are **massive** and **timelike** while all **divergent** terms including counter terms become **finite**.

No particles with **different energies** can have the **same zero mass** and move at the **same speed** c. **DE fact** (each sort of particles has different energies) and **IR** (infrared) **fact** have disproved **massless** particles once for all.

All photons and neutrinos are massive and timelike. Their speeds are less than c, frame-variant, and frequency/energy dependent. Since photon mass is extremely small, the speeds of all **v-measured** photons (with the same frequencies as those that have been tested for speed) squeeze in a very narrow interval $(u_{\gamma LB}, c)$ containing empirical value of the speed of light. Here $c - u_{\gamma LB}$ is too small to measure. Ether-drag experiments can be explained without squeezing entire velocity spectrum of photons into a singleton $\{c\}$ and claiming the same zero mass and the same speed c for all photons. The unique frame-invariant speed c is just the limiting speed all particles can approach but cannot reach. Velocity addition law remains Einstein's for $f_S(V_{S'S})$ cancels in $d\boldsymbol{x}'/dt'$.

Consistency condition is obtained. The **isotropic** (**preferred**) reference space $\{K\}$ is unique. The anisotropy of $f_S(V)$ in terms of any earth system S and the 'absolute' velocity V_{SK} are hidden for $f_S(V)$ is too close to isotropic constant 1 for all $\gamma < 10^{1\sim2}$ phenomena in any earth system S.

The empirical and theoretical **isotropic** atomic spectrum of any source at rest and observed in any **anisotropic** earth-system is due to isotropic energy spectrum $E = m\lambda\gamma c^2$ associated with the differences between quantum levels independent of the moving directions of emitted photons. The anisotropic (direction-dependent) speeds and γ-values of the emitted photons compensate the anisotropy of λ and yield isotropic product $\gamma\!\!\!/ \equiv \lambda\gamma$. Their very large γ-values make the corresponding direction-dependence of their speeds practically unobservable. Even very large **LV** (**Lorentz violation**), say $\lambda - 1 = 99$ and the severe anisotropy of $\lambda(\boldsymbol{u})$ in any anisotropic earth-system at super-large $\gamma\!\!\!/$- values of the photons in visible light is in perfect agreement with the empirical isotropic atomic spectrum. In fact, $\lambda\gamma \sim 10^{7+2Z}$ for a 1 eV photon if

$$m_\gamma \sim 10^{-27-z} eV/c^2 = 1.78\times10^{-60-z} g.$$

Here $z = \pm\Delta m_{err}$ serves as the error parameter of the experiment. Consequently, for the 1 eV photon, a very large difference in γ-value such as $\Delta\gamma = 10^{27} - 10^{25}$ may be caused by the difference between $\lambda = 100$ at $\gamma = 10^{25}$ in new physics and $\lambda \equiv 1$ at any γ-value including $\gamma = 10^{27}$ in relativity, or by the difference between two λ-values in opposite moving directions of the 1 eV photon emitted from the rest source in any anisotropic earth-system. Unfortunately, such large departure of γ-value only causes $\Delta\beta \approx .5\cdot10^{-50}$, too tiny to test experimentally.

If K is the reference space where *cosmic background radiation* is exactly isotropic, **anisotropy** is expected due to Doppler shift of the CBR observed from any frame in absolute motion such as the earth.

For large-γ phenomena, very large deviation from $\gamma = E/mc^2$ and from λ=1 causes deviation from $\beta = \sqrt{1 - m^2c^4/E^2}$ that is too small for experiment to verify. Even for electrons, protons, and ions in large accelerators with $10 < \gamma < 10^4$, possible small LV might have been ignored because the measurement of their speeds is not accurate enough to check LV. Charged particles in a uniform magnetic field move along circles or spirals with $R = mc\lambda\gamma u / qB \sin\theta$, which cannot determine speed u accurately without knowing λ in the formula. The LV can be detected iff the high speeds can be detected directly with high precision. This is the very difficulty of modern experimental technique. We need a very smart innovation to carry a crucial test.

It is unknown how λ enters the formula of lifetime of decaying particles and whether it cancels the λ in the formula of time expansion to makes the flying distances insensitive to LV even at $\gamma = 100$. Even if the flying distances in the experiment have proved Einstein time expansion with zero LV for the fast decaying particles with $\gamma < 10^2$, it may not true for faster decaying particles with $\gamma > 10^{2\sim3}$.

Ultraviolet **cutoffs** caused by limiting LV as $\gamma \to \infty$ yield the **minimum spacetime sizes** of quantum particles via uncertainty relation. **IR** *catastrophe* counter terms fail to cancel is removed by genuine **IR cutoffs** and **convergent factors** associated with the removal of catastrophic IR departure of inverse potentials ($\propto 1/r$) from **square-integrable local** inverse potentials with **YFs** (Yukawa factors) as **common** solutions of both **classical** equations of *massive* vector fields and **quantum** equations for *massive* spin 1 particles. All gauge symmetries are undone by massive L_{LY} (**Lee-Yang**'s [2] original and alike), supported by DE fact, IR fact, and the nonexistence of gauge byproducts (Higgs particles and gluon balls).

Massive vector theory for all non-gravitational forces without gauge invariance and Higgs mechanism (HM) is **renormalizable**. The reasons are seen below.

Weinberg's [3] **guess** is right for the conservation of lepton-quark numbers requires **conserved currents**. The force carriers do not carry these numbers though W-particles do carry electric charges. The charge conservation is sufficient but not necessary for probability 4-currents to conserve.

In **Sterman**'s [4] **propagator**, $k^{\beta}k^{\gamma}/m_V^2$ term is nullified if the gauge fixing parameter in original L_{LY} is chosen to yield quantum equations also for free massive vector particles. It is also nullified in finite-N massive theory, where Sterman-type propagator (deduced from L_{LY} similar to original) automatically takes the form that contains no $k^{\beta}k^{\gamma}/m_V^2$ term.

Massive vector fields are automatically fixed: universal Lorentz condition $\partial^{\mu}V_{\mu}=0$ is a simple property of the solutions of classical new (non-Lorentz invariant) **Proca equations**, called **massive Maxwell equations** in honor of Maxwell, in a simple unified massive non-gauge vector theory without GS (gauge-self) and Higgs couplings. There is e.m. (electromagnetic) interaction between two electrically charged W-particles. But it is not an example of GS couplings. Vector forces differ by interaction charges (coupling constants) and coupling styles. The merit of the standard model can be recovered with SU(3)⊗SU(2)⊗U(1) **algebraic** coupling styles in non-gauge coupling terms, leaving masses of leptons and quarks to be explained via future composite models.

Gluons found in jet events are virtual and 'overdrawn' like W-particles in β-decays. Virtual pions between double virtual gluon lines (with opposite spins) and/or u/d quarks in loops extend strong force range from very short range of gluon force (exchanging single heavy gluons only without loop quarks and/or virtual pions in the middle) and the charge radii of small quark cores to hadrons' sizes with reduced coupling $\underline{g}_S^2=\eta e_S^2$. In **LH** (**large** strong coupling constant e_S^2 and heavy gluon mass M) model, the pair (e_S^2, M) and the functional form of λ are adjustable. They determine the shape of the **strong PE** (**potential energy**) and explain empirical smooth (no divergent pole) running of effective strong coupling at distances larger than small quark cores.

The equation of motion $F_{(\mu)}=dp^{\mu}/dt\equiv d(m\lambda\gamma u^{\mu})/dt$ and the force splitting $F_{(\mu)}=I_{(\mu)}+S_{(\mu)}$ with $S_{(\mu)}\equiv m(d\lambda/dt)\gamma u^{\mu}$ yield the **ET** (**energy theorem**)

$$dE/dt = \boldsymbol{I}\cdot\boldsymbol{u} + mc^2\gamma d\lambda/dt\,.$$

Many non-constant λ make $\int mc^2\gamma d\lambda/d\gamma d\gamma$ integrable and

$$E+k_{eff}q\phi-mc^2\int\gamma d\lambda/d\gamma d\gamma=c_0$$

is **solvable** to yield $E\equiv mc^2\lambda\gamma=E(k_{eff}q\varphi,c_0)$. It implies

$$\underline{U}(r)=-E(k_{eff}q\phi,c_0)+\underline{C}\,.$$

The strong PE that explains experimental facts now have a radical rather than phenomenological origin that does not affect atomic physics and QED numerical results.

Non-monotone model λ (20) yields **finite** limits $\lim_{u\to c}\lambda\gamma$ but features a **velocity shell space (VSS)** near the characteristic γ_0 ($10^5 \le \gamma_0 << N$) with very large ***positive*** values of $d\lambda/d\gamma$. This makes $mc^2\gamma d\lambda/dt$ term ***negative*** for any ***escaping*** quarks within small quark cores with large γ-values and being ***decelerated*** ($d\gamma/dt < 0$) in the escaping and joining forces with ***negative*** $\boldsymbol{I\cdot u}$ term to decelerate escaping quarks and confine them. It also makes a wide velocity space $\gamma_0 < \gamma < N/10$ with large values of λ for most v-measured photons and energetic neutrinos to provide extra λ^5-enhancement of their effective gravity masses. For neutrinos and photons, $T^{\mu\nu}$ without $0\cdot\infty$ make them sources of gravitational fields. In III, one can see that

$$T^{\mu\nu} = \rho^* U^\mu U^\nu = \lambda^5 \rho\gamma u^\mu u^\nu$$

features **γ-enhancement** of the effective gravity masses of non-soft photons and neutrinos in both general relativity ($\lambda \equiv 1$) and new physics with LV ($\lambda \neq 1$). In new physics with LV there is an extra-large **λ^5-enhancement** of the effective gravity masses of most non-soft photons and energetic neutrinos via model λ (20). Both general relativity ($\lambda \equiv 1$) and the new theory ($\lambda \neq 1$) tell that the **total energy (TE)** of **high-speed** particles ($\gamma >> 1$), not their total mass determines their contribution to gravitational field. **Dark matter** can be explained especially for model λ such as (20) with the said λ^5-enhancement in addition to the γ-enhancement. The quasi-vertical piece of $\lambda(\gamma)$ at and near γ_0 yields quasi-vertical pieces of PEs for all interaction forces at and near characteristic distances (CDs), proportional to effective couplings. They are the walls of deep PWs (potential wells). For $\gamma < \gamma_0$, $\lambda(\gamma)$ is almost or approximately constant 1 and all PEs between the CDs and the ranges of forces are almost or approximately inverse PEs ($\propto 1/r$). The large positive derivative $d\lambda/d\gamma > 0$ in VSS also produces powerful boosting for future space vehicles to reach upper layers of the VSS to get large extra λ-enhancement of time expansion and powerful extra braking for the vehicle to return to the lowest layer of the VSS. The VSS provides not only **spacetime bags** to confine quarks but also **spacetime tunnels** for future intergalaxy voyages.

Quark confinement, the absence of gluon spectrum, and the empirical smooth running of effective couplings without divergent pole can be explained for large variety of the values of the pair (M, e_s^2) and of the forms of λ. Accurate estimation of them is impossible due to the absence of gluon spectrum and radially excited states of hadrons. However, small quark-core model yields radially excited states beyond the accessed energy scale and the spin 3/2 baryons relate to the fact that $r^{-1}d\underline{U}^{st}/dr$ in spin-orbit coupling term is very small for a quasi-constant piece of the strong PE (covering the distances in the main probability regions for bound quarks). It corresponds to and is due to a quasi-inverse piece $\lambda(\gamma) = \lambda_0/\sqrt{a+\gamma^2}$ for the large γ-values ($\gamma_2 > \gamma \ge \gamma_1 \sim \gamma_0$) of the bound quarks in small cores. The mass splitting in baryon octet as ground states of composite baryons is still given by the **algebraic** SU(3)-symmetry breaking.

A general **CPT** theorem tells: CPT invariance is independent of the choice of f. If a Lorentz invariant quantum theory is CPT invariant then its counterpart with c-postulate and LV (Lorentz violation) is also CPT invariant. If c-postulate is exactly true, the rareness of antimatter is either 'local' (in a space equal to or larger than the observed part of universe) or owing to initial conditions at the beginning.

Frank Q Sun
Chicago, Illinois
January 2005

Contents

Note: In the text, I refers to Chapter 1, II refers to Chapter 2, and so on.

Highlights

(Many Articles and Copulas Omitted)

(I)

LV and **Massive** Carriers of **NG**-Forces

LV (**L**orentz **V**iolation) and $E-\boldsymbol{p}$ Relation

General Linear Transformations Leaving Finite Speed c Invariant Obey

(H1a) $$dS'^2 = f_S(\vec{V}_{S'S})dS^2, \ \forall \text{ IFs (Inertial Frames) } S, S'.$$

Let S^* Be Instantaneous Rest Frame of a Timelike Particle and $\vec{u} = \vec{V}_{S^*S}$ Be its Velocity Observed in IF S. Then

(H1b) $$d\tau = \sqrt{-dS^{*2}/c^2} = dt/\lambda\gamma, \quad \lambda \equiv f_S^{-1/2}(\vec{u}).$$

LV Factor Function λ Enters **Eqs** (**Eq**uations) of Laws of Physics, Covariant under New Transformations. In Particular,

(H1c) $$p^\mu = m\lambda\gamma u^\mu, \quad c^2\vec{p}^2 - E^2 = -m^2\lambda^2c^4.$$

$\lambda-1$ Measuring LV: Zero/Positive/Negative LV Means $\lambda-1=0/>0/<0$.
Zero/Small/Large LV Refers to Zero/Small/Large Value of $|\lambda-1|$.
Limiting LV Relates to Theoretical Results of Limit $N \equiv \lim \lambda\gamma$ (as $u \to c$ and $\gamma \to \infty$) and Relevant Empirical Evidences and Facts.
Non-Limiting LV Relates to Theoretical Results of Deviation $\lambda-1$ at All Energy Scales Accelerators Have Reached and will Reach in Future & Related Facts and Evidences Experimenters Have Obtained and will Obtain.

Limiting **LV** with **Finite** Limit $N \equiv \lim \lambda\gamma$ (as $u \to c$, $\gamma \to \infty$)

Coleman-**G**lashow [1] ***E-p*** Relation Gives Same Finite Limit with Setting $\varepsilon = -N^{-2}$.
Limiting LV Ultraviolet **C**utoff or Simply **UC** (**U**ltraviolet **C**utoff) and **MSS** (**M**inimum **S**pacetime **S**izes) in the Sense of **Uncertainty**.
Non-Existence of **Massless** Particles and **0-Mass By-Products**, such as Every **GI** (**G**auge **I**nvariance), Gauge-Self Couplings in Non-Abelian Cases, HPs (Higgs Particles), and Gluon Balls. (Details about HPs on Pages 156, 252, 253).

- **Massive Carriers** of All **NG** (**N**on-**G**ravitational) Forces and **Infrared Cutoffs**.
- **Square-Integrability** with **Yukawa Factors** and without Infrared Catastrophe.
- **IR** (**I**nfra**r**ed) **Convergent Factor** Manifests Itself in Massive Propagator as We Rewrite It in Terms of Massless Propagator:

$$\frac{g^2}{m^2+q^2}=\frac{(1+m^2/q^2)^{-1}g^2}{q^2}$$

Empirical Foundation for **Zero** LV ($|\lambda-1|\equiv 0, \lambda\equiv 1$) Collapses

- Statement "Zero LV Has Been Proved Experimentally with High Precision and It Has Empirical Foundation" Now is Proved Untrue in Large-γ Sectors with Simple Facts and Reasoning:

Example 1

Electrons with $\gamma=10^5$ and $\lambda=10^2$ OR $\gamma=10^7$ and $\lambda=1$ Have Same Energy but Differ by $\Delta\gamma=10^7-10^5=.99\times10^7$ that Only Causes $\Delta\beta< 5\cdot10^{-11}$, beyond all Experimenters' Ability to Measure and Judge.

Example 2

1 eV Photons ($m_\gamma\sim10^{-27}eV/c^2$) with $\gamma\sim10^{27}$, $\lambda=1$ OR $\gamma\sim10^{25}$, $\lambda\sim100$ Differ by Super-Large γ-Departure $\Delta\gamma=10^{27}-10^{25}=0.99\cdot10^{27}$ that Only Yields

$$\Delta\beta<0.5\times10^{-50},$$

Too Tiny To Test Experimentally.

- It Requires ***Accurate*** ***Empirical*** γ-Values to Test Every Theoretical ***Assertion*** about LV (Zero/Tiny/Small/Large) in All *Large-*γ Phenomena. It Needs ***Super-Accurate*** Speed Measurement, Far **Beyond** Experimenters' Ability. This is Called **STD** (**S**peed **T**est **D**isability). We Don't Know How Many Centuries Mankind Has to Wait to See STD Changing to **STA** (**S**peed **T**est **A**bility). But All Facts We See Every Day from Extremely Low to Extremely High Energy Regions Prove Limiting LV with Limit $N\equiv lim\,\lambda\gamma$ Finite as We will See Soon.
- QED Numerical Results Not Sensitive to Departure at $\gamma>10^{2\sim3}$: Bethe's [5] Low Ultraviolet Cutoff ($=m_ec^2$) Yielded Good Numerical Results in QED.
- **DE** (**D**ifferent **E**nergy) **Fact** tells that Each Sort of Lightlike Particles Have Different Energies. Here 'Lightlike' Particles Refer to Particles Moving with Same or Almost Same High Speeds as Photons.
- **IR Fact** is that All Physical Quantities Associated with Exchanging Soft Force-Carrier or Long Distance Behaviors Permanently Proved to Be Finite in All Performed Experiment without Single Exception.
- **DE Fact** & **IR Fact** Have Proved **Massive** Photons and Neutrinos Once and for All. No Particles with Same Zero Mass and Traveling at Same Speed c Could Have DEs. If Force Carriers were **Massless** and Infrared Cutoffs or Infrared

Regulators were **Fictitious**, No Process of ***Only*** Emitting (Exchanging) Soft Force Carriers with Energies ***Greater*** than a ***Positive*** Value, No Matter How Tiny, Could Ever Exist in Physical World.

- Photons are Rigorously Proved to Be Regular Members of Particle-Family Just Like Electrons: It Has Been Proved Rigorously Since 2005 that Same Formulas of Particle's Momentum-Energy, when Used for Photons, Lead to Exactly Same Formula of Doppler Effect Deduced from Transformation Rule of Wave Vectors. Such Perfect Match Not Only Further Proves Wave-Particle Duality, but Also Proves Once for All that Photons & Electrons Obey Exactly Same Formulas for Momentum-Energy, which are Also Obeyed by All Quantum Particles as well as by All Classical Particles and Heavy Bodies. See Details on Pages 117, 181, 182.
- DE with Same 0-Mass/Same Speed c Cannot Be Justified by Claiming Photons as Different Physical Objects that Do Not Obey Formulas Others Do.
- **Direct** Experimental **Evidence** of Limiting **LV** will **Never Exist** because any Experiment Proved LV (Zero/Small/Large) at Any Energy Scale, Accessed in Any Future, will Not Provide Scientific Guarantee for Such LV to Continue to Be Like That beyond Accessed Energy Scale.
- **Direct** Experiment Data about LV at Any Scale Accessed in Any Future Will Be Totally **Useless** and **Irrelevant** to the Issue of Limiting LV.
- Opposite to Limiting LV with Finite $N \equiv lim\,\lambda\gamma$, Infinite N as By-Product of 0-LV Has Produced Many By-Products Radically Opposite to By-Products of Finite N. All Directly Relate to Well-Known Empirical Facts about Things We Can See Every Day from Extremely Low to Extremely High Energy Phenomena.
- **Overwhelming Direct** Experimental **Evidences** that Disprove All By-Products of Infinite Limit and Prove By-Products of Finite N are Found and Pointed out in This Book. Every Such Evidence Serves as Indirect Experimental Evidence of Limiting LV with Finite Limit N.

Non-Limiting **LV** (Theoretical Results/Empirical **Evidences**)

- We Call **GS** (**G**amma-**S**ector or γ-Sector) $10^{n-1} \le \gamma < 10^{n}$ ($n = 1,2,3,...$) nth GS. We're Quite Sure: LV is Zero or Almost Zero in Lower Part of 1st GS. STD May Put Uncertainty on LV in Each Higher GS after 1st or 2nd GS.
- Once LV Factor Function λ Enters almost all Eqs of Laws of Physics, One Can Always Find Many <u>Non-Speed</u> λ <u>-Depending</u> Physical Quantities Not Only <u>Measurable</u> Everyday but also with Values <u>Sensitive</u> to <u>Large Departure</u> of λ from 1 at Large γ-Sectors. Quantities, whose Empirical Values Enable Us to Make Decisive Conclusion in Spite of STD, Can Be Called **LV-Quantities**.

- Although Non-Limiting LV, its Theoretical Outcomes and Empirical Evidences are Useless for Us to Figure Out Limiting-LV and its Theoretical Outcomes and Empirical Evidences, Limiting-LV Does Guarantee Non-Limiting LV that will Produce Corresponding Theoretical Results Measurable at some unknown Large γ-Sectors. But Limiting-LV with All its Theoretical Outcomes and All Known Empirical Evidences Have Absolutely No Way to Tell Us at Which GS Non-

Limiting LV will Be Practically Measurable. But Experiment May Have Shown LV's Effects. In Fact, (H1c) and **Eqn** (**Eq**uation) of Motion $F^{\mu} = dp^{\mu} / d\tau$ Yield

(H1d) $$F_{(\mu)} \equiv F^{\mu} / \lambda\gamma = dP^{\mu} / dt = m\lambda d(\gamma u^{\mu}) / dt + m\gamma u^{\mu} d\lambda / dt = I_{(\mu)} + S_{(\mu)}$$

- Interaction Force $I_{(\mu)}$ Determined by Force Formula. Universal Space Force $S_{(\mu)}$ Matters as $d\lambda/dt$ Not Negligible. It Changes Way How Interaction Force Causes Dynamic Effects. Solution for **2B** (**2 B**ody) System in Collision Proves Violation of Momentum Conservation as $d\lambda/dt$ Not Negligible. Evidences are Bizarre Jets with Nothing to Balance in Opposite Directions and Reaction $\gamma p \to n\pi^+\pi^0$ where π^+ Didn't Move on Reaction Plane where $\gamma n\pi^0$ Moved.
- Astronomically Large Number of Photons and Neutrinos in Universe with Large γ-Values far beyond what Quarks & HE-Particles in Accelerators Have Reached Produce Cosmological Phenomena and LV-Quantity, Such as Fast Orbiting Seen at Large Cosmological Scales.
- Once LV Factor Function λ Enters *Field Eqs*, λ^4-*<u>Enhancement</u>* of 4-Momenta of Photons and Neutrinos is Seen. **DM** (**D**ark **M**atter) is Immediately & Trivially Explained (see chapter 8 for details). It Also Explains why No Supersymmetry Particles with LV and Masses of Photons and Neutrinos Ignored Has Ever Been Found Since Beginning of Particle Physics.
- **CHSPs** (**C**osmic **H**igh-**S**peed **P**articles) Don't Co-Move with Receding Galaxies. A CHSP-Theorem Shows: One Can Easily Prove Field Eqs with 4-Momenta of CSHPs Forming ***<u>Inconsistent</u>*** System of Differential Eqs unless Coefficient of **CT** (Cosmological Term) Not Pre-Assumed to Be Constant (0 or Nonzero). Formula for CT is Part of Solution while Formula for $R(t)$ Showing Expansion is another Part of Solution. CT-Formula (72) Gives (87) in a Parabola Model with Extremely Small Values since 5 Billion yrs. ago. Schwarzschild Metric with 0-CT is Justified for Solar Problems. Thus, Gravitation and Cosmology with LV Trivially Recover All 3 Major Successes of Einstein's **GR** (General Relativity) with 0-LV. Three Tests of GR Unable to Tell which is Right, 0-LV or Nonzero-LV. For Instance, Condition $d\tau = 0$ is Required in Explanation of Bending of Light Rays Passing by Sun. Due to Photon's Extremely Small Mass, $d\tau = 0$ too Accurate even for Massive Photons in Visible Lights.
- Due to DM, Non-Limiting LV Becomes Certain in Large γ-Sectors beyond Man Made Accelerators in Coming 500 to 50,000 Years, such as $\gamma \sim 10^{10}$ to 10^{47}. SDL (See Details Later)

Massive Lagrangian of **Lee-Yang**'s [2] Type without **GI**

- **Massive** <u>Gluons</u> Seen in **Timelike** <u>Jets</u> Only with Total 4-Momentum of Final State Particles of Each Such Jet Being Timelike.
- Massless Gluons Cannot Explain Short Range of St Force.
- All Carriers of NG-Forces (Photons/Z-Bosons/Gluons) are Massive.
- DE-Fact and IR-Fact Have Proved Massive Photons/Neutrinos Once and for All.

- Massive Carriers of All **NG** (**N**on-**G**ravitational) Forces Now Become Reality. They Indicate: Idea in LY (Lee-Yang) Massive Lagrangian without GI May Easily Become Corner Stone to Build Unified Theory of All NG-Forces if We Make Extension to Include Massive Gluons and Massive Photons.
- Common Solutions of both C and Q Eqs for Massive Vector Fields
- Common Solutions of Massive Maxwell **Eqs** (Eqs) & Quantum Eqs for Massive Spin 1 Particles Possess Very Simple **Property** $\partial^{\mu}V_{\mu}=0$ that is neither a Gauge Condition Nor a Gauge Choice. From Now on We Call well-Known **LC** (Lorentz Condition) $\partial^{\mu}V_{\mu}=0$ **Lorentz Property**. Massive Vector Fields Now Quantizable at Very Beginning and in Any Inertial Frame because of such Property.
- Classical and Quantum ***Proca*** Eqs are <u>*Lorentz Invariant*</u> Eqs for Massive Vector Particles, Embedded in Such Basic Theory where Ultraviolet Divergence Can Never Become Finite: As Speed of ***Proca-Type*** Massive Photon or of any ***Proca-Type*** Massive Particle Approaches Frame-Invariant Speed *c* (as *Limiting Speed* of all Massive Particles Predicted by *both* Lorentz-Invariant Theory and New Physics with LV, **KE** (**K**inetic **E**nergy) Approaches Infinity.
- Eqs in New Physics with LV Covariant under **New Transformations** (More General than Lorentz Transformations) that Leave Same Speed *c* Frame-Invariant and Lead to Finite Limit $N \equiv lim\,\lambda\gamma$ as $u \to c$ and $\gamma \to \infty$.
- No Gauge Arbitrariness/Choice/Condition for Massive Vector Fields.
- Finite Range of a Force is a Physical Reality, Not Allowing Us to Change It in any Fixed Frame by Performing Gauge Transformation to Change Mass Term. Arbitrary Gauge Fields Not Physical Quantities.
- Masslessness of Photons and its By-Product U(1) Gauge Invariance Work in **CED** (Classical Electrodynamics) and Maxwell's Theory Not because Photon's Mass too Small to Consider Theoretically, Numerically, and Empirically but because Classical Theories Do Not Deal with **Q-Facts** (**Q**uantum **Facts**) so that Corresponding Quantum Wave Eqn for Massless Spin-1 Particle & Catastrophic Results Out of Sight.

  Note: Decisive Evidence of **Wk** (Weak) Charge & Wk **MM** (**M**agnetic **M**oment) is Observed Preferred Spin Direction of Anti-Neutrino Emitted in beta Decay in **EMF** (**E**xternal **M**agnetic **F**ield). Every Electrically Charged Elementary Particle Carries Wk Charge too. Every Magnetic Field Accompanied by Wk Magnetic-like Field. Fictitious Neutrino with 0-Mass Could Not Have Finite Wk MM to Respond to Wk Magnetic-like Field so Gently and Surely.

LH-Model (**L**arge Strong Coupling and **H**eavy Gluons)

- Every Jet where Gluon was Found was **Timelike**. Total 4-Momentum of All Final State Particles of Each Such Jet was Timelike. This Undeniable Evidence of Massive Gluons, if Found Ones were Real, Not Virtual. If They are Virtual, Why Real Ones Never found now that They are Massless? In Any Possible Case Gluon Cannot Be Massless.
- Short Range of Strong Force is Another Decisive Evidence of Massive Gluons.
- Smooth Running of Effective **St** (**St**rong) Coupling Caused by Heaviness of

Gluons Do Not Lead to Any Divergent Pole at Any Energy Scale.

- In LH-Model, Gluons Much Heavier than Pions. Virtual Pion in Middle of Virtual Gluon Line Extends Range of Gluon Force to Observed Range. As Such Higher Order Feynman Diagram's Treated as 2nd Order Feynman Diagram with Virtual Massless Gluon Only without Virtual Pion, Effective Strong Coupling Constant in Terms of Massless Propagator is Seen Running Smoothly without Divergent Pole.
- If We Focus on Such Short Distances where Exchanging Single Virtual Gluon without Virtual Pions in Middle Plays Important Role and Re-Write Propagator in Terms of Massless Propagator, One Can See a Terribly **Non-Perturbative** Feature of **ES** (Elastic Scattering) (see New Def of ES soon) 2nd Order Feynman Diagrams for t-Channels as long as Transferred 4-Momenta Become Very Large:

$$\frac{g_s^2}{M_{gluon}^2 + K^2} = \frac{(1 + M_{gluon}^2 / K^2)^{-1} g_s^2}{K^2}$$

As Energy Scale Becomes Large Enough and M_{gluon}^2 / K^2 No Longer Very Large, Very Large g_s^2 in LH-Model would Cause Terribly Non-Perturbative Feature. It Can Be Turned off if All Records of Collision Events in Cosmic Showers and Accelerators Prove Existence of **NLVUC** (**N**on-**LV** **U**ltraviolet **C**utoff) of 4-Momenta Transferred in ES. See < NLVUC > soon below.

- If 3-Jet and 4-Jet Events Really Mean that Gluons were Found, then They Must Be Virtual because There were No Tracks of Gluons. For **Fundamental** Force, its Real Carriers Living too Short to Leave Tracks would Be too Eerie to Believe. This Implies Real Gluons Heavier than What was Apparently Shown in Those Jet-Events. NLVUC, if Lower than Rest Energy of Real Gluon, Forbids Free Real Gluons to Appear in Any Event and in Any Future, No Matter Impact Energy in Collision is 9 TeV or 10^{12} GeV. Billions or Nearly a Billion of Photos that Recorded Events in Cosmic Showers and in Accelerators are Enough to Prove or to Disprove Appearance of any Free Real Gluon.

< NLVUC >

- Elementary Fermion (u-Quark, Electron, or any One of Their Antiparticles), Free or Bound, Gets Struck in Collision with 4-Momentum Changed but It Remains Same Sort of Particle before and after Collision. That's **ES** (**E**lastic **S**cattering) Event in Particle Physics. ES Can Only Be Sub-Event in **DI** (**D**eep **I**nelastic) Collision, Co-Existing with Particle-Production. Such Most Generally Defied ES Even Include Classical ES, that Produces Soft and Low-Frequency Photons Only, and Classical ES in Rigid Body Mechanics with Nothing Produced, as Special Cases. In Classical Scattering Seemly Continuous and Gradual 4-Momentum Transfer Formed by, in View of Quantization of NG Forces, Large Number of Quantum 4-Momentum Transfers with 4-Momentum Transferred in Each Single Quantum Transfer very or extremely Small. DI Collision Causes Large Transfer.
- Nonzero Masses of NG-Force Carriers Yield Nonzero Lower Limit of Energies Transferred. **VI** (**V**acuum **I**nelasticity) Leads to Finite Upper Limit, i.e., NLVUC, of Energies Transferred in All Quantized NG-Interactions. To Verify NLVUC

via **HE** (**H**igh **E**nergy) DI Collisions, We Must Focus on Each ES among Furious Particle Productions. Unbound u-Quarks Not Available. Focusing on Electrons with Visible Tracks after Collisions & Ignoring Collision Produced Particles that Did Not Exit before Collisions, One Should Be Able to Verify NLVUC.

- In about 80~90 Years, All Photos That Recorded Cosmic Showers and HE DI Collisions in Accelerators (Probably Billions of Photos) Should Have Shown Value of NLVUC: ***Largest Energy of Non-Incident Electrons Flying Away from Centers of Collisions***.
- Non-Existence of beyond Upper Bound Elastic Scatterings or NLVUC Not Only Can Make Terrible HE Non-Perturbative Feature of LH-Model Never a Reality but also Required if **QC** (**Q**uark **C**onfinement) is Correct Explanation of Ironic **FQA** (**F**ree **Q**uark **A**bsence) Fact.

  Note: DS is Able to Explain Ironic FQA Fact Even if Quarks are Not Confined, **QKoE** (**Q**uark **K**nock**o**ut **E**nergy) Lower or Much Lower than NLVUC and than Energy Scales of Today's Large Accelerators, and Quarks are Knocked Out Everyday Millions of Times. SDL
- **VI** (**V**acuum **I**nelasticity), as Reason of Existence of NLVUC, Seen Every Day in DI Scattering/Collision. Energy-Consuming in **PAPP** (**P**article-**A**ntiparticle **P**air **P**roduction) Breaks down Collision Process into Many Sub-Processes of Lower Energy Scales that Make Effective Couplings in LH-Model Perturbative, at Least Not Terribly Non-Perturbative. If Cosmic Shower Events and HE Collisions in Accelerators Have Proved NLVUC for 4-Momentum Transfers btw (between) 2 ***Real*** Particles Unchanged before and after 4-Momentum Transfer, That NLVUC Must Be Reality for ***Virtual*** Force Carriers Responsible for any t-Channel 2nd-Order Feynman Diagram, whether or Not It is Part of Higher-Order Feynman Diagram. If NLVUC Exists, then It Forbids Energy Transfer Larger than it in Each Single Quantum Energy Transfer. If it's 1 *GeV*, Repeated and Gradual Energy Transfers Taking Place in Accelerators Can Easily Produce HE Particles with KEs beyond or far beyond NLVUC. If Universe Does Not Have Super-Accelerators, Where Did Super-HE Protons in Prime Cosmic Rays Come from?

< Origin of Super-HE Protons in Prime Cosmic Rays >

NLVUC May Range from 0.5 to 50 *GeV*. But NLVUC Never Prohibits Particles from Gaining Cosmic KEs as Large as 10^{11} Times of NLVUC. Supper Energy Gains were Completed during Explosions right after Gravitational Collapses of Stars via Series of Quantum Energy Transfer Processes with Energy of Each Quantum Transfer Absolutely No More than NLVUC. Common Phenomenon of all Explosions is Out-Flying Pieces of Matter with Lower **TE** (total energy) of Left-Over System to Obey Energy Conservation Law. Supernova Means Super-Explosion with Super Gravitational Collapse and Birth of Core Much Denser than before. KEs of Out-Flying Particles Came from What were Released in Collapses of Dense Cores into Much Denser Cores. In Dense Condition/Curved Space, Quantum Level Differences Unknown. Gravitational Collapse of Dense Core, in View of Quantum Theory, is Collective and Quick Stepping down from High Levels to Ground Level. Since Original Dense Core Contained Too Many Nucleons, Such Supper **MB** (**M**any-**B**ody) System Expected to Have Extremely Large Number of Quantum Levels with Large Differences in Energy. On

Contrary, 2B Atomic Quantum Levels Only Contain 5 to 10 Quantum Levels (Most Close to Ground Level and Called Atomic Major Quantum Levels) that Have Atomically Large Differences btw Neighboring Levels, while All Rest Levels Squeezed btw 0-Energy Level and Near-0 Level. Suppose Average Difference btw Neighboring Major Quantum Levels of Dense Core is about 1 MeV. Quantum Transition btw Neighboring Levels is Done in about 10^{-18} Seconds (according to Uncertainty Relation). Suppose it Takes 10^{-6} seconds, for Example, for Dense Core to Finish Collapse. A Total of $10^{-6}/10^{-18}=10^{12}$ as Many Quantum Energy Transfers Done One after Another. Then 10^{9} *GeV* would be Average Energy of Cosmic Protons Produced by Supernova. Here We Assume: Energies Released in Collapse Take Form of Photons and Virtual Gluons which Then Hit Nucleons in Out-Layer of Dense Core. Such Nucleons Form Out-Flying Pieces of Matter in Explosion with Super-High KEs Coming from above Photoelectric Encounters. Dense Condition Necessary for One-after-Another Quantum Energy Transitions and Photoelectric Encounters to Occur in Very Short Time Period of Explosion. This Just Rough Description. No Real Gluons Produced for They are much Heavier than 1 *Mev*, 10 *Mev*, or 1 *Gev*.

Quantum Eqs in Curved Spacetime for MB-System Not Solvable. However, Observations Able to Tell whether or not Protons in Cosmic Rays and with Extremely High Energies Did indeed Come from Explosions of Stars.

Renormalizability of **Massive** Unified Theory of NG Forces

- **DS** (**D**eeper **S**tructure) of Quarks & Leptons Proves **CC** (**C**urrent **C**onservation) and Correctness of *Weinberg* [3] Guess. (see pages 112, 290, 291)
- Massive Vector Fields, as Common Solutions of Massive Maxwell Eqs & Quantum Wave Eqs for Massive Spin 1 Particles, are Fixed Automatically at $\xi=1$ Leading to **Nullified** $k^{\beta}k^{\gamma}/m_V{}^2$ Term in Sterman's [4] Propagator that is Based on Lee-Yang's Massive Lagrangian without Gauge Invariance.

(II)

Gravitation and Cosmology

EF(**E**instein-**F**riedman)-Setting and Failures of EF-Cosmology

EF-Setting (No CHSP, 0-LV, Constant G Extending G-Constancy from only Few Hundred Years since Kepler/Newton to Billions of Years and to the Beginning) Yields Iconic Formula $|dR/dt|=(const\cdot G\cdot R^{-1}-1)^{1/2}c$ with ∞/Near ∞ Expansion Speed at Singular/Nonsingular Beginning, Near c Receding Later, and Extremely Large Red Shifts Never Seen, Producing Age much Less than Observed Largest Red Shifts (see Age Theorem in VIII). Age Failure Made Einstein [7] Giving Us One of His Lifetime Conclusions: "I See No Reasonable Solution".

Dirac Suggested Nonconstant G but Never Completed a Theory to Fix EF-

Failures.

Gravitation and Cosmology in **Most General** Form

Gravitation and Cosmology in Most General Form Has Settings Undetermined, Allowing EF-Setting as well as Other Settings to Show Clearly Their Theoretical Results so We Can Compare and Eventually Prove that All Problems Can Be Solved including Dark Matter/Dark Energy if Undoing EF-Settings. Field Eqs Written in **Same Tensor Form**, as Seen in Einstein's GR, Actually Differ by Settings: **w/o** (**w**ith OR with**o**ut) LV, CHSPs, *G*-Constancy, and Inflation.

- Existence of CHSPs, such as Photons and Neutrinos, is Reality. GR and EF-Type Cosmology Ignore CHSPs. It was OK/Not OK in Solar System/at Large Scales. For any Spherical Space with Certain Density ρ of Photons and Neutrinos, the Larger the Space Size, the Larger the Magnitude of Gravity Force Exerted on a Unit Mass Outside by All Photons and Neutrinos inside the Sphere. ($f \sim k\rho r^3/r^2 = k\rho r$)
- Generally (See Details in VIII),

(H2) $$T^{\mu\nu} = \rho^* U^\mu U^\nu = \lambda^5 \rho\gamma u^\mu u^\nu = \lambda^4 \varepsilon u^\mu u^\nu / c^2, \; \varepsilon \equiv \rho\lambda\gamma c^2$$

- CHSPs Not Co-Move with Receding Galaxies and the CHSP Part of $T^{\mu\nu}$ Takes Very Special Form in Co-Moving Coordinate System (e.g., $T^{ii} \neq 0$, $i = 1, 2, 3$).
- **CHSP-Theorem** Proves: As 4-Momenta of CHSPs Enter $T^{\mu\nu}$, Field Eqs with ***Constant*** (Zero or Nonzero) Coefficient of **Cosmological Term (CT)** Form an ***Inconsistent*** System of Eqs.
- Formula of CT's Coefficient that Makes Field Eqs in Most General Form, w/o CHSP/LV/*G*-Constancy/Inflation, Form Consistent System of Differential Eqs is Deduced in **CT Theorem**. Only Under **EF-**Settings, It Gives Friedman's 0-CT. Since Einstein Accepted Friedman's Cosmology without CT by Openly Claiming Nonzero CT as His '***Biggest Blunder***', He Had Made Another Mistake: CT itself Not Mistake. Mistake was that Both He & Friedman Ignored CHSPs and Didn't Deduce Consistency Condition of Field Eqs with CHSPs. Friedman Considered Consistency but His 0-CT Yields Consistency Only when CHSPs Fully Ignored.
- Proof of CHSP Theorem/CT Theorem Given in VIII. Merits of Einstein's GR in 3 Major Tests in Solar System (Bending of Light Rays Passing by Sun/Motion of Perihelion of Mercury/Gravitational Red Shift) Trivially and Fully Recovered.

DM (**D**ark **M**atter)

- Source Tensor Contains Large Valued λ^4-Enhancement of Energy Density ε of Photons/Neutrinos, a Key to Explain DM Phenomena. Large Values of λ^4 for Massive Photons/Neutrinos with $10^{38} > \gamma > 10^{5\pm\sigma}$ Do Make Them Non-Baryonic DM in Spite of Their Extremely/Very Small Masses. *Newton*: Mass & Only Rest Mass Determines **GF** (Gravitational Field). *Einstein*: Field Eqs in GR Tell that For Moving Matter, 4-Momentum, Not Rest Mass, where Rest Mass Receives γ-

Enhancement, Determines GF. *New Physics with LV*: λ^4-Enhanced 4-Momentum Determines GF, where rest mass receives both γ- & λ^5-Enhancement, leading to λ^4-Enhancement of 4-Momentum as seen in eqn (H2) on page 9.

- In Universe, Photons and Neutrinos that Do Not Arrive at Earth are Transparent, Dark, and Resistance-less as Seen in *Bullet Cluster*. (SDL in VIII) They Dark Simply Because Never Arriving at Earth. Even for Closest Star Sun, Less than $2.46 \cdot 10^{-8}\%$ of its Radiations Could Ever Arrive at Earth.
- Large LV at Large γ-Sectors of Photons/Neutrinos (beyond γ-Values Electrons, Ions, Protons in Accelerators ever Reached/will Reach in 2500) Produces Large λ^4-Enhancement of their Energy Density $\varepsilon \equiv \rho\lambda\gamma c^2$ Making them Negligible, Significant, Large, and Dominating Sources of *G*-Fields (Gravitational Fields) at Different Scales Ranging from Luminous Galaxies, Vicinities of Spheres which Enclose Luminous Galaxies, to Clusters and Super-Clusters. The Larger the Scale, the Larger the Contribution of CHSPs as Non-Baryonic DM that Occupies Largest Space with Density More Uniform at Large Scales than all Baryonic Matter. Solar Constant & Average Radiation Rate of Stars Shows Negligible Gravity Force by CHSP Non-Baryonic DM within Luminous Galaxy and Observable Fast Motion of Lonely Stars outside Galactic Sphere that encloses Luminous Galactic Disc with their Distances from Center of Galaxy about Two to Five Times Longer than Radius of Luminous Disc.
- Empirical Velocity Curve of Stars on Luminous Disc of Milky Way Has Also Been Explained as We Explain DM. All Explanations are in Good Agreement with Solar Constant and Its Clue to Average Rate of Radiation of Stars in a Galaxy or in Universe.

Dark Energy

***G*-Theorem**'s Proved to Yield *G*-Eqn:

(H3)
$$G = \frac{\pi c^2}{2M_{total}B} R\left(1 - \frac{R^2}{c^2}\frac{dh}{dt}\right), \quad B \equiv 1 + \frac{k+3}{3}\overline{\lambda_d^4}\frac{\varepsilon_d}{\varepsilon}.$$

- Here ε_d is Energy Density of CHSPs, Not Co-Moving with Receding Galaxies, $\overline{\lambda_d^4}$ is Average λ^4-Enhancement of CHSPs, ε is Energy Density of Matter Co-Moving with Galaxies including All Non-CHSP Dark Matter, and M_{total} Total Mass of All Non-CHSP Particles at Rest in Well-Known Chosen Cosmological Coordinate System That's Co-Moving with Receding Galaxies.
- EF Settings $G = Const$, $\lambda \equiv 1$ (Zero LV), $\varepsilon_d \equiv 0$ (No CHSP), c_1=0 (No Inflation) Lead to EF Universe and Some Severe Problems. SDL in VIII.
- Formula $|dR/dt| = (const \cdot G \cdot R^{-1} - 1)^{1/2} c$ in EF-Cosmology and Modified Formula with LV/CHSPs Causing Similar Failures (if *G*=*const*) Shown Below.
- Constant G with **Any Amounts** of LV ($|\overline{\lambda_d^4} - 1| = 0, <<, \leq, >, >> 1$) and of High-Speed Non-Baryonic DM ($\varepsilon_d / \varepsilon = 0, <<, \leq, >, >> 1$) Makes Speed of Recession Ranging from Infinity/Near-Infinity at Singular/Nonsingular Beginning to 0 at

Turning Point, Near-c Speed of Recession Sometimes btw Causing Extremely Large Red Shifts Never Seen, Inducing Age Less than Observed Large Red Shifts (see **Age-Theorem** in VIII) So that Einstein [7] Made One of His Lifetime Conclusions: "*I See No Reasonable Solution*."

- **Non-Constant** *G* (Attributed by Weinberg [6] to Dirac), Spontaneous Breaking of *G*-Symmetry, Running *G*, and Mach's Principle with Running Mach Factor All Cosmologically Equivalent. They Automatically Explain Genuine Behavior of Many Fundamental Cosmological Functions such as $R(t)$, $h(t)$, dh/dt and all Their Future Empirical Curves as *G*-Eqn in *G*-Theorem is Satisfied. In Absence of Such Empirical Curves, Sample Models're Proposed with Less than *c* Speed of Recession at All Times, Normal Supernovas within 5 Billion Light Years, Dimmer Supernovas beyond, with Relative Rate of Change $\Delta G/G \sim 10^{-8}$ in 400 Years, Less than 16% in Past Three Billion Years.
- Hawking's Space Trip Has Showed No Observable Impact of Change in Gravity on Human Life. A Very Simple/Clear/Well-Known Fact in Physics/Chemistry/Biology is the Domination of **e.m.** (**E**lectro**m**agnetic) Force Exerted on Atoms in Molecules. No Part of Any DNA Molecule Could Be Sensitive to Negligible Gravity Force on Earth. Extremely Small Ratio ($<10^{-16\sim-20}$) of Earth Gravity against Valence Force Exerted on Same Atoms Well-Known Fact that Can Be Easily Understood by any One with College Physics Training. Such Small Ratio Remains Small No Matter *G* Decreases or Increases Ten or More Times. To DNA Structures, Earth and Place Nearby Black Hole with Gravity 10^8 Times Stronger would be Same Kind of Environment. Bacteria are Much Larger, More Complicated than DNA. They Easily Survive in Centrifuges Spinning at High Speed. The Centrifugal Force there is Equivalent to Very Heavy Gravity. Highly Developed Life Forms such as Fish, Birds, and Men, with Their Brains, Body-Organs, and Blood Circulations Unable to Survive in Harsh Environment of Hypothetic Very Heavy Gravity on Earth nor in Centrifuges Spinning at High Speed, Could Have Easily Survived since They Appeared on Earth Because the Past Changes in *G*-Value were That Small.

(III)

SA Physics with **DS**

Brief Introduction to **DS**

- Among All Quarks and Antiquarks, Only $u\bar{u}$ are Elementary.
- Every Other Quark than **u-Q** (u-Quark) Composite Containing One u-Q & Some Stable **DSL**s (**DS-L**eptons), for Example, $e^{-}\bar{\nu}_e$. Every Antiquark Other than $\bar{u}$ Composite Containing One $\bar{u}$ and Some Stable **DSL**s, for Example, $e^{+}\nu_e$. By DSLs We Mean Leptons Bound or Quasi-Bound within **SA** (Sub-Atomic) Small Spaces.

- Thus, Every Meson Contains **HMC** (**H**adronic **M**eson **C**ore) $u\bar{u}$, Every Baryon (including Nucleon) Contains **BC** (**B**aryon **C**ore) uuu, and Every Antibaryon Contains **AntiBC** $\bar{u}\bar{u}\bar{u}$. Each One of HMC, BC, and AntiBC is Called **Q-Core** (Quark-Core).
- Among All Composite Quarks/Antiquarks, **d-Q** (d-Quark) and $\bar{d}$ Only Ones in **Ground** States.
- Each One of DS-Electrons/DS-Positrons in Hadrons is Bound by Resultant Force of **MF** (**M**agneton **F**orce) and Coulombian Force with its **OAM** (**O**rbital **A**ngular **M**omentum) Opposite to its Spin to Have Both Spin and Orbital Electric Currents Attracted by Net MM of Same Q-Core that Binds it. **OSOD** (**O**pposite **S**pin-**O**rbital **D**irections) Reconcile Small Empirical Net MMs of Composite Nucleons or Baryons with Large Spin MM of Elementary Electrons under the Condition that Baryons Do Contain DS-Electrons. Here Baryons Include Nucleons. For any DS-Electron in Baryon, its Quantum Orbital Number is Not Zero but **Positive** in **Ground** State because MF is *Non-Spherosymmetric*. It Forbids S-States. S-States of Atomic e^- Exist for MF Almost Vanishes at Atomic Distances. Formulas of MFs, (H11) through (H19), Given on Pages 47-54.
- Every Meson Contains HMC $u\bar{u}$ that Must Annihilate Quickly, before or after Some Other Fast DS-Process. It Immediately Explains Why All Mesons Live Short **without Single Exception**. It Also Explains Why in All Events of Decaying Mesons Only Leptons and/or Photons Appear as Final State Particles **without Single Exception**.
- Composite Quarks are d, s, c, b, t Quarks. Only One in ***Ground State*** is d-Quark ($d = ue^-\bar{\nu}_e$). This Immediately Explains why Non-Nucleon Baryons All Have Very Short Lifetimes **without Single Exception**: SA De-Excitations Always Happen very Quickly **without Single Exception**.
- Solutions of Dirac Eqn in **IPE** (**I**nverse **P**otential **E**nergy) Fields (e.m. or Wk Filed) Prove Large Sizes of **MPR** (**M**ost **P**robable **R**egion) for Bound Electrons & **NFMs** (**N**eutrino-**F**amily **M**embers) due to their Light Masses. The Legendary Solutions Prove Permanently that Other Types of Forces, such as MFs Must Exist in SA World so that DSLs Can Be Bound within SA Small Spaces.
- Quantum Eqs with MF-Binding Not Solved and May Be Unsolvable. DS Unable to Explain Mass Spectrum, Orbital Quantum Number, Net MM, and Lifetimes Quantitatively. But DS Has Shown its Power to Explain Huge Number of Facts Qualitatively and Half-Quantitatively. Many of Them Never Explained before.
- Example of **PCQM** (**P**erfect **C**lassical-**Q**uantum **M**atch) Shows Importance of **CI** (**C**lassical **I**mitation): Solution of Dirac Eqn Gives Ground State Energy & Wave Function for Atomic Electron in Hydrogen. Average Values of $1/r$ and **PE** (**P**otential **E**nergy) $U = -e^2/r$ as well as **DBR** (**D**irac **B**ohr **R**adius) Can All Be Deduced from Ground State Wave Function. Ground State Energy and Ground State Average PE $\bar{U}$ Determine Average KE and $\bar{\gamma}$. But We May also Use Circular Motion CI and See Ground State Electron Bound in Hydrogen as Classical Particle Moving along a Circle in Classical Coulomb Field. By Solving Classical Eqn of Motion in Relativistic Mechanics, We Obtain Infinitely Many Classical Circular Motion Solutions Associated with Corresponding **IC**s (**I**nitial **C**onditions). Classical Theory itself Unable to Get Quantized Orbital Sizes and

Identify which One among Infinitely Many Circular Motions Best Represents Quantum Motion of Atomic Electron in Hydrogen and in Ground State. But if We Choose Classical Circular Motion with Same Size as DBR, Corresponding γ Value is Exactly Same as $\bar{\gamma}$ We Got from Relativistic QM. This is Example of PCQM. It Encourage Us to Use CI Approach to Penetrate Dense Fog Covering SA World's Secrets. Detailed Calculations in PCQM will Be Seen Later.

- Milestone-Type Solutions in QM Fully Based on Central Potential Functions and Curves in Classical Theory of Forces that Obey Inverse Square Law. MF's Awful θ-Depending Non-Spherosymmetric Feature May Make Even Classical Systems Unsolvable if ICs Not Special. In CI with Classical Planar Motion of Magneton (e.g. d-Electron or u-Quark) on Plane Perpendicular to Source MM, MF Binding with awful θ-Depending Non-Spherosymmetric Feature Simplified. To Further Simplify Dynamics and See Effects Caused by MF We Ignore LV in GSs of Quarks and DSLs. Then Resultant Force is MF + Coulomb Force for d-Electron and St Coulombian Force + MF for u-Quark and Anti-u-Quark. St MF May Be Included. Circular Motions on above Said Plane Has Simple Potential Function that Depends on r Only, where r is Radius of Circular Motion. Solving Eqn of Motion in SR Rigorously for Classical Circular Motion in Said Resultant Force Field Leads to **TMET** (**T**otal **M**echanical **E**nergy **T**heorem) and TME Function that Takes Minimum Value (Negative) at Critical Distance r_c. See Definition of TME in (H5) on Page 32. As r Increases from r_c to 10^{-4} cm, for instance, TME Increases Monotonically to Near 0. But as r Increases from 10^{-24} cm, for instance, to r_c TME Decreases Monotonically from very Large Positive Value to 0 at **0TMED** (0TME Distance where TME=0) $r_{0TME}(<r_c)$ and to the Minimum Value at r_c. Namely, at r_c TME-r Curve Has Lowest Point and Goes Up in Both Right and Left Direction from $r=r_c$. PE $U(r)$ Still Negative at All Distances, Monotone Decreasing as r Decreases. But around Critical Distance r_c and Nearby, TME-r Curve Has V-Shape. This May Imply Small Quantum Level Differences of Near-r_c Bound States, Having Something to Do with Small Mass Differences btw η' and ϕ, ρ^0 and ω^0, Σ^0 and Λ^0, and some N*-resonances. We Leave Possible Near-r_c Bound States Open for Future Study. As We Focus on **Ground** State of HMC $u\bar{u}$ in π^0 that Has Lowest Value of Negative TME and Smallest Mass in Whole Family of Mesons, TMET Gives Hint and Puts Light on Such State of Lowest Negative TME: If MFs Suppressed Greatly in **OSTP** (**O**rbit-**S**pin-**T**ilting-**P**recession), r_c Shifts to Distance r_c' Much Shorter than DBR and St Coulombian Force now Dominates Forbidding Near-r_c Bound States. New TME-r Curve Changes in Shape Greatly Dropping down from Previous Lowest Point to Much Lower Places and Reach New Lowest Point at New Critical Distance r_c'. This May Yield Ground State with TME Much Less than that of Near-r_c Bound States. Now It's Possible to Explain Low Mass of π^0. Its Mass is Naturally Low, Not 'Accidentally' Low. (see details of OSTP later) TMET Has Rigorously Proved Bizarre **Positive ME** (**M**echanical **E**nergy) of MF-**Bound** Particles at Distances Shorter than 0TMED $r_{0TME}(<r_c)$. It's True for any Bound State Once Binding Force Proportional to r^{-4} such as MF.

  Examples of Negative and Positive TME Given Later.

- **1BEx** (**1 B**ody **Ex**traction) of MB-Problem Focuses on One Body in Resultant

Force Field (**RFF**) Produced by Other Constituent Particles in Same MB-System. See 1BEx, Time-Varying ME of any Individual Constituent Particle that is in Time-Varying RFF Produced by Others (**MB-Theorem**), and Time-Independent TME of MB System (**MB-Postulate**) with Overwhelming Evidences Later.

- All-Time Negative ME of a Constituent Particle Closes/Forbids **QTC** (**Q**uantum **T**unnel **C**hannel) for it to Become Free via QT. Sometime or All-Time Positive ME of Any Individual Bound Constituent Particle Opens QTC for it to Become Free via QT.
- Negative TME Necessary for 2B or MB System to Be Stable.
- Positive TME Sufficient for 2B or MB System to Be Unstable.
- **FQA** Fact Requires All-Time **Negative** ME for Every u-Quark and Every Anti-u-Quark Bound in any Q-Core so that They Have No any Chance to Flee to Free Region through QT. FQA Fact under HE and Super HE Knocking Requires other Things (**4TDSP** and NLVUC) to Explain. SDL
- All-Time Negative ME of Each One in HMC $u\bar{u}$ Can Be Realized in Different Possible Ways:
  1) OSTP Suppresses MFs Greatly Guaranteeing Negative ME at All Times.
  2) HMC $u\bar{u}$ without OSTP and with MFs in Full Scales are Bound at Distances Near Critical Distance r_c Leading to Negative ME at All Times.

  Note: One Major Difference btw above 2 Possible Cases is Found that $u\bar{u}$ with OSTP Have All-Time Antiparallel Spins and Parallel MMs Leading to Net MMs of HMC $u\bar{u}$ that May Bind DSLs while $u\bar{u}$ with Parallel Spins without OSTP Have 0-Net MM. Net MM Binding Leads to Non-Existence of $d\bar{d}$ while 0-Net MM of $u\bar{u}$ Leads to Existence of $d\bar{d}$ where d-Electron & $\bar{d}$ -Positron Have Parallel Spins MF-Binding Each Other with Fast e^+e^--Annihilation & Very Short Lives. For Ground State, We Focus on OSTP. We Leave 2nd Possibility for Future Study. At Present, We See a Possibility that Near-r_c Binding of Small Level-Differences without OSTP Corresponds to Neutral Mesons Heavier than Zero-Pion and with Smaller Mass-Differences. All Excited States of HMC $u\bar{u}$ May Have OSTP States with a Series of OSTP Angles and MF-Suppressions Ranging from Complete/nearly Complete to Zero/Nearly Zero-Suppression. All Quantum States without OSTP are Characterized by Zero-Suppression of MFs. DS & MFs Naturally Explain Why Experimenters soon Easily Discovered So Many New Mesons and New Baryons far beyond Original Meson-Octet, Baryon-Octet, and Baryon-Decuplet and Why Experimenters've Failed to Find Particles Predicted by Supersymmetry Theory. Relation btw Structure and Symmetry from Newton, Einstein, to Founders of Quantum Theory Did Not Serve as True Mathematical Principle of Natural Philosophy. But Quark Contents in SM Work everywhere if DS and MFs are Taken into Account. Reason why Quantitative Solutions Can't Be Obtained Has Been Crystal Clearly Revealed by DS with MFs.

- DBR Inversely Proportional to Mass of Bound Particle. Heavy u-Quark (see details below) Can Lead to Small Binding Size with St Coulombian Force Only without any MF. Then It's Possible to Have Small Binding Size with MF Greatly Suppressed.
  Quantum MB-Systems Unsolvable. Even Equal Mass Quantum 2B-System such

as HMC $u\bar{u}$ Not Solved and May Be Unsolvable too due to MF-Binding. At Same Time, however, DS Can Explain Many Things in Absence of Quantitative Solutions. We'll See How DS and MFs Explain Facts after Facts.

- Classical MB-Systems Unsolvable too. But CI Can Be Used to Get a Series of Rigorous Classical Solutions for Quantum Equal-Mass 2B-System such as HMC $u\bar{u}$, with a Series of Different Settings of Basic Unknown Quantities Related to Undetermined Value of u-Mass and Give Precious Clue for Us to Identify Best Solutions that Agree with Accurate Quantitative Empirical Data and Facts. Once $u\bar{u}$ Equal-Mass 2B-System with MFs Suppressed Greatly by OSTP is Solved in CI, We Found that HMC $u\bar{u}$ 100 Times Smaller than 1fm and May Be Even Smaller than that due to Heavier u-Quarks. (see Rutherford Issue on P. 30) This Tiny Heavy Core Allows us to Treat MB-Systems of Charged Pions that Contain DSLs as 1-Body in MF Field Produced by Net MM of $u\bar{u}$ with Small DS NFM Contribution to π-Mass Omitted. This MF-Binding Leads to Positive ME of DSLs in Charged Pions so that Not Only Small π^0-Mass is Explained under Heavy u-Mass Setting that is Required by Simultaneous Explanation of Masses and MMs of Nucleons p and n and of Stable Free Protons, but also $\pi^\pm$-Mass is Explained. (see π^0-mass and $\pi^\pm$-mass eqs later) Note: eqs refers to equations.
- QTC Allows Some Constituent Particle(s) of *Nonnegative* ME in Resultant Force Field Produced by the Rest of All Constituent Particles in Same Bound System to Penetrate **PW** (**P**otential **W**ell) and Flee to Free Region through **QT** (**Q**uantum **T**unnel). Here PW Refers to Region with Negative PE. Its Shape May Not Be Well-like. For Particle in Coulomb Field or Coulomb-type Field without MF, Positive ME Means Unbound with No Need of QT. Stable Free Proton Has Negative TME Yielding **HQMT** (**H**eavy **Q**uark **M**ass **T**heorem): u-Quark Mass Larger or Much larger than 1/3 of Proton's Mass. It Explains MMs of Nucleons as Shown by D.H. Perkins in His Book "Introduction to High Energy Physics". He Didn't Consider DS. It May Be Ok if d-Electron's OSOD Yields Very Small Net MM of d-Electron. HQMT Only Provides Inequality for u-Mass. OAM of d-Electron Empirically Immeasurable/Theoretically Unknown. Uncertain u-Mass & Unknown Value of d-Electron's OAM Tell Us: We're Not in Position yet to Quantitatively Explain MMs of Nucleons. But HQMT, DS with MF-Bound d-Electron Having OSOD, and the Method Perkins Has Used Join Forces to Have Explained MMs of Nucleons Half-Quantitatively. Moreover, DS with MFs Crystal Clearly Shows and Proves Impossibility to Get Quantitative Explanation to Many Quantitative SA Facts.
- Ironic FQA Fact in All Collisions in Accelerators/Cosmic Showers with Incident Particles' Energies Ranging from 0.1 to 10^{12} GeV will Be Explained soon.

In Circular Motion CI for Quantum 2B-System HMC $u\bar{u}$, It's Found that MF btw $u\bar{u}$, when Suppressed Greatly by OSTP, Can Reconcile Heavy u-Mass with $m_u > or >> m_p/3$ and Small π^0-Mass with $m_{\pi^0} << 2m_p/3$ so that Masses of Pions and Nucleons, MMs of Nucleons, and Stable Free Protons All Can Be Explained Simultaneously.

See how OSTP can greatly suppress MFs btw $u\bar{u}$ soon.

See how greatly suppressed MFs btw $u\bar{u}$ yield π^0-mass eqn & $\pi^\pm$-mass eqn via CI to explain masses of pions in Last Section (page 75-83).

- TMET Shows that as MF-type Force Gets Suppressed and Decreases, 0TMED Decreases too and as MF-type Force Approaches 0, 0TMED Approaches 0 too. Namely, Negative ME or TME is **IPEBF** (**IPE B**inding **F**eature) for All Binding Sizes. As Coulomb-type Force Approaches 0, 0TMED Approaches ∞. It Means Positive ME or TME is Pure **MFBF** (**MF B**inding **F**eature) for all Binding Sizes.
- **LMC** (**L**eptonic **M**eson **C**ore) e^+e^- Contained in Composite Muon is Born to Have Size Smaller than 0TMED with Large Positive TME. It Makes Muon Heavy and Opens QTC for Muon-Decay: $\mu^+(e^+e^-e^+\nu_e\bar{\nu}_\mu) \to e^+e^-e^+\nu_e\bar{\nu}_\mu$. See How DS of Muons Explains All Confirmed Decay Modes of Muons as well as Heavy Muon-Mass Later. Also See How DS of Muons with MF-Binding Miraculously Explains Lifetime Inequalities with Lifetimes of Muon/3 States of Positronium e^+e^- Later. DS-Reason for Slower Decay of Charged Pions than Neutral Pion is Similar to What is Seen in Slower Annihilation of LMC e^+e^- in Muons than Annihilation of Positronium e^+e^- in p-P_S & o-P_S States. Composite Muons with Lifetime $2.19 \cdot 10^{-6}$s Leaves 6.57m to 65.7m Long Tracks as their Speeds Reach $0.01c$ to $0.1c$. Such Long Tracks Make Them Look Like Elementary Particles.
- Tau Contains LMC e^+e^- in Higher MF-Excitation and Much Heavier. From Now on Context will Determine which Do We Mean by LMC e^+e^-, or by e^+e^-.
- PCQM Achieved by Inputting DBR into Classical Eqn in Relativistic Mechanics while Classical Theory itself Cannot Determine Binding Size. Rigorous classical solutions in CI of Equal-Mass 2B-System Give a Series of CI-Outcomes via a Series of Inputted Binding Sizes and Other Undetermined Quantities such as u-Mass. They Provide Precious Clue and We Can Easily Compare CI-Outcomes with Accurate Empirical Data and Facts to Identify Some Good Combinations of These Fundamental SA Quantities under Condition of Unsolvable/Unsolved Eqs in SA QM for Said Equal-Mass 2B-System. BC 3B-System totally Unsolvable. Even Classical Solution for Equal-Mass 3B-System Not Available if MF and MM-Correlations are Considered. We Still Unable to Determine Value of u-Mass Completely.
- MF Can be Turned off or Greatly Suppressed by Magnetons' Special Quantum State, Called OSTP because OSTP Keeps Their **ZL**s (**Z**one-**L**ocations) being R0 at Each Other. (see three regions R1, R2, R0 in MF field and OSTP later)
- Particle Bound with its ME ***<u>Negative</u>*** **<u>at All Times</u>** Has No Chance to Penetrate PW and Flee to Free Region through QT. <u>Examples</u>:
  Every u-Quark and Every $\bar{u}$.
  Single Pair of d-Ls $e^-\bar{\nu}_e$ in Proton without Identical Ones and Repulsion from them. It Explains **<u>Bizarre Uniqueness</u>** of **<u>Stable</u>** Free Protons while All Other Hadrons including Neutrons Not Stable.
  HMC $u\bar{u}$ in π^0 as Equal Mass 2B-System Bound by St Coulomb-Type Force with MF Greatly Suppressed by OSTP. SDL
- Particle Bound with its ME ***<u>Positive</u>*** **<u>at All Times or Sometimes</u>** Has Chance to Penetrate PW and Flee to Free Region through QT. This is QTC of Decay.
  Example 1: One More Pair of d-Ls Step in Proton Turning it into Neutron. Later Comers Have Time-Varying ME Negative at Most Times, Positive Sometimes in a Variety of Times Later. Repulsion from Identical d-Ls Originally in Proton Now in Neutron Causes **ENM** (**E**agle **N**est **M**echanism) to Slowly Push New

Comers Out. Without Fast SA De-Excitation and with Slow ENM, We Now Understand **<u>Bizarre Uniqueness</u>** of Neutrons, among All Unstable SA Particles, with Average Lifetime about 10^8 Times Longer than Slowest Decaying Hadrons other than Neutrons and than Other Slowest Decaying Unstable SA Particles.

Note, In this Book, *SA Particles* Refer to *Hadrons & Leptons* in Particle Physics. Example 2: Charged Pions 4.59 MeV Heavier than Neutral Pion. DS of Pions ($\pi^0 = u\bar{u}$ / $\pi^+ = u\bar{d} = u\bar{u}e^+\nu_e$ / $\pi^- = \bar{u}d = \bar{u}ue^-\bar{\nu}_e$) Shows Positive ME of DSLs in Charged Pions. It Opens QTC for Decay: $\pi^+ \to \pi^0 e^+ \nu_e$ / $\pi^- \to \pi^0 e^- \bar{\nu}_e$. Here DSLs Bound by MFs from HMC $u\bar{u}$ Have Positive ME Allowing Them to Penetrate MF-Coulomb PW Via QT Going to Free Region outside HMC $u\bar{u}$. HMC $u\bar{u}$ in Charged Pion Becomes π^0 ($= u\bar{u}$) after DSLs Leave.

- **SGE** (**S**tern-**G**erlach **E**xperiment) Shows: There're Two and Only Two Possible Directions for Electron's (Free or Bound in Atoms) MM in EMF: Either in Same Direction as or in Opposite Direction to EMF.
- Formulas of MFs Seen in (H11) through (H19). They Can Be Wk MF or St MF if Charges q_1q_2 are Extended from Electric to Wk or St Charges.
- Magnetic Field of any SA Source Magneton Highly Local, Non-Uniform, Non-Spherosymmetric, θ -ZL Depending, Inversely Proportional to r^4 in all θ -Directions but almost Ceases to Exist at Distances Much Longer than **MD** (**M**agneton **D**istance), Dominates Coulomb-type Field at Distances Shorter than MD, <u>too Complicated</u> <u>to Make Even Classical Motion of any Other Magneton Nearby Solvable</u> (unless Special IC Leads to Simple Solution) No Matter Source Magneton at Rest (too Heavy to Move) or Not and No Matter Source Magneton Effective One with Effective Net MM Produced by More than One Elementary Magnetons Crowded in Tiny Space or Not. Inclination θ Well-Known Spherical Coordinate. MF btw 2 Magnetons $\vec{\mu}_1$ & $\vec{\mu}_2$ Depends on Distance r , ZL Parameter θ, and MM-Correlation $\vec{\mu}_1 \cdot \vec{\mu}_2$ or Spin-Correlation. Sign of any MF Determines if It's Attractive or Repulsive. Both Magnitude and Sign of any MF btw 2 MMs of Same Kind Depend on θ & Directions of MMs. MF Field Axially Symmetric wrt (with respect to) Source MM-Line, as Zenith Axis of Spherical Coordinate System, so that MF is ϕ-Independent.
- Let MM $\vec{\mu}_1$ be at Origin, Pointing at Positive Zenith Direction. Formulas of MF btw 2 Magnetons $\vec{\mu}_1$ and $\vec{\mu}_2$ with Parallel or Antiparallel MMs Indicates Three Regions with their Particular Properties:

R1 with $79.5^0 < \theta < 100.5^0$ like a Horizontal Disc with Thickness Increasing as r Increasing.
R2 with $0^0 \le \theta < 20.4^0$ and $159.6^0 < \theta \le 180^0$, Two Vertical Circular Cones.
R0 with θ Equal or Close to Two **0R-MF** Angles about 54.7^0 and 125.3^0 .
0R-MF Angle Such θ -Angle where Radial Component of MF btw above 2 MMs Vanishes.

Three Regions Have Following Properties:
In R1 and R2, Radial Component of MF Dominates its Non-Radial Component.
In R0, Radial Component of MF Vanishes or is Very Small.

For Parallel $\vec{\mu}_1$ and $\vec{\mu}_2$, R1/R2 is Repulsive/Attractive Region respectively.

For Antiparallel $\vec{\mu}_1$ and $\vec{\mu}_2$, R1/R2 Attractive/Repulsive Region respectively. If $\vec{\mu}_2$ is in R1, R2, or R0 with $\vec{\mu}_1$ at Origin, We Say that $\vec{\mu}_2$ is in R1, R2, or R0 at $\vec{\mu}_1$. One Can Check that This ZL is Mutual Relation as long as Two MMs either Parallel or Antiparallel.

- Radial Component of MF btw two Magnetons with Parallel or Antiparallel MMs Vanishes/Very Small At/Near 2 Specific θ-Angles, Called **0R-MF** Angles. It Allows Some Particular Motions of Two Equal Mass Magnetons of Same Kind and Their **2B** (**2-B**ody) System to Yield Domination of Coulombian or Coulomb Type Force that Obeys Inverse Square Law with MF and/or St MF Turned off or Suppressed Greatly. Such Particular Motion of 2 Magnetons is Called OSTP.
- If We Only Consider Two Possible Quantized Directions of Spin-Correlation: Parallel or Antiparallel with $\vec{\mu}_1 \cdot \vec{\mu}_2$ Having Opposite Signs, MF-Formula Has Simpler Theoretical Results. We'll Use These Results to Give DS-Explanations to Many ***Lifetime Details*** of SA Composite Particles. They Serve as Sample Explanations. However, We Do Have ***Many Other Empirical Facts*** That Can Be Explained without Knowing Details of Dynamic Effects of MFs.
- Every Baryon Contains *uuu* **BC** (**B**aryon **C**ore) Either in Ground or in Excited State Never Changing its *uuu* Content. Different Colors Do Not Make any two Quarks Annihilate. This Leads to Conservation of Baryon Number.
- To Fully Explain Baryon Number Conservation in All HE Collisions, We Need Something More: To Simultaneously Explain Meson Number Nonconservation and Ironic FQA Fact. SDL

- **d-Q** (**d-Q**uark) Formed by $ue^-\bar{v}_e$ in Ground State without Excitation.
- Neutron Unstable, Decays Slowly for Identical Leptons in Two d-Quarks inside Neutron Repel Causing ENM while Proton Stable for Single d-Quark in Proton Does Not Have 2 Identical d-Leptons to Create ENM. Atomic Electrons Repel each other with Repulsion Much Weaker, Compared with Attraction of Nucleus (See Details on Page 63). While Repulsion btw 2 DS-Electrons in SA Small Space Significant, Compared with Centripetal Force Binding Them.
- DS's Explained Continuous Spectrum of β-Particles/Preferred Spin Directions of d-Leptons Emitted in beta Decay. Parity Violation/SM-Handedness/DS/MFs Join Forces to Have Explained all Details of β-Decay in EMF. See Pages 93-96.

- Idea of W-Bosons and Decay Modes $d \rightarrow W^- + u$ with $W^- \rightarrow e^-\bar{v}_e$ Don't Agree with Simple Facts that d-Quark in Unbound Proton Stable and No Double Electrons Ever Found Flying Out from One Neutron so that We Could See Baryon *uuu* of Label *J=1/2*. Elements on Stability Island Disproves W-Idea too. All Facts Disproved W-Idea at Beginning when Quark Model Revealed said Discrepancy Caused by W-Idea.
- Decay Modes $u \rightarrow W^+ + d$ & $W^+ \rightarrow e^+v_e$ in SM Have Been Disproved by Stable u-Quarks in Unbound Nucleons Once For All. DS Explains β^+-Decay without W-Scenario and Proves Exactly why β^+-Decay Forbidden if Available Energy Less than $2m_ec^2$. At Same Time DS Explains e^--Capture So Naturally and Proves that e^--Capture Not Forbidden Even if Available Energy Less than $2m_ec^2$.

- Certain SA Excitation, Called $d \to s$ Excitation of Hadron Containing d-Quark Turns d-Quark into s-Quark. Since MF-Binding Leads to Complicated/Unknown Details of Excitations, $d \to s$ Excitation, Containing LE (Leptonic Excitation), May or May Not Contain Q-Core Excitation.
- Excitation of **MBS** (**M**any-**B**ody-**S**ystem) with More than One Kinds of Forces Involved Complicated with Unknown Details.
- **MFE** (**MF E**xcitation) Refers to Excitation of any System Bound by any Kind of MF. MFE Details Unknown due to Insolvability of MB-System. MF-Bound 2B-Systems Not Solved yet and May Be Unsolvable too.
- Higher Excitation Turns s-Quark into b-Quark. Quark-Excitation May Occur.
- c-Quark Contains $ue^-\bar{v}_e e^+ v_e$ in Excitation. Asymmetric/θ-Depending MF Field Capable to Simultaneously Stick Particles with Opposite Charges in Different θ-ZLs and/or Having Right Spin-Correlations with u-Quark and with Each Other.
- MF Ceases to Play Significant Role at Atomic Distances but Dominates Coulombian Force at Distances Shorter than MD. Coulombian Repulsion btw u and e^+ Overcome by MF Attraction inside c-Quark.
- Among All Leptons/Antileptons Other than NFMs, Only e^+ & e^- Elementary. DS of Muons is Written $\mu^+ = e^+e^-e^+v_e\bar{v}_\mu$ and $\mu^- = e^+e^-e^-\bar{v}_e v_\mu$, with Heavy LMC e^+e^- Bound by MF and Born to Have Large ***Positive*** TME with Large γ-Value (see TME theorem and muon-mass eqn later). Such DS Explains All Confirmed Decay Modes of Muons Naturally while Predicts Impossibility for Some Unconfirmed Decay Modes. For Example, $\mu^+ \to e^+ + \gamma$ Can Never Happen, for μ^+ Not Excited State of e^+.
- To Produce Heavy LMC in Muon, Needed Energy May Come from Collision such as $e^+ + e^- \to \mu^+ + \mu^-$ or from $u\bar{u}$-Annihilation such as Seen in Pion-Decay $\pi^+ \to \mu^+ + v_\mu$: $u\bar{u}$-Annihilation in $\pi^+(u\bar{d} = u\bar{u}e^+v_e)$ Produces 2 Pairs (Heavy LMC e^+e^- and $v_\mu\bar{v}_\mu$) with $\bar{v}_\mu$ and Relics e^+v_e (Left Over after $u\bar{u}$-Annihilation) Captured by LMC e^+e^- to Form μ^+ and v_μ Flying away. LMC e^+e^- May Annihilate Quickly as Seen in Decay Mode: $\mu^+ \to e^+v_e\bar{v}_\mu$. Here Heavy LMC e^+e^- Annihilate into Virtual γ and/or Virtual Z-Bosons for Other Leptons in Muon to Absorb and Gain KEs Fleeing Out. e^+e^- in LMC ***<u>Bound</u>*** but Possess Large ***<u>Positive</u>*** TME. They Can Penetrate Deep **PW** (**P**otential **W**ell) and Flee Out to Free Region with Same Positive TME through **QT** (**Q**uantum **T**unnel). It Predicts Decay Mode $\mu^+ \to e^+e^-e^+v_e\bar{v}_\mu$. Experiment Proves it. It's QTC of Decay.
- π^- Contains HMC $u\bar{u}$ plus d-Ls $e^-\bar{v}_e$ while π^0 Contains HMC $u\bar{u}$ Only.

  Note: OSTP Plays Key Role to Explain Small Messes of Pions under Condition of Heavy u-Mass that Simultaneously Explains Masses of Nucleons, MMs of Nucleons, and Stability of Free Proton. OSTP Suppresses MF Greatly, Causing Small HMC $u\bar{u}$ with Antiparallel Spins and a Net MM, which Unable to Bind DS-Electron and DS-Positron Simultaneously to Form any Particle with Content $d\bar{d}$. Due to OSTP, $d\bar{d}$ Must Live too Short to Be Called a Particle. Without

OSTP, $d\bar{d}$ Can Exist as a Particle or Part of a Particle. But MF btw d-Electron and $\bar{d}$ -Positron will Cause Fast e^+e^--Annihilation and Very Short Lifetime of $d\bar{d}$ 2B-System. DS Reliably and Easily Explains Very Short Lifetimes of All DS-Symmetric Mesons no matter they contain mixture or not. SDL

If OSTP and Non-Existence of $d\bar{d}$ are True, π^0 is the Only Example of 2B-System in Set of All Hadrons.

- Mass Difference btw π^- and π^0 Proves Positive **ME** (**M**echanical **E**nergy) of $e^-\bar{\nu}_e$ inside π^-, Making **QTC** Open for $e^-\bar{\nu}_e$ to Flee. Experiment Proves This Decay Mode: $\pi^- \to \pi^0 e^- \bar{\nu}_e (QTC, 1.036 \cdot 10^{-6})$. It's Greatly Suppressed/Dominated by HMC $u\bar{u}$ Annihilation Channels that Lead to Decay Modes:

$$\pi^- \to \mu^- \bar{\nu}_\mu (PPC, \;\; 98.98770\%),$$
$$\pi^- \to \mu^- \bar{\nu}_\mu \gamma (PPC + PR, \;\; 0.02\%),$$
$$\pi^- \to e^- \bar{\nu}_e (KC, \;\; 0.0123\%).$$

More Details of **PPC** (**P**article **P**roduction **C**hannel), **KC** (**K**inetic **C**hannel), and QTC Seen on Pages 64-67.

< OSTP >

- We'll Postulate Special Classical Circular Motion in CI, Called OSTP to Imitate Hypothetic Quantum Ground State of HMC $u\bar{u}$ with MMs of $u\bar{u}$ Parallel at All Times, Their θ -ZLs Being R0 at Each Other and Very Close to 0R-MF Angle at All Times, and MF btw them Attractive but Greatly Suppressed at All Times. Such Hypothetic Quantum State Allows Quantitative Calculation of Circular Motion CI on Fixed Orbit Plane. Two Spin MMs Tilted Against Orbit and Change Directions in Precession to Maintain MM-Correlation and Relative θ -ZLs Near 0R-MF Angle. In Circular Motion CI, Two Magnetons $u\bar{u}$ Move along Same Circle on Same Orbit Plane, Revolving about their Mass Center with Same Angular Velocity. At Every Instant Time, Center of Circular Orbit is Middle Point of Segment Joining Two Magnetons. Such Special Motion is Simplest Classical Solution of any Equal Mass 2B-System. OSTP Means that Each Magneton's Spin Vector Changes its Direction Constantly in such way that It Keeps Forming Same Angle with Orbit Plane with its Projection Line on Orbit Plane Passing through Center of Circular Motion at All Times. When a Skater Makes Circular Motion on Ice Floor with Body always Stretched, Not Vertical, always Forming Same Angle with Floor in the Way that Projection of Body-Line on Floor at Every Instant Time Passes through Center of the Circle. Let 2 Skaters Represent $u\bar{u}$ in OSTP. Body-Vectors from Feet to Head Represent 2 MM-Vectors of $u\bar{u}$. Their Body-Lines are Not Vertical but Parallel at All Times as They Move alone Same Circle on Ice Floor Opposite to Each Other, Forming Same Angle with Floor-Plane at All Times. Projection Lines of 2 Parallel Body-Lines Coincide on Floor-Plane, Rotating about Center of Circle and Passing through that Center at All Times. From This Picture We May See Exactly what

Does It Mean by 'OSTP'. As We Use Relativistic Mechanics to Rigorously Calculate Classical Circular Motion in CI for HMC $u\bar{u}$ 2B-System in Ground State, Existence of Such 2 MMs is Represented by Suppressed Attractive MF. Calculation Explains Mass of π^0 under Heavy u-Mass Setting, which is Needed to Simultaneously Explain Nucleons' Masses, their MMs, and Stability of Free Proton. See a Bunch of Possible Tiny Sizes of HMC $u\bar{u}$ in Ground State (< $2.6 \cdot 10^{-15} cm$) on Page 79. The Heavier the u-Mass, the Less the Size.

Note: Precession in OSTP is Needed for Spin MMs to Change their directions in Order to Maintain Orbit-Spin Tilting Near 0R-MF Angle ZL Simultaneously. Orbit-Spin Tilting alone Cannot Fix two Magnetons in Same Near 0R-MF Angle ZL all the Times. In OSTP, Two MMs with Directions Changing All the Times Remain Parallel at all Times so that Spins of Two Magnetons Remain Antiparallel at all Times to **Obey Conservation of Total Angular Momentum of Spins**.

- TMET Proves: Negative TME **BF** (Binding Feature) of Coulombian/Coulomb-Type Force when it Dominates or No Other Type of Force Exists while Positive TME BF of MFs when MFs Dominate and IC Puts Binding Size Less than 0TMED.
- OSTP Cannot Happen to LMC e^+e^- 2B-Bound System. As OSTP Suppresses MF btw e^+e^- Greatly, Coulombian Force Unable to Bind e^+e^- within SA Small Space due to Their Small Mass. It then Becomes Positronium e^+e^- of Atomic Size with MF at such Atomic Distances too Weak to Cause Significant Orbital Motion for e^+e^- to Depart S-State.
- Without OSTP and with MF-Binding Domination and its Positive TME Feature, Muon-Mass Eqn is Deduced to Give Explanation of Empirical Value of Heavy Muon-Mass Compared with Much Smaller Total Mass of All its Leptonic DS-Constituents. (see pages 84-85)
- Even 2B-System Bound by Non-Spherosymmetric and θ -Depending MF Not Solved and May Be Unsolvable. To Explain Many Facts Only Means to Explain Half-Quantitatively or Qualitatively due to Above Said Difficulty & Unsolvable **MBD** (**M**any **B**ody **D**ifficulty).
- DS, Powerful and Penetrating, Has Explained One after another Empirical Fact. Some Never Explained before. With Very Few Kinds of Elementary Particles (So Stable and Never Decay) and Only 4 Kinds of Basic DS Processes, Called **4TDSP** (**4 T**ypes of **DS P**rocesses), DS Has Demonstrated its Ability to Explain Hundreds of Kinds of Middle Stage Particles and Thousands of Decay Modes on Most Basic and Most Fundamental Platform. ….

Unknowns, Spin **Misconception**, TAM **Disability,** and **4TDSP**

- ***Composite Particles Do Not Spin*** for Constituent Elementary Particles Usually Do Not Spin about Same Axis and Do Not Spin with Same Angular Velocity. They Do Not Have, Usually, Same Charge-Mass Ratio.
- Two Elementary Fermions with Same Spin ½ Even Do Not Spin with Same Angular Velocity if Their Masses Different.

- J-Label of ***<u>Composite</u>*** Particle Should Be **TAM** (**T**otal **A**ngular **M**omentum) as Vector Sum of Spins and Orbital Angular Momenta of ***<u>Constituent</u>*** Elementary Particles in **CMF** (**C**enter of **M**ass **F**rame).
- For Any ***<u>Unstable Particle</u>*** Decaying into Final State Particles, its Correct Spin-Label (if Believed to Spin) or J-Label Must Be **TAM** of ***<u>Final State Particles</u>***, Measured in CMF.
- ***TAM*** of Constituent Particles of Composite Particle *Empirically* ***<u>Immeasurable</u>*** Since Tiny Departures of Momentum-Lines of Final-State Particles from Exact Mass-Center Empirically Immeasurable.
- ***TAM*** of All Constituent Particles of Composite Particle *Theoretically* ***<u>Unknown</u>*** because Their Total **OAM** (**O**rbital **A**ngular **M**omentum) Cannot Be Known Theoretically due to Unsolvable MB-Problems and Unsolvable or Unsolved 2B-Systems with **Mag** (**Mag**netic) Forces Involved. Non-Spherosymmetric MFs Tell us: S-State with 0-Orbital Number (l=0) Does Not Exist in SA World.
- Spin/J-Values for Composite Particles in **SM** (**S**tandard **M**odel) Only **Labels**. Even TAM of Constituent Particles May Not Equal SM Spin/J-Value of the Composite Particle They Form While Exact Empirical Value of Sum of Spin-Vectors of All Final-State Particles Does Not Scientifically Prove Any Value of Total OAM nor TAM.
- Spin/TAM Misconception/Disability in SA Physics Refer to Two Simple Facts:

  Fact 1. ***Composite Particles Do Not Spin***.
  Fact 2. **OAM/TAM** of each/all Final State Particles and **OAM/TAM** of each/all Constituent Particles ***Empirically Immeasurable*** and ***Theoretically Unknown***.

Example 1

Even in Rest Frame of Decaying Particle, a Very Small Deviation of a Final State Particle's Momentum Line from Accurate **MCtr-MCtr** (**M**ass **C**enter to **M**ass **C**enter) Line Would Invalidate a TAM-Measurement where Sum of Only Their Spin-Vectors is Treated as TAM with OAM Uncounted. For a **FSP** (**F**inal **S**tate **P**article), MCtr-MCtr Line Means that 1st MCtr its MCtr and 2nd MCtr MCtr of All Final State Particles. It is Easy to Calculate and See that Flying Out 1 MeV Electron's OAM is $\hbar/2$ if Said Deviation is $1.123\cdot10^{-11}cm$. Once Future super Advanced Tech in 200 to 800 years Enables Experimenters to Measure Deviation as Small as that Lots of Important Basic Facts of SA Structure will Manifest themselves via Direct Measurement of Fundamental Physical Properties of Final State Particles. OAM's Not Been Measured since Beginning of Particles Physics.

Example 2

Although S-States Don't Exist in SA World with Existence of Asymmetric **MF**s, Theoretical Values of Nonzero Quantum Orbital Numbers of Constituent Particles Unknown for Quantum Eqs for **2B**-System and MB-System in any MF Field and/or Time-Dependent Non-Central Non-Uniform Magnetic Field, Never Solved, Unsolvable, or May Be Unsolvable at Least at Present.

- Universe at Large Scales Reveals Many Cosmological Features Much Clearer than SA World. We See Stars/Galaxies after Stars/Galaxies Everywhere, with Red Shifts Revealing Stunning Similarity of Atoms/Ions/Chemical Elements, all Over Universe and in All Directions.

- Newton Was Able to Propose and Verify His Great Laws (Laws of Motions and Law of Universal Gravitation) because Tycho and Kepler Could Observe Details of Solar System and Motions of Solar Planets and then Found/Verified Laws of Planetary Motion to Provide Crystal Clear Picture and Quantitative Observation Data for Newton to Think. But No One Could Get Unbound Quarks, Observe Their Motions, Find Empirical Laws of Forces btw Them. Moreover, There is No Way to Observe and See Anything Inside SA Particles. No One Knows How Long Mankind Has to Wait Until Unknown Future High-Tech Photos May Reveal whether Electrons and Quarks Have Been inside Hadrons.
- Experimental Ways to Measure Masses and Speeds of Quarks Do Not Exist.
- Direct Measurement of e.m. Forces btw Electric Charges and Currents/Universal Gravitational Force btw Heavy Bodies Done by Coulomb, Ampere, Faraday, and Cavendish. No Similar Ones Can Be Down in SA Physics due to Unavailability of Free and Unbound Quarks and Short Range of St Force. Quark Model then is Immature & Uncertain, Containing More than 1 Possibilities in 2 Major Aspects: Constituent Particles (Structure) and Forces (Dynamics). It Never Means that Quark Model's Bad. On the Contrary, Quark Model One of the Most Important Milestones on the Way to Ultimate Truth of SA World, Just Like Newton's and Einstein's Theories that're 2 Most Important Milestones on the Way to Ultimate Truth of Fundamental Laws of Physics.
- There will be Little Chance that Some One or Some Group of Physicists Will Be Able to Get Certain and Mature Model of Mysterious SA World in This Century. But Major Changes that Advance Our Understanding are More than Likely to Happen.
- Forces btw Moving Protons Provide Little Clues to Forces btw Moving Quarks since any 2 Quarks inside Different Protons are Separated by Distances Close or Longer than Range of St Force. Such Skin-Deep Experiment with Unbroken and Unchanged Protons Cannot Penetrate Distances from 1 to 10^{-2} fermions. Such Distances are Crucial to **QD** (**Q**uark **D**ynamics) if Quark-Cores and Hadrons Have Same Sizes. Moreover, It Cannot Cover Distances from 10^{-2} to 10^{-8} fermions, Crucial to QD if Quark-Cores Have Smaller or Much Smaller Sizes than Hadrons.
- Simplicity of Dynamics Seen and Revealed by Great Old Masters of Physics Does Not Appear in SA World.
- To Experimentally Prove/Measure Force btw Electric Charges/Currents is Easy. To Experimentally Measure Force btw u-Quarks is at Least Almost Impossible. All Theories/Models in SA Physics are Proposed in Absence of Experimental Data about Forces btw u-Quarks. What Theory/Model is Powerful to Explain Experimental Facts?
- Experimenters' Attempts to Probe Deeper into Hadrons and to Produce HE e^+e^-, pp, and Heavy-Ions Head-On Collisions Always Produce Enormous Phenomena <u>So Complicated</u> that *Hundreds* of Kinds of Middle Stage Particles & *Thousands* of Decay Modes Have Been Seen while <u>So Simple</u> that All Can Be Explained in Terms of DS with Only Few Kinds of Elementary Particles and Only 4 Types of **DS P**rocesses, Called **4TDSP**.
- In View of DS, ***Original*** Pair Productions Consume Energies Released from HE

Collisions, SA De-Excitations, and Annihilations. Each Original Pair Production Takes Form of One or More Elementary **PAPs** (**P**article-**A**ntiparticle **P**airs). Elementary **PAPs** Refer to $e^+e^- / \nu_e\bar{\nu}_e / u\bar{u}...$, Not $\nu_\mu\bar{\nu}_e / c\bar{s} / u\bar{d}$ .... Many PAPs, $d\bar{d} / s\bar{s} / \mu^+\mu^- / p\bar{p} / ....$ are Not Elementary PAPs.

- Every Meson Contains Elementary **PAP** (Particle-Antiparticle Pair) $u\bar{u}$ (HMC) and Muons/Taus Contain Elementary PAP e^+e^- (LMC). Their Annihilations are DS or SA Annihilations, Not Annihilations of Free PAPs in Collisions. DS-Annihilation of $u\bar{u}$ and Most DS-De-Excitations Release Enough Energies to Produce One or More PAPs, Of Course, Elementary PAPs. But Hundreds of Kinds of Middle Stage Particles in Collision & Decay Events Much Complicated than Elementary PAPs plus Initial State Particles due to Simple and Only Four Types of DS-Processes (4TDSP):

< **4TDSP** >

1. DSE (**DS-E**xcitation) and **DSDE** (**DS-De-E**xcitation)

DS-Leptons, Sometimes uuu -**BC** (**B**aryon **C**ore), $\bar{u}\bar{u}\bar{u}$ -**AntiBC**, $u\bar{u}$ -HMC, or e^+e^- -LMC Get Excited, Usually due to Collisions.

DSE and DSDE Always Happen Quickly **without Single Exception**.

Examples:

Target Nucleon Gets Excited Turning Itself into a Short-Lived Baryon then SA DS-De-Excitation Quickly Happens. Neutron Slowly Decays for d-Quarks and Neutron itself Not in DS-Excitation. (see how DS explains facts after facts later including bizarre uniqueness of stable free proton)

DSLs as DS-Contents Add Leptonic Excitations to Entire Collection of SA Excitations. MFs Non-Spherosymmetric and θ -Depending. DSLs MF-Bound. Complicated Excitations of MF-Bound MB-Systems with Unknown Details Qualitatively Explain Empirical Large Collection of Hadrons that is beyond SM without DS and with Big Roles of MFs Ignored.

2. PAPP (**PAP P**roduction) and **PAPA** (**PAP A**nnihilation)

PAPP Consumes Energy Released in *Collision*, SA *De-Excitation*, or *PAPA* while SA PAPA such as Annihilations of HMC $u\bar{u}$ and LMC e^+e^-, either in Ground or Excited States, May Produce Real Photons, Virtual Carriers of Forces Absorbed by Nearby DSLs and Hadrons to Gain KEs and Flee away, or PAP(s), Leading to Radiative Decay Modes, **KC** (**K**inetic **C**hannel) of Decay, and **PPC** (**P**article **P**roduction **C**hannel) of Decay. (see details of KC, PPC, and QTC on pages 64-67)

Examples:

e^+e^- Collision Produces PAPs Needed to Form $\mu^+\mu^-$ or $\tau^+\tau^-$.

Mesons, Muons, and Taus Appear as Middle-Stage Particles in Events of HE Collisions due to Same Type of DS-Processes: PAPP with HMC $u\bar{u}$ and/or Heavy LMC e^+e^- Produced, w/o Non-Heavy PAPs e^+e^- $\nu_e\bar{\nu}_e$ $\nu_\mu\bar{\nu}_\mu$ and so forth. All Non-Baryonic Middle-Stage Particles (Mesons, Muons, & Taus) are Made

Out of Produced Elementary PAPs via 3rd Type of DS-Processes Described soon below. De-Excitation of Excited d-Leptons in Composite s-Quark in Baryon Σ^+ Produces Bound PAP $u\bar{u}$ (HMC) to Form π^0 with s-Quark Returning to d-Quark and Σ^+ (uus) to Proton (uud). This is DS-Way Decay Mode $\Sigma^+ \to p\pi^0$ is Going On. It's So Natural/Simple.

Annihilation of HMC $u\bar{u}$ in π^+ May Produce Two Elementary PAPs: Heavy LMC e^+e^- and $\nu_\mu\bar{\nu}_\mu$, Followed by Other DS-Processes Described before and below to Cause Decay Mode $\pi^+ \to \mu^+ + \nu_\mu$.

Annihilation of Excited LMC e^+e^- in Tau May Produce One or More HMC $u\bar{u}$, Leading to Production of Meson(s) in Hadronic Decay Modes.

3. PQTR (**P**article **Q**uantum **T**unnel **R**elocating) with SA Particle(s) nearby in Same SA Small Space to Capture Relocating SA Elementary Particle(s).

PQTR Creates **MSPs** (**M**iddle **S**tage **P**articles) and FSPs Out from **IPs** (**I**nitial **P**articles) & Produced Elementary PAPs so that Set of MSPs and Set of FSPs Usually Not Initial Particles plus Produced Elementary PAP(s).

Examples:

d-Leptons Emitted from Neutron Relocate into Proton Nearby Bound in Same Nucleus, Changing p/n into Each Other and Causing Stability of all Elements on Stability Island. Many Stable Elements Contain more Neutrons than Protons for Neutrons Do Not Decay At Same Time.

HMC $u\bar{u}$ in π^+ Annihilate Leaving Relics $e^+\nu_e$ & Producing LMC e^+e^- plus another PAP $\nu_\mu\bar{\nu}_\mu$ then $e^+\nu_e$ and $\bar{\nu}_\mu$ Relocate in LMC e^+e^- via QT to Form μ^+ ($= e^+e^-\ e^+\nu_e\ \bar{\nu}_\mu$), with ν_μ Flying away. $\pi^+ \to \mu^+ + \nu_\mu$ is Said to Occur.

4. SA **PR** (**P**hoton **R**adiation) Followed by SA Photoelectric-Encounters

SA PR an Important Channel for DSDE and PAPA to Release Energy. Spectrum-less SA PR and Atomic PR with Spectrum Bring us to a Unified Picture for **QS** (**Q**uantum **S**cattering) of Photon by Electric Charge. (see topics < Unified QS of Photon by Charged Fermion > on page 27)

Examples:

$J/\psi(c\bar{c}/J=1) \to \eta_c(c\bar{c}/J=0)+\gamma$. Small Difference in Mass btw η_c and J/ψ Not Enough to Produce HMC $u\bar{u}$ or π^0. As long as η_c Appears as Middle Stage Particle in J/ψ Decay, No Other Meson Can Appear together with η_c. Too Many Other Examples of PR after DSDE and PAPA Seen among Observed Thousands of Decay Modes. They are Called ***Radiative*** Decay Modes.

Note: Some DSLs among FSPs Originally were Constituent Particles of Initial SA Composite Particles while some of them were Produced in Middle Stages as Members of Produced PAPs. Some were Originally Bound, Even with Negative MEs. To Change to Free FSPs with Positive MEs (or with Larger Positive MEs than Initial Positive Values, Proper Energies Must Be Received. Some Needed

Energies Come from SA QS with photons in SA PR. Photon in PR Not Virtual Photon. Real Photons in SA PR Born within Small SA Spaces Loaded with u-Qs and DSLs. QS Inevitable within Crowded SA Spaces. But Needed Energies May also Be Gained by Absorbing Virtual Force Carriers that are Produced in PAPA as Described below:

AVC (**A**nnihilation **V**irtual **C**hannel) May Be another Important Channel for SA PAPA to Release Energy and Transfer Energy to Nearby Constituent Particles. AVC is Seen when SA PAP Annihilates into **Virtual** γ and/or **Virtual** Z-Boson to Be Absorbed by SA Elementary Particle(s) Nearby to Gain KE and Flee away, where Virtual Z-Boson the Only One NFM within Same Small SA Space Can Absorb. See Examples of AVC Later.

Note: 2nd, 3rd, and PR after PAPA in 4TDSP Create Three Types of Non-De-Excitation Decay Channels: **KC**, **PPC**, **QTC** Shown on Pages 64-67.

Energy Released in SA De-Excitation May Be Enough to Produce 1, 2, or 3 Bound HMCs $u\bar{u}$ plus Some Other PAP(s). PAPP Channel in **ERM** (**E**nergy **R**elease **M**echanism) May Suppress PR Channel Leading to Small Branch Ratios of Radiative Decay Modes. ERM Claims: *energy released in any classical transition or quantum transition from higher to lower level or in any particle-antiparticle (not valence pair) annihilation may take form of one or many* ***particle-antiparticle*** *pairs, called* ***PAPs****, as long as permitted by energy conservation.* (Classical Transition Occurs when free incident particles lose or gain KEs as they collide). In SA World, PAPPs May Suppress PR-Channel due to VI. In Atomic Physics, ERM Seen in Atomic De-Excitations Seems to Have PR-Channel Only. PAPs $u\bar{u}$ & e^+e^- Can't Be Born in Atomic De-Excitations due to Conservation of Energy. But Unified ERM Does Predict Radiation of Low-Energy NFMs in De-Excitations of Atoms. Low-Energy NFMs Elude Detection. Two Hundred Years Later, Experiment May Be Able to Tell if Unified ERM Covers Both Atomic and SA World. If Modern Tech Allows Experimenters to Confirm Single Photon Emission or Absence of Photon Emission from Single Hydrogen Hit by Single Incident Photon Only with No Chance to Get Excited from Thermo-Collision with Other Particles than Single Incident Photon, Then Atomic Emission of Low Energy NFM Can Be Indirectly Proved or Disproved in Ten Years or Later in This Century.

Decay Mode $\mu^+(e^+e^-e^+\nu_e\bar{\nu}_\mu) \rightarrow e^+\nu_e\bar{\nu}_\mu\gamma$ and DS Tell: Energy Consumed in PR here Released in PAPA (Here, Annihilation of LMC e^+e^-). DS Asymmetry Causes **Asymmetric DS-Photoelectric Effect**: Only 1 of 2 Photons Produced in LMC e^+e^--Annihilation Absorbed for FSP Positron to Gain KE. If Photon Survives after Photoelectric Encounter, It May Elude Detection due to its Low Energy. Possibility to Empirically Find Soft Photons and Neutrinos is Exactly 0, if Their Low Energies Even Less than Experimental Errors to Measure Energy.

Energies Released in All SA PAPAs and in All SA De-Excitations are Not Large Enough to Produce Radiations of Very Heavy Real Gluons and Real Z-Bosons. Experimenters've Proved Their Absence among FSPs of Decays. If Non-SA PAP Formed by Free Particle/Antiparticle in Head-On Collision, PAPA May Produce Real or Virtual Gluons in 2-Jet and 3-Jet Events. Photons Never

Decay. If Real Gluons/Real Z-Bosons Confirmed to Decay, It Would Be too Bizarre and We Might Need to Rethink Essence of Strong and Weak Force. If All Collision Events Recorded in Past 100 Years are Indeed Showing NLVUC that is Lower than Energy Needed to Produce Real Gluons, then Real Gluons will Never Show up. (see definition and details of NLVUC on page 6-7)
SA PR May Be Followed Immediately by QS btw Radiated Real Photon and Nearby Elementary Constituent Particle(s). (see QS details after remark below)

<u>Examples of AVC</u>:

In Decay $\pi^+(u\bar{d}=u\bar{u}e^+\nu_e)\rightarrow e^+\nu_e$ (0.0123 %), $u\bar{u}$ Annihilate into Virtual Force Carrier(s) to Be Absorbed by $e^+\nu_e$ to Gain Energy and Flee away. Due to VI, This KC is Greatly Suppressed by Following PPC:

$$\pi^+(u\bar{d}=u\bar{u}e^+\nu_e)\rightarrow e^+e^-\nu_\mu\bar{\nu}_\mu e^+\nu_e\rightarrow \mu^+(e^+e^-e^+\nu_e\bar{\nu}_\mu)+\nu_\mu \quad (99.98770\ \%)$$

In Decay $\mu^+(e^+e^-e^+\nu_e\bar{\nu}_\mu)\rightarrow e^+\nu_e\bar{\nu}_\mu$ (~100 %), LMC e^+e^- Annihilate into Virtual Force Carrier(s) to Be Absorbed by above Leptons Nearby. This KC Could Dominate for PPC (LMC $e^+e^-\rightarrow$HMC $u\bar{u}$) is Forbidden in Muon Decay by Energy Conservation.

We'll Soon See Examples of How to Use 4TDSP to Explain SA Events.

Remark: Within SA Composite Particles, <u>MB-Interactions</u> in Crowded Tiny Spaces May Lead to Production of Many Virtual Force Carriers in Single DS PAPA or Single DS De-Excitation. Energy-Momentum Distribution of FSPs Obtained via Large Number of Decay Events of **<u>Same Mode</u>** May Indicate Such Production of Multiple Virtual Force Carriers. Please note, SA PR Events May Look like AVC Events if Radiated Photons Lose too Much of Energies Eluding Detection after SA Photoelectric Encounter.

< **Unified QS** of Photon by Charged Fermion >

<u>Compton</u> Scattering May Have Shown Clue to Possible Unified Physical Picture of Interactions btw Photon & Charged Fermion, Electron for Instance. Classical Scattering of e.m. Wave by Free Electron is just Classical Approximation and Limit of Compton Scattering: When Photon Energy Much Less than Rest Energy of Electron Compton Formula Indicates Negligible Change in Frequency. Compton Scattering is in Class of QS for Classical e.m. Wave is Treated as Formed by Particles Called Photons. Electron in Original Compton Scattering is, however, Unbound and Free. Once Electron is in Bound State, such as Bound in Metal Body or Bound in Atom, Does Photon Have to Disappear after Hitting and Scattering? Experimenters Unable to Detect Soft Photons. It May Be Wrong to Think that Photon Must Be Completely Absorbed and Totally Disappear after Knocking Out Electron in Photoelectric Effect Experiment and after Hitting Atomic Electron and Sending it into Higher Level. Unified Picture of Collision, Interaction, and Scattering btw Photon and Charged Elementary Particle May Provide One of Reasons for Final State Particles to Have So Many Different KEs

Even for Decay Events of Same Mode. Another Reason Seen in AVC, Especially Final NFMs among Final State Particles May Gain KEs from Virtual Z-Bosons as One of Products of DS PAPA. Real Photons May Become Soft after Losing Lots of Energies in SA QS, Eluding Detection, or May 'Survive' Such Scattering Flying away as Final State Particle Observable. SA Constituent Particles in Same SA Small Spaces where Photons are Born after SA PAPAs or SA De-Excitations May Gain Large amounts of 4-Momenta, Fleeing Out from Original SA Composite Particles with High Speeds. Needed 4-Momemta Come from New Born Photons via SA Photoelectric Encounter and Collisions. Virtual Force Carrier Produced during SA PAPA is for Nearby SA Elementary Particles to Absorb and to Gain Energy.

Soft Photons 100 % Surely Elude Detection. SA QS Does Not Guarantee Real Photons Produced in SA PR to Be Seen among Final State Particles. In SA QS, Real Photon is Produced So Close to Particle to Be Hit that Very Large Amount of Energy Can Be Transferred from Photon to Particle It Hits and Spectrum of Transferred Energy May Be Continuous Making 'Surviving' Photons Appearing among Final State Particles Spectrum-less. While Photons Radiated by Atomic Electrons, when Stepping down to Ground or Lower Levels, Have No Anything Nearby within Short SA Distances to Hit. They Fly Out without Scattering. Thus They Nicely Preserve Atomic Spectrum. In SA Physics, Photons Produced in DS PR Processes, Could Hardly Become Simple Messengers. But They Might Be ***Sophisticated*** and ***Subtle*** **Messengers** Like Other FSPs. Original Compton Scattering is 2 Body Collision. But SA QS or SA Compton Scattering is Usually MB Interaction Process. For Example, In Radiative Decay Channel of J/ψ ($J/\psi \to \eta_c + \gamma$), Photon is Born Right within SA Small Space HMC $u\bar{u}$ and DS-Electron/DS-Positron are Crowded in. MB-Scattering Lead to Difficulty of Quantitative Calculation.

- Many Composite Particles Live too Short to Leave Tracks of Visible Lengths. Stern-Gerlach Method is Unable to Measure Their Spin MMs. Even if Some Composite Particles Leave Long Enough Tracks and Have Accurate Empirical Values of Spin MMs, Such Values are Unable to Determine Their 'Spin' and TAMs of Their Constituent Elementary Particles. **Spin Misconception & TAM Disability** are Clearly Explained Below.
- Stern-Gerlach Experiment Directly Measures Spin e.m. MM of Electron, Not Spin. To Know Spin from Empirical Value of Spin MM, Connection Formula is Needed. We're Unable to Verify Connection Formulas and to Determine Value of g-Factor, unless Spin is Determined via Completely Different Combination of Theoretical and Empirical Methods. Such Needed Combination is Shown below:
- Good Agreement btw Solution of Dirac Eqn for Atomic Electron in Proton's Coulomb Field and Experimental Spectrum of Hydrogen Atom Can Be Called **Dirac Success** that Indirectly Convinces us of Theoretical Value of Electron's Spin, Deduced from Abstract Quantum Operator-Representations of Spin/OAM and Conservation of TAM. It Proved ½ to Be Electron's Spin, So-Called **Dirac Spin** for Dirac Eqn Contains Spin-Orbital Coupling Term but Schrödinger Eqn Contains No Such Term. Empirical Value of Electron's Spin MM Measured in Stern-Gerlach Experiment and Dirac Success/Spin before **R**adiative **C**orrections Prove Valid Formula of Quantum Connection that Yields Electron's Spin MM in

Same Order of Magnitude as Predicted by Classical Connection with Dirac Spin Inputting. They Differ by a Factor 2. **RC**s (**R**adiative **C**orrections) Further Modified Dirac Spin-Orbital Coupling, Giving More Accurate Theoretical g-Factor. But This Triumph's Irrelevant to whether such Formulas of Connection Remain Right or at Least Approximately Right in SA Physics as SA Composite Particles such as Hadrons are Studied.

Composite Particle Does Not Spin and its J-Label as TAM of All its Constituent Particles Empirically Immeasurable and Theoretically Unknown. But Electron's Spin Story Provides Outstanding Example How Physicists Confirm its Spin and Spin Value 1/2. The Criterion is Called **S1/2abc**.

< S1/2abc >

Three Things, Called **S½abc**, are Needed to Determine Spin of Electron as an Elementary Particle:

a. TAM Conservation and Its Quantum Operator-Representation
b. Solution of Dirac Eqn, which Works for Fermions Only, for Atomic Electron
c. Experimental Spectrum of Atom in Good Agreement with First 2 Things

MTH (**M**uon-**T**au **H**ypothesis) Claims That Their Spins Equal ½ and They Not Composite Particles. Since Composite Particles Do Not Spin, S½abc Become Only Criterion for any Particle to Be Elementary and to Have ½ Spin. As One Thinks about Unstable Muons and Taus, He/She Can Recognize How Difficult It is to Use S½abc to Prove or Disprove MTH. Tau's Lifetime Too Short to Leave Any Track and Form Atom with Electron Replaced by Tau. Muon's Lifetime is 'Long' enough to Leave Long Track But Not 'Long' enough to Form Such a Muonic Atom that Lasts Long enough to Allow Experimenters to Record its Light Spectrum as Accurately and Reliably as They Record Light Spectrum of Hydrogen.

- Empirical Disability to Measure TAM of All FSPs in Muon-Decay Events and in Other Unstable SA Particles' Decay Events Makes MTH absolutely Not Provable until OAM of each FSP will be Measured in 200 to 800 Years.
- Dirac Success and Dirac Spin for Electron are Not Only Based on But Also Proving **UGH** (**U**hlenbeck-**G**oudsmit **H**ypothesis) and Importance of Genius Pauli Matrices.
- UGH Further Proved by Numerical Success of QED with **RC** (**R**adiative **C**orrection). RC Maximizes Dirac Success and UGH's Success. However, If S½abc Cannot Be Done for Muon and Tau, Dirac Success and Dirac Spin Cannot Be Repeated for Muon and Tau Even if One Believes Them to Be Elementary.
In Spite of Blocked Way to Prove or Disprove MTH via QM and RC Approach due to Short Lifetimes of Muon and Tau, DS of Muon and Tau Shows its Power to Explain All Facts about Them.
- MB-Systems/**2B**-Systems of Composite SA Particles with Unsolvable/Unsolved **PE**s (**P**otential **E**nergies) are Unable to Repeat Dirac Success in SA Physics.
- Classical Connection Formula that Connects Empirical Spin MM of Electron and Its Spin is Deduced by Assuming: Every Part of Electron Revolves about Same

Axis, with Same Angular Velocity, and with Same Charge-Mass Ratio. Such Hypothesis Does Not Match Reality of Composite Particles.

- Spin Values of Composite Particles in SM Can Be Treated as Statistics Labels, Not Spin, Not Even Necessarily **TAMs**.

< **Rutherford** with MF and **Sping Crisis** with OAM/TAM Disability >

- Even if Rutherford and Mott Type Experiment and Method Can Determine 'Impact Parameter' and Estimate Distance or Even Size of Nucleus's 'Surface' (such as $r < 2.7 \cdot 10^{-12} cm$) where Deviation from Inverse Potential Began by Measuring Cross Section and 4-Momentum of Every Initial or Final State Particle, 'Turning Impact Parameter' of Deviation Never Means Size of Nucleus or Nucleon. Formula for MF Explicitly Proves Deviation at Short Distance, No Matter if Incident Particle Has Hit Surface of Struck Particle or Not.

- Attempts of Experimenters to Probe Deeper into Nuclei or Nucleons Have Always Caused Particle Productions in Deep Inelastic Scattering/Collision. This is Empirical Foundation of Inelasticity of SA Vacuum. Information of Impact Parameters and Nucleon-Sizes Hidden in Deep Inelastic Scatterings with Said Productions of MB-Systems.
- Even Very Small Deviation from MCtr-MCtr Collision Means Large Initial OAM in So-Called HE *Head-On* Collision. Inelastic & Deep Inelastic Collisions Absorb Part of Collision Energy, Linear & Angular Momenta, and then Release Them in Form of PAPs. Thus, '***Spin Crisis***' Emerges as Uncontrollable and Unknown Deviations from Exact MCtr-MCtr Collisions Ignored. Head-On Collisions Have Been Going On in Largest Accelerators Repeatedly All Over the World for Forty Years. They Usually Not Exact MCtr-MCtr Collisions.

- J-Label in SM without DS as TAM of all Constituent Elementary Particles Not Real TAM. Accurate Value of OAM of Each Constituent Elementary Particle, Hence TAM or J-Value of Composite Particle, Empirically Immeasurable and Theoretically Unknown. Example: In 1950's Wu's *Simultaneous* Measurement of Super Low *Temperature* and *Angle* Distribution of Particles Emitted in β-Decay Stunning. But Empirical Disability to Measure OAM of 2B-System formed by Emitted $e^{-}\bar{\nu}_e$ Still there 50 Years Later. Such Disability Can Be Easily Recognized if Experimenters Try to Measure Deviation of beta Particle's Momentum Line from Mass Center of a Single Isolated Free Neutron from which that beta Particle Flies away. When Linear Momentum of beta Particle is 1 MeV/c and Said Deviation is $10^{-11} cm$, then It Had Initial OAM inside that Neutron Approximately Equal to $\hbar/2$ before Decay Took Place. In Fact,

(H4) $$\hbar/2 = 1.93 \cdot 10^{-11} cm \cdot m_e c = 0.986 \cdot 10^{-11} cm \cdot Mev/c$$

- Empirical Disability to Measure OAM and TAM of FSPs Puts a Question Mark after Each Decaying SA Particle's J-Label no matter It's Postulated as Composite or Elementary in a Model. Then We Don't Know which's Radial Excitation and which's Non-Radial Excitation as We Read those Empirical Mass Spectra.

< **Spin**-Related **Statistics** >

SM without DS and DS that Reserves Some Components of SM Have **Same Statistics**: Composite **Boson** Contains **Even** Number of Elementary Fermions and Composite **Fermion** Contains **Odd** Number of Elementary Fermions, No Matter DS Considered or Ignored. Every Composite Quark and Every Composite Lepton Contains Odd Number of Elementary Fermions, according to DS. Exchanging Such 2 Identical Composite Particles Means Exchanging Odd Number of Spin ½ Identical Elementary Fermions. Wave Function Has to Change its Sign. This is Why Nucleons are Fermions No Matter if They Spin or Not and What are Their TAMs. Besides, Mesons are Bosons No Matter whether They Spin or Not and What are Their TAMs. But Still, Vector Mesons May Be Excited States of Corresponding Mesons with Zero J-Label for They Heavier and Shorter-Lived. All Non-Radial Excitations Due to Spin-Flipping and/or Larger Orbital Quantum Number l . Although l Not Empirically Measurable and Its Value Theoretically Unknown, Larger Norm of Vector Sum of Spin Vectors of Final State Particles Might Be Statistically True when any Composite Particle with Higher J-Label Decays. Composite Particles' J-Labels in SM without DS May Qualitatively or Half-Quantitatively Right as Used to Estimate TAM of Constituent Particles.

To Be Fermions, Nucleons Do Not Have to Have Spin ½ Nor Any Spin. Also, TAM of All Constituent Particles in a Composite Fermion Does not Have to be ½ or k/2 (k=1,3,5,…). To Be Bosons, Composite Mesons Do Not Have to Spin or Have ANY Spin Values. TAM of All Constituent Particles in a Composite Boson Does Not Have to be a Whole Number such as 0, 1, 2, or 3.

- Higher J-Labels of Composite Particles May Have Some Empirical Meaning if Experimenters Can Identify Statistically Larger Norm of Vector Sum of Final State Particles' Spin Vectors in Decay Events of Such Composite Particles.
- Spin Direction of Composite Particle is Correct Direction Label **iff** (if and only if) It's Same Direction as TAM of All its Constituent Elementary Fermions. Now that TAM of Composite Particle Immeasurable/Unknown, Spin Direction of any Composite Particle Not Meaningful, Even Serving as Direction of TAM. Only Total MM or Net MM Vector of Composite Particle with Long Enough Lifetime Can Be Labeled because Both Its Direction and Magnitude (Norm) Measurable. If We Keep Calling it Spin MM, then It Just Refers to MM Measured without Local Orbital Motion of Whole Composite Particle (though its Constituent Particles Have Local Orbital Motions inside Composite Particle). For Composite Particle, Term 'Spin MM' Bears Stamp of Old Spin Misconception. But its Empirical Meaning of No Confusion. We All Know What MMs Experimenters Have Measured in All Experiments of Stern-Gerlach Type.
- Spin MM and Spin Vector of Free Electron Can Be Measured and Determined Empirically and Theoretically. One Can Put Arrow Label to Indicate its Spin Direction. But Once It Becomes Constituent inside a Composite Particle, its Spin Direction and Orbital Angular Momentum Turn to Be Unknown. Still, a Few Things Can Help to Make Sample Spin Arrangement inside Composite Particles:

1. Permanent/Temporary MF-Binding Demand Right Spin Directions to Yield Attractive Net MFs.

2. Spin Directions Also Determine Location Regions of d-Electrons, which Further Determine How Fast Composite Particles Decay. Thus, Lifetime of a Composite Particle Provides Clue for Possible Sample Arrangement for Directions of d-Electrons' Spin Vectors.
3. Preferred Spin Direction of Final State Particles in EMF.

- Above Three Things will Lead to Sample Arrangement (R1 E_0, R2 E_1, R1 E_2) for Bound d-Electron's Spin Direction, States of Ground and Excitation.
- **MMD** (**M**any-Body/**M**ag-Force **D**ifficulties) and **STD** Yield and Reveal More Unknowns. Unknowns/Spin Misconception/TAM Disability Cause Uncertainty as One Tries to Explain Mass Spectrum of Hadrons, Making SA Physics a 'Hot Pot'. No One Can Touch It without Getting 'Scald'. No Theory in SA Physics Could Be Developed in Process without Speculative Component and without Making Any Mistake. Generally and Basically, Basic Theory of Fundamental Physics without Mistakes and without Speculative Components Did Not Exist, Does Not Exist, and will Not Exist. SA Physics before and after Quark Model, Has Been So and Most Challenging. We are Eagerly Looking for **Better** Theory for We Want to Explain Unexplained Empirical Facts Badly.
- Some Mistakes in any Better Theory May Be Invisible, Uncertain, Unclear, but will Be Seen in Future by Some One. This is Our Scientific Religion, Leading to No Superstition but Endless Effort of Mankind to Pursue Ultimate Truth of Physical World, whether or Not One Believes Such World was Created by God. Mistakes in Mistaken Theories are Already Seen by More and More Scientists. The True Merits of Each Basic Theory also Seen after Mistakes are Recognized. One Mistake Author Made is that He Had Ignored the Reality of MF and its Crucial and Big Effects in Dynamic Foundation of Composite Structure of SA Particles for 30 years until Recent Three Years. As a Consequence, Another Mistake We All Made was that We Had Believed an Assumption for too Long. That Assumption Claims Zero Possibility for Electrons to Be Contained in SA Composite Particles.

CME (**C**omposition **M**ass **Eqn**) and **TMET**

- TME Crucial in Physics after MFs Enter Fundamental Dynamics.
- Each Composite Particle is 2B or MB-System. Its PE U, Its Mass m, and Masses of Its Constituent Particles are Related in Following CME:

(H5) $$mc^2 = \Sigma E_i + U = \Sigma m_i c^2 + \Sigma E_{ki} + U = \Sigma m_i c^2 + TME\ ,$$

$$E_i = m_i \lambda_i \gamma_i c^2\ ,\ E_{ki} = m_i c^2 (\lambda_i \gamma_i - 1)\ ,\ TME = \Sigma E_{ki} + U$$

If TME is ***Negative***, We Have **CMI** (**C**omposition **M**ass **I**nequality):

(H5a) $$m < or << \Sigma m_i$$

If TME is ***Positive***, We Have **RCMI** (**R**eversed **CMI**):

(H5b) $$m > or >> \Sigma m_i$$

TME Issue is of Vital Importance in SA Physics.

Remark: Both Classical and Quantum MB-Systems Not Solvable and Quantum 2B-System with MF Involved Not Solved and May Be Unsolvable. Quantitative Explanation of Mass Spectrum Can't Be Made. In Qualitative/Half-Quantitative Explanations, **CI** (**C**lassical **I**mitation) is the Only Theoretical Approach to Find Out TME of 2B-System such as HMC $u\bar{u}$ and LMC e^+e^-. DS Tells Every Quark/Antiquark Contains $u\,/\,\bar{u}$ respectively. Then Every Meson Contains One HMC $u\bar{u}$. Every Muon and Every Tau Contains One LMC e^+e^- with Very Large Positive TME. (see TMET, pion mass eqn with OSTP and muon mass eqn without OSTP later) Empirical Facts Shown in Masses, Stability, Decay Modes of Composite SA Particles Touchstones of TME Features for All SA Composite Particles.

< TMET >

We Have Described TMET and TME-r Curve before. Detailed Statements're Given on Page 71 in Topics < Three Parts of TMET with Proof >.
Solution of Quantum Eqs for Atomic Electron in Coulomb Field Shows: Higher Excitation Means Larger ME and TME and Larger Binding Size. Classical TME -r Curve Shows Monotone Increasing TME as r Increases if Binding is Pure Coulombian or Coulomb-type. Quantum Eqs in MF-Fields plus Coulomb or Coulomb-type Field Not Solved and May Be Unsolvable. But TMET Puts Light on and Shows Brand New Type of TME-r Curve for such *MF-Coulomb Bound System*: as r Increases from near-zero to 0TMED, TME-r Curve Drops from Super Large Positive Value to 0 and as r Increases from 0TMED to Critical Distance r_c, TME-r Curve Continues Dropping down from Zero to a Lowest Point Reaching Negative Minimum Value of TME, and then as r Continues Increasing after Critical Distance r_c, TME-r Curve Now Climbs up and Approaches r-Axis as r Approaches infinity. The Curve is below r-Axis btw r = 0TMED and any r Larger than 0TMED and is above r-Axis btw r = 0TMED and any r Shorter than 0TMED.
Classical TMET Implies **Smaller** Binding Size for ***Quantum*** MF-Bound 2B-System in **Higher** Quantum Level Once Binding Size Quite Less than 0TMED, **Larger** Binding Size for **Higher** Quantum Level Once Binding Size is Quite Larger than Critical Distance r_c. Near-r_c Binding May Have Small Difference in Energy btw Two Quantum Levels. Near-r_c Binding and its Role and Position in Mass Spectrum of HMC $u\bar{u}$ and Mesons is Open, Left for Future Study. The Excited Bound States where Smaller Binding Size Corresponds to Higher Level are Called ***Inward Excited States*** (**IES**s). Bound States with Binding Sizes Close to r_c are Called **NCB** (Near-Critical Bound) States. NCB States Have Small Energy Differences.

Note: "MF-Bound" Does Not Always Mean Only MF Participates in Binding. MF-Binding May Appear in SA World together with Other Types of Forces. MF

May Dominate Sometimes. Sometimes MF Plays Significant Role Even if It Does Not Dominate. Sometimes It Means Pure MF-Binding or MF-Binding Dominating other Types of Forces. Context Determines what Does It Mean by 'MF-Bound'.

< Examples of ***Negative*** TME >

Stable Proton Proves **Negative** TME for TME Conservation Forbids Proton's Constituents to Flee Bound State via **QT** (**Q**uantum **T**unnel) Only in Case TME is Negative. For any ***Unbound*** System, TME's ***Non-Negative***. Negative TME Proves HQMT via DS and CME with CMI: *Mass of u-Quark Larger than 1/3 of Proton's Mass*. DS Shows: d-Quark Contains an Electron Called d-Electron. Single d-Electron in Free Proton Bound and Very Light in Comparison with u-Quark so that its Existence Does Not Change Conclusion in HQMT.

At Atomic Distances, MF Acting on Each Atomic Electron Almost Vanishes. With Suppressed Effects of its Existence, Effects of Its Non-Spherosymmetry Also Almost Vanish. (see formulas of MF later) Thus, Atomic QM and Atomic QED with MF Omitted Could Be So Successful and S-States Exist in Atomic Physics.

Mass of π^0 with $u\bar{u}$ Content then Less than u-Mass. It Just Proves Negative TME of $u\bar{u}$ Bound System inside π^0, via CME (H5), Implying Inverse Square Law Binding Dominating Suppressed MF-Binding. DS and MF Nicely Explain This with OSTP: MF in (H12) and (H15) Depends on Both r and θ. (H12) Tells that MF's ***Radial*** Component Vanishes at

$$\theta = \theta_{zero1} \equiv \cos^{-1} 1/\sqrt{3} \text{ and } \theta = \theta_{zero2} \equiv \pi - \cos^{-1} 1/\sqrt{3} .$$

We Call Them **0R-MF Angles**. **OSTP** May Turn off or Suppress MF-Attraction as its Radial Component, Causing Domination of Coulomb-Type St Force and Negative TME Feature. SA Small Binding Size due to Heavy u-Mass and **DBR** (**D**irac **B**ohr **R**adius) Shorter than **SBR** (**S**chrödinger **B**ohr **R**adius). (see formula of general DBR later)

Necessary Condition for Composite Particle to Be *Stable* is Bound State of Each of its Constituent Particles. Original Examples and Concept of Bound States Can Be Seen in Newton's Theory of 1B-System with One Heavy Body Moving in Central Gravitational Field. Circular Motions Simplest Classical Examples of Bound States with ME of Said 1 Body Proved Long Time Ago to Be Negative at All Times. Electric Charge Moving/Bound in Central Coulomb Field was another Classical Example. CED Proves its All Time Negative ME for Every Circular Motion. Since Source Body Producing Central Gravitational or Central Coulomb Field is at Rest, ME for 1B-System Same as TME of 2B-System. Then Connection btw **Bound** State & **Negative** TME Established. QM further Proves that Atomic Electron Bound in Hydrogen Has Negative ME. Since Proton is too Heavy to Move in CMF with Enough KE, Hydrogen Atom as Bound 2B-System is Proved to Have Negative TME.

Connection btw **Bound** State & **Negative** TME for Unsolvable MB-System Can Only Be Made as Postulate. But 1B-Treatment of Solar System and of Atoms of 1<Z<3 is So Accurate Approximation and Yields Negative TME, We

Have very Good Reason to Postulate Negative TME for Bound MB-System. But We Need More than Just Negative TME to Explain Bizarre **Unique** Example of **Stable** Bound **Composite** System in SA Physics, which is Stable Free Proton.

Negative TME is Necessary for Proton to Be Stable but Not Sufficient. If Proton p with Negative TME Emitted an Electron and Became a Neutral Particle p' while the TME of p' Reduced to Compensate Electron's KE such that TME of Entire System { p' e^- } were Maintaining Same Negative Value as p Had before So That Negative TME & Conservation of TME Would Remain True. To Fully Explain Stable Free Protons, Especially to Explain Unstable Neutrons with Continuous Spectrum of beta-Particles, **1BEx** and **MB-Theorem** Needed Badly.

< **MBT (M**any **B**ody **T**heorem) >

1BEx (1-Body Extraction): Each MB-System is 1 Body in Resultant Force Field Produced by Others in That MB-System. 1BEx Proves **MB-Theorem**: **ME** of Each Individual Particle in MB-System ***Time-Varying*** and ***Non-Discrete*** as long as Resultant Force Field Time-Varying. (See unsolvable eqs of 1BEx on page 61)

Proof: Any Stationary Wave Function with Constant ME E Only Satisfies Quantum Eqs with Time-Independent PE U. Time-Varying Force Fields Do Not Yield any Time-Independent PE U. MB-Theorem is then Proved Rigorously.

Note: In SA Composite Particles, Resultant Force Fields Time-Varying because Others in Rapid Motions are Nearby. Atomic Electrons far away from Nuclei so that Quarks & DS-Electrons inside Nuclei Do Not Produce Time-Varying Force Fields in Most Probable Regions where Atomic Electrons Stay. Quarks and DS Electrons inside Nuclei with Rapid Orbital and Spin Motions Produce Coulomb Fields and MF Fields. Local Orbital Motion of Charge Produces Force Field. Its Effect outside like what Caused by MM, called Orbital MM. Charged Particle in Spin Motion Makes Spin MM. Rapid Local Orbital Motion Produces Time-Varying Force Fields at Places Not Far Away from Orbit. At Places Far Away from Orbit, Force Fields, both Coulomb and MF Fields, Become Stationary.

Insolvability of 1BEx Means: Accurate Math Expression of Time-Varying ME Cannot Be Obtained. But MB-Theorem's Been Proved Rigorously.

MB-Theorem Explains Bound d-Electron and Stable d-Quark in Proton, One Quasi-Bound d-Electron and One Unstable d-Quark in Neutron, and Continuous Energy Spectrum of beta-Particles. Individual Particle in MB-System Has Time-Dependent ME with its Value Changing Continuously. It's All-Time Negative if Permanently Bound as long as No Knocking with Enough Energy Received. Some Individual Particles in MB-Systems Quasi-Bound. Their ME Can Change from Negative to Positive so that QT-Escaping that Makes It Flee into Free Region Outside May Happen. Inside Neutron, Wd-e's ME May Change from Negative to Positive, Triggering beta-Decay. (Wd-e Refers to One of the two d-Electrons Contained in two Composite d-Quarks in Neutron) Another is Sd-e. Pauli's Principle Forbids them in Same State. Wd-e Further away from Q-Core and Bound Weaklier. Wd-e/Sd-e Represent 2 States of 2 Identical Particles. SDL

< **MBP** (**M**any-**B**ody **P**ostulate) >

MBP: Isolated MB-System's TME ***Constant*** (***Time-Independent***) Determined by ICs. **Bound** Quantum MB-System Has **Discrete** Spectrum of TME with Constant Values Determined by ICs. A constituent Particle's Time Varying ME Can Be a Continuous Function of Time t .

Overwhelming Evidences: Empirical Masses of Each Sort of Composite Particles in Ground & Excited States Produced in Collisions form Discrete Spectrum and Beta-Particles' Energies Indeed Form Continuous Spectrum.

< Example of ***Positive*** TME >

TMET Shows Existence of MF-***Bound*** States with ***Positive*** TME. Conservation Law of TME Allows Change from Bound States with Positive TME to Free States with Same Positive TME via QT. LMC e^+e^- in Muon is an Example. Positive TME Feature of MF-Binding Opens QTC for Bound State to Decay, Making MF-Bound System with Positive TME Unstable.

In Muon-Decay, QTC is Dominated by **KC** (**K**inetic **C**hannel) for LMC e^+e^- in Muon to Annihilate. DS Nicely Explain All Decay Modes of Pions and Muons with Different Branch Ratios. SDL

Insolvability in SA Physics and **CI** (**C**lassical **I**mitation)

- **SA** Dynamics Has Many Unsolved Problems such as MB-Problems and **1BEq** (**1**-**B**ody **Eq**uivalence) for 2B-System $u\bar{u}$ as **MF**s and Other **Mag-Forces** btw Charges with High Speeds Involved. Right Now We're Not Sure if HMC $u\bar{u}$ and LMC e^+e^- Quantum 2B-Systems with MFs Solvable. Other SA Composite Systems Surely Unsolvable MB-Systems. We Cannot Get Accurate Quantitative Solutions of Quantum Wave Eqs in SA Physics. Impossible to Quantitatively Explain Mass Spectrum, MMs, Lifetimes, and Branch Ratios of SA Composite Particles.

  Remark: Many Other SA Basic Facts than Mass Spectrum, MMs, and Lifetimes of SA Composite Particles Can Be Explained via DS Quantitatively, Even Exactly. For Example, DS Even Explains Exactly why β^+ -Decay Forbidden if Available Energy is Less than $2m_ec^2$ and why e-Capture Not Forbidden Even if Available Energy Less than $2m_ec^2$. DS Explains why One d-Quark in Neutron Dooms to Decay while d-Quark in any Free Proton Never Found to Decay. DS Also Explains why One u-Quark in Proton Can Be Changed to d-Quark in β^+ -Decay or Electron-Capture while No any u-Quark in Free Isolated Proton Can Ever Be Changed to d-Quark so that Such Free Isolated Proton Can Be Changed to Neutron. Such Power to Explain Oldest Experimental Facts Never Explained by SM without DS Proves Once and for All that SM Modified by DS Better than SM without DS.

  EqM (Equal-Mass) 2B-System HMC $u\bar{u}$ and LMC e^+e^- with Perturbations from Other Particles nearby Ignored, is Solvable ***Classically*** via CI for Each Individual Body/Particle at Least under Some Special IC such as Circular Motion about Mass Center, Even if Non-Spherosymmetric MF is Involved.

- Before Seeing Details of DS/CI/TMET/Mass-Eqs for π^0 and $\mu^\pm$ /Simultaneous Explanations of Nucleons' Masses, π^0 -Mass, $\pi^\pm$ -Mass, MMs of Nucleons, Muon-Mass, and Tau-Mass, We Now Reveal an Example of Perfect Match btw QM and CI. As 1st Step, We Deduce Formula for General DBR and then We'll Show and Prove Example of Perfect Match btw QM and CI.

< General DBR >

Consider Dirac Eqn for Any Particle with Mass $m = k_m m_e$ ($0 < k_m < \infty$) and in any Coulomb-Type Force Field with IPE

$$U = -k_C e^2 / r = -k_C \alpha\hbar c / r = -\alpha_C \hbar c / r \quad (0 < k_C < 137,\ 0 < \alpha_C < 1)$$

(α is fine structure constant equal to 1/137.035999679). Solution of such Dirac Eqn Gives **Ground** State Wave Function and Energy, Written as

(H6)
$$\psi_0 = \begin{pmatrix} b_0 a^s r^s e^{-ar} \chi_0 / r \\ d_0 a^s r^s e^{-ar} \chi_0 / r \end{pmatrix}, \quad s = \sqrt{1-\alpha_C^2}\ , \quad E_0 = mc^2\sqrt{1-\alpha_C^2}\ ,$$

$$a = \frac{\sqrt{m^2c^4 - E_0^2}}{\hbar c} = \frac{mc^2\alpha_C}{\hbar c} = \frac{mc^2 k_C e^2}{\hbar^2 c^2} = \frac{mk_C e^2}{\hbar^2} = \frac{1}{r_b^{Sch}}\ ,$$

$$r_b^{Sch} = \frac{\hbar^2}{mk_C e^2} = \frac{r_b}{k_m k_C} \quad (r_b \equiv \frac{\hbar^2}{m_e e^2} = 0.529141 \cdot 10^{-8} cm\ ).$$

Note: Bohr Radius $r_b \equiv \hbar^2 / m_e e^2$ is Special SBR r_b^{Sch} for Atomic Electron with Special Settings: $m = m_e$, $k_C = 1$, and $\alpha_C = \alpha \approx 1/137$.

Ground State in Spherosymmetric Potential Field is S-State, Indicating Constant χ_0 . Then Normalization Gives

$$1 = (b_0^2 + d_0^2)\chi_0^2 a^{2s} \int_0^\infty 4\pi e^{-2ar} r^{2s} dr = \frac{4\pi s\Gamma(2s)}{2^{2s} a}(b_0^2 + d_0^2)\chi_0^2$$

Therefore,

$$\frac{1}{r_b^{Dirac}} = \overline{\left(\frac{1}{r}\right)}_{Dirac} = (b_0^2 + d_0^2)\chi_0^2 a^{2s} \int_0^\infty 4\pi e^{-2ar} r^{2s-1} dr =$$

$$= \frac{4\pi\Gamma(2s)(b_0^2 + d_0^2)\chi_0^2}{2^{2s}} = \frac{4\pi s\Gamma(2s)(b_0^2 + d_0^2)\chi_0^2}{2^{2s} a} \cdot \frac{a}{s} = \frac{a}{s}\ ,$$

With $a = 1/r_b^{Sch}$ and $r_b^{Sch} = r_b / k_m k_C$, Obtained in (H6), We Get

(H7)
$$r_b^{Dirac} = \frac{s}{a} = s r_b^{Sch} = \sqrt{1-\alpha_C^2}\, r_b^{Sch} = \frac{\sqrt{1-k_C^2\alpha^2}}{k_m k_C} r_b$$

One Can Easily Check that This DBR Also the Most Probable Radius.
For ***Atomic*** Electron in Hydrogen, $m = m_e$, $k_m = 1$, $k_C = 1$, $\alpha_C = \alpha$. Then

(H7a)
$$r_b^{Sch} = \frac{\hbar^2}{m k_C e^2} = \frac{\hbar^2}{m_e e^2} = r_b,$$

$$r_b^{Dirac} = \sqrt{1-\alpha^2}\, r_b.$$

For ***Atomic*** Electron in Hydrogen, DBR r_b^{Dirac} Almost Same as Bohr Radius r_b because α^2 too Small. But from General Formula $r_b^{Dirac} = \sqrt{1-\alpha_C^2}\, r_b^{Sch}$ in (H7) We See that for HMC $u\bar{u}$ in π^0, St Force with $\alpha_C^2 = \alpha_s^2 >> \alpha^2$ Makes DBR r_b^{Dirac} in (H7) Significantly Less or Much Less than $r_b^{Sch} = r_b / k_m k_C$. Please Note, this SBR for Bound u-Quark itself is Much Less than Bohr Radius r_b for Bound Atomic Electron in Hydrogen due to St Force with $k_C >> 1$ and Heavy u-Quark with $k_m >> 1$.

< Example of **Perfect Match** btw **QM** and **CI** >

Solution of Dirac Eqn for Electron in Coulomb Field with PE $U = -e^2/r$ Yields Ground State Energy

(H8)
$$E_0 = m_e c^2 \sqrt{1-\alpha^2} = m_e c^2 \gamma + U = m_e c^2 \bar{\gamma} + \bar{U}$$

Ground State Wave Function Gives Average Value of $1/r$ and $\bar{U}$:

(H9)
$$\frac{1}{r_b^{Dirac}} = \overline{\left(\frac{1}{r}\right)}_{Dirac} = \frac{1}{\sqrt{1-\alpha^2}\, r_b^{Sch}} = \frac{1}{\sqrt{1-\alpha^2}\, r_b}$$

$$\bar{U} = -e^2 / r_b^{Dirac} = -\frac{e^2}{\sqrt{1-\alpha^2}} \cdot \frac{m_e e^2}{\hbar^2} = -\frac{\alpha^2}{\sqrt{1-\alpha^2}} m_e c^2$$

(H8) and (H9) Lead to

(H9a)
$$\bar{\gamma} = \sqrt{1-\alpha^2} + \alpha^2 / \sqrt{1-\alpha^2} = 1/\sqrt{1-\alpha^2}$$

We Now Prove that $\bar{\gamma}$ in (H9a) is Exactly Same as What We Can Obtain from

Classical Mechanics for Circular Motion CI. For $\gamma < 10^3$ We Assume $\lambda = 1$ and $d\lambda / dt = 0$. Note, Quantum Ground State Energy in (H8) is also Obtained under 0-LV Approximation Condition $\lambda = 1$. Uniform Circular Motion's Speed is Constant, with $d\gamma / dt = 0$ and $\vec{u} \cdot d\vec{u} / dt = 0$. Centripetal Acceleration's Written as $\vec{a} = d\vec{u} / dt = -r^{-2} u^2 \vec{r}$, $a = u^2 / r$. One Can Check

(H10) $$f = e^2 / r^2 = m_e \gamma u^2 / r \Rightarrow \beta^2 = k\gamma^{-1},$$

$$k \equiv e^2 / m_e c^2 r \Rightarrow \gamma = (\sqrt{k^2 + 4} + k)/2$$

Setting $r = r_b^{Dirac} = \sqrt{1-\alpha^2} r_b$ Yields $k = \alpha^2 / \sqrt{1-\alpha^2}$, $\gamma = 1/\sqrt{1-\alpha^2}$. Such γ Drawn from Relativistic Classical Mechanics plus Setting $r = r_b^{Dirac}$ is Exactly Same as $\bar{\gamma}$ Drawn from (H8) and (H9), which is Obtained via Relativistic QM. This is Example of **Perfect Match** btw QM and CI, Called **PCQM** (**P**erfect **C**lassical-**Q**uantum **M**atch).

Note: PCQM for Excited States Open for Future Study. Orbital Excited States're Not S-States. May Be, Elliptical Classical Motions are Needed in CI to Achieve PCQM.

- PCQM is Obtained in 1-Body Problem Where Source Particle of Coulomb or Coulomb-Type Force Field is Treated as Rest Particle without Magnetic or Magnetic-like Field. Surely, Rest Charge Produces No Magnetic or Magnetic-like Field. HMC $u\bar{u}$ & LMC e^+e^- are EqM 2B Systems with Every Member Moving at Near-c Speed. We Don't Know Quantitative Effect of Magnetic Fields and of Possible Magnetic-like Fields and How to Get Perfect Match. Still, We've Been Inspired by above PCQM and Power of CI. It Immediately Yields TMET, Putting Lights on Crucial TME Issue, Leading to Simultaneous Half-Quantitative Explanations of Masses of Nucleons, Nucleons' MMs, Masses of Pions, Mass of Muon, and Mass of Tau.
- PCQM is ***Natural*** for QM Uses Exactly ***Same*** Formulas of Energy, Momentum, and $E-p$ Relation as Used in Classical Mechanics for Heavy Bodies.

Remark: Perfect Match Cannot Be Determined/Proved in SA Physics because No Theoretical Value of Binding Size r Can Be Used to Do Inputting due to Absence of Quantum Solution. Power of CI is that It Uses Rigorous Calculation in SR Mechanics with LV Omitted in 1st, 2nd, and 3rd GS and Leads to Mass Eqn for EqM 2B-System, Serving as Quantitative Connection-Formula to Connect 3 Corner-Stone Quantities: Mass of Composite Particle (Mass of HMC $u\bar{u}$ in Pion or LMC e^+e^- in Muon) as EqM 2B-System, Mass of Constituent Particle (Mass of u-Quark or Electron), and Radius of Circular Motion in CI Serving as Binding Size to Imitate **MPR** (**M**ost **P**robable **R**egion) of Relevant Quantum 2B-System. π^0 is Treated as EqM 2B-System $u\bar{u}$ Even if it Contains d-Leptons for d-Leptons Make Little Contributions to Pion-Mass. But as We Pointed out before that to Reconcile Small Mass of Pion and Heavy Mass of u-Quark, HMC $u\bar{u}$ 2B-System Must Have very Negative TME that Needs OSTP to Suppress MF Greatly. Then

It's Found that HMC $u\bar{u}$ Has Very Small Possible Sizes and Effectively Appears as Single Particle with Net MM. It Cannot Bind 2 d-Electrons to Form Particle $d\bar{d}$. Thus, π^0 is EqM 2B-System $u\bar{u}$ without Mixture and DSLs if HMC $u\bar{u}$ in Zero-Pion is in OSTP State. Muon Also Treated as EqM 2B-System (LMC e^+e^-) for Other Constituent Stable Leptons in Muon Make Little Contributions to Muon-Mass while Heavy LMC e^+e^- Contributes Most to Muon-Mass. Binding Size r Appears in Mass Eqn. Once One Inputs Muon-Mass into Muon-Mass Eqn, Binding Size r is Deduced from Muon-Mass Eqn. But in π^0-Mass Eqn, We Must Input both π^0-Mass & u-Mass to Get Binding Size r. Heavy u-Mass Required by Small MMs of Nucleons then is Giving Simultaneous Explanation to both π^0-Mass and Nucleons' MMs. SDL

DS of **Quarks** with Bound and/or Quasi-Bound **Stable Leptons**

- DS Reduces Sorts of Elementary Particles to a Few: $e^+e^-u\bar{u}\nu_e\bar{\nu}_e\nu_\mu\bar{\nu}_\mu\nu_\tau\bar{\nu}_\tau$, γ, Z, and g (Gluon). Structure of Muon/Tau Neutrinos Open. u-Qs Have 3 Colors. Also, Their 3 Antiquarks Have three Different Colors with Their Natural Color identifications depending on definition of 'anti-color'.
- d-Quarks and All ***Unstable*** Particles (s, c, b, t Quarks/Muons/Taus) ***Composite***.
- Basic DS with MFs Seen before in "Brief Introduction to DS"
- All **SA Leptons** Can Be Called **DS-Leptons** or **DSLs**.
- 1st **Excitation** Turns d-Q into s-Q. Higher Excitation Turns s-Q into b-Quark.
- c-Quark Contains $ue^-\bar{\nu}_e e^+\nu_e$ in Excitation. Such DS Naturally Explains Decay Modes Involving c-Quark. Both e^+e^- Can Be Bound to u-Q if in Right Zone Locations at u-Q and with Right Spin-Correlations. SDL

Now We List Some Facts We Can Explain in Terms of DS and MFs.

< Experimental **Facts DS** with **MF**s Can Explain >

- DS Can Explain Many Details of **Lifetimes** Half-Quantitatively or Qualitatively. DS Simultaneously Explains *Unstable Neutrons* with Very Slow Decay (Bizarre Uniqueness of Long Life among All Unstable SA Particles) via ENM, *Stable Free Protons*, All Stable Nuclei on *Stability Island* with One or More Neutrons (Even More than Protons), Nucleons' *Anomalous MMs*, Many *Lifetime Details* such as Much Longer Lifetime of Charged Pion than that of Neutral Pion, Longer Lifetime of Charged Pions than Charged Kaons though Pion and Kaon with Same Charge Have Exactly Same Elementary Constituents (s-Quark Contains Same DS-Content as d-Quark but in Excitation with d-Electron in Different ZL in View of DS so They May Decay into Same Set of Final State Particles), Life-Inequalities $1.1\mu s > 1.42 \cdot 10^{-7}s > 1.244 \cdot 10^{-10}s$ for Ps-2s-State, *o*-Ps, and *p*-Ps, Longer Lifetime of Muon than Positronium Ps, All Mesons of **DS-Asymmetry** and J=0 or without J-Label (Except DS-Asymmetric B*-Mesons of No J-Label and Unmeasured Lifetimes), Such as Charged Pions and All Charged as well as Neutral K/D/B Mesons Living Much **Longer** than **DS-Symmetric** (Hence

Neutral) Mesons of Label J=0 ($\pi^0, \eta, \eta', \eta_c$), Always Very Short Lifetimes of Mesons of J=1 Label, Such as $\rho^{\pm}, \rho^0, \omega, \phi, J/\psi, \Upsilon$, All K*- & D*-Mesons, Life Difference btw So-Called K_S^0 and K_L^0, Bizarrely Having Same Quark-Content as well as Same DS-Content while Decaying into Same Set of FSPs.

Note: According to PDG (Particle Data Group) Particle Summary Posted on World Web, Lifetimes of η_b and B*-Mesons Not Measured, while J-Label of B-Mesons ($B^{\pm}$, B^0, B_s^0, $B_c^{\pm}$) and Corresponding B*-Mesons Not Obtained.
DS itself Makes It Clear that All DS-Symmetric Mesons are Neutral without any Exception. It's Also Clear that Some Neutral Mesons such as Neutral K, D, B Mesons are DS-Asymmetric while All Charged Mesons are DS-Asymmetric.

- DS Exactly Explains Why β^+-Decay ***Forbidden*** if Available Energy is Less than $2m_e c^2$, and Why We See Stable u-Quarks in All Free Nucleons.
- DS Exactly Explains Why e^--Capture ***Not Forbidden*** Even if Available Energy is Less than $2m_e c^2$.
- DS Explains ***Continuous*** Energy Spectrum & ***Preferred*** Spin Direction of β-Particles, Parallel to EMF with Spin MM Bizarrely Opposite EMF's Direction.

Remark: If *Handedness* in *SM* Explains R-Handed $\bar{\nu}_e$ and L-Handed Electron in beta-Decay while *Parity Violation* Explains Existence of Preferred Moving Direction of beta Particles, then DS with MF-Dynamics & MF Spin-Correlation Further Explains Why beta-Particles' Preferred Moving Direction Must Be Opposite to Direction of EMF. If beta Particles' Preferred Moving Direction were Same as EMF with Opposite Spin Direction and Correspondently Anti-Neutrinos' both Moving and Spin Directions were Reversed, Parity Violation and Handedness in SM Would Remain True. DS Helps to Explain such Bizarre Phenomenon. Namely, to Satisfy Parity Violation and Handedness in SM, Spin Difference btw Parent and Daughter Nuclei Can be Either Positive Or Negative. If Ignoring Dynamical Details, We Unable to Fully Explain Experimental Facts. Parity Violation, Handedness in SM, and DS with MF Join Forces to finish Breakthrough. At Last, Every Detail Seen in beta-Decay Experiment Explained.

- DS Undoes Change from One Elementary Particle to Another One of Different Kind. All Particle That Decay are Composite. Every Elementary Particle Stable (Not always Vice Versa). Every Stable SA Particle is Elementary except Proton. Elementary Particles Can Only Be Created in PAPs and Disappear/Annihilate in PAPs. All Events Naturally Preserve All Physical Numbers Every Elementary Particle Carries. DS Immediately Reveals that All Physical Processes in HE Non-Elastic Collisions and in SA Decays belong to Only 4 Types of Processes, So-Called 4TDSP: **DSE** or **DSDE**, **PAPP** or **PAPA**, **PQTR**, and **PR** Followed by SA Photoelectric Encounters, That Enable DS to Explain Not Only Hundreds of Kinds of Middle Stage Particles and Thousands of Decay Modes, but also Ironic FQA Fact in both Two Possible Cases:

1. Collisions in Our Accelerators and in Cosmic Showers Do Knock Out u-Quarks,

Even Millions of Times Every Minute.

2. Quarks are Permanently Bound and Confined for Some Reason Even if HE Particles in Prime Cosmic Rays Hit Nucleons in Air with Energies Larger than $10^5 \sim 10^{10}$ GeV.

- Largest **KkE**s (**K**nock **E**nergies) in Primary Cosmic Rays Far beyond Reach of Man Made Accelerators in Coming 300 Years, Making Confinement Scenario without NLVUC Impossible. DS Able to Explain FQA in both 2 Possible Cases:

1. Quark Has Never Been Knocked Out due to Existence of NLVUC that is Lower than **QKoE** (**Q**uark-**K**nock**o**ut **E**nergy).
2. Quarks are Knocked Out Millions of Times Every Minute because QKoE's below or Far below Energy Scales of Today's Largest Accelerators and also below NLVUC.

- We Call 1st Case above **QC** and 2nd Case **QNC** (**Q**uark **N**on-**C**onfinement). At this Moment, Author Not Clear which's Right. But One Thing is Clear Now: DS Able to Explain Ironic FQA Fact in both Two Cases.

DS-Explanation of Ironic **FQA** Fact in Case 1**: QNC** with **AVI**:

- **PAPP** is Evidence of **SAVI** (**SA** **V**acuum **I**nelasticity) or Simply **VI** (**V**acuum **I**nelasticity). SA Vacuum Place where PAPs Produced. In Order for HE Incident Particle to Hit Bound u-Quark Hardly, Very Large Amount of Energy Must Be Transferred to that u-Quark. **AVI** (**A**bsolute **VI**) Means that PAPP Has Absolute Priority to Consume Impact Energies, if Larger than Rest Energies of Bound or Free PAPs, of Incident Particles, in All HE Collisions Seen in Cosmic Shower Events and Large Accelerators in Past 90 Years, in the Sense that PAPP Must Occur Right before any Knockout Event. Just because of **4TDSP**, Free Quark, as Naked Quark, Never Appears among FSPs. Right before a u-Quark Knocked out from Hadron, if Ever Happens, a $u\bar{u}$ Pair Production Must Occur, Ready to QT-Relocate Somewhere Nearby or Annihilate. Knocked Out u-Quark, if any, would Leave a Hole in Hadron while Immediately Capture $\bar{u}$ from Produced $u\bar{u}$ with Hole Left behind Being Filled by u-Quark in that Produced PAP $u\bar{u}$. QT in SA Spaces Can Easily Make Relocating and/or Repairing Happen. DS and SAVI Naturally Explain FQA Fact in All Events, No Matter It Means **C**onfinement or Not, No Matter All Quarks are Deeply or Non-Deeply Bound, and No Matter **QKoE** Below or Higher than Energy Scale of Today's Largest Accelerators. In ***Non-Deep*** Binding Model like this, u-Quark Knockout Events Always Preceded by PAP Productions Making Naked u-Quark Impossible to Appear among Final State Particles Even after Being Knocked Out from Original Hadron.
- One Can Use Beam of HE Electrons to Bombard Rest or Low-Speed Electrons. Any Clean Record Showing Non-Existence of Single Elastic Collision Event, in which Large Amount of Energy Transferred from HE Electron to Low-Speed Electron while Particle Production Did Not Happen, May Serve as Evidence to Validate AVI. If Elastic Collision with Large Amount of Energy Transferred but without Particle Production Ever Exists, It May Serve as Evidence to Invalidate AVI iff Transferred Energy Has No Upper Limit. Large-Angle Elastic Scattering Events Not Eligible to Test AVI if Energy Does Not Change a Lot No Matter

How Largely Linear Momentum Changes.

DS-Explanation of Ironic **FQA** Fact in Case 2**: QC** with **NLVUC**:

If u-Quarks ***Permanently*** Confined in Sense that u-Quarks Cannot Be Knocked Out from any Hadron but QKoE ***Below*** Impact Energies in Cosmic Showers or Even below Impact Energies Produced in Largest Accelerators, then there May Be Upper Limit of Energy Transferred in All ***Elastic Parts*** of HE Collisions. We Call That Upper Limit **NLVUC**. An Elastic Part in Collision Refers to Unbound Particle Receiving KkE to Gain KE or Bound Particle Receiving KkE to Jump to Higher Quantum Level (Including to Be Knocked Out). Experimenters May Check to See if there is such Upper Limit of Energies Transferred in All Elastic Parts of Collisions Even if Too Many Particle Production Sub-Events Happen in Collisions. We Expect Experimenters to Check Collision Data to Make NLVUC Issue Crystal Clear.

- Confirmed Existence of ***Heaviest*** Excitation States of Baryon and Mesons Far below Energy Scales of Largest Accelerators May Imply Said Upper Limit if QKoE below Reached Energy Scales. Said Upper Limit Forbids HE Non-PAPP Energy Transfer to Take Place if it's Larger than the Limit. Beyond the Upper Limit, Energy Transfer that Pushes Bound u-Quark to Higher Energy Level Can Not Happen and Consumption of Impact Energy Can Only Go through PAPP Channels. If Such Limit Does Exist and is below QKoE, Quarks are Bound Permanently without Possibility to be Knocked Out.

  Remark: In 2005, Author Thought that FQA Fact Implied QC **and STB** (**S**pace**t**ime **B**ag) could Explain QC. It's **Wrong**. So Far We Unable to Prove QC by Disproving QNC-AVI Scenario. FQA Fact Doesn't Necessarily Mean QC while Super KkEs Carried by Super-High Energy Particles in Prime Cosmic Rays Do Make any Bag a 'Toy', Impossible to Confine Quarks under such Knocking if NLVUC Does Not Exist or It's Higher than QKoE.

- In View of SM without DS, Every Meson's **2B-System**. In View of DS, All Hadrons except $\pi^0 = u\bar{u}$ are Unsolvable **MB-Systems**. This is 1st Reason Why We've Been Unable to Explain Mass Spectra of Various Hadrons Quantitatively with Good Accuracies as We Can Do for Atomic Spectra. 2nd Reason: None of Many Important or Possibly Important Forces such as MF, Possible St MF, Possible Significant St Magnetic-like Force btw Quarks with Relativistic High Speeds Calculated, No Matter LV Space Force in SA Dynamics is Negligible or Not (details of space force $S_{(\mu)} \equiv m\gamma u^{\mu} d\lambda / dt$ seen in II). Even for Pure $u\bar{u}$ Content, Any Quantitative Agreement with its Mass in a Model without DS and **MT-Force** (**M**agne**t**on-**Force**) Can Only Be Accidental Coincidence in Single Isolated Case. Discrepancies would 100% Surely Occur in Large Number of Other Facts if DS and MF are Not under Consideration.

- Every Neutron, Bound or Free, Unstable due to Inner ENM, which Refers to Mechanism of **Sd-e** ('**S**tronger' **d-E**lectron) Slowly Pushing Out **Wd-e** in Same Neutron. Here Wd-e Means '**W**eaker' **d-E**lectron. Sd-e Bound Closer to u-Q or u-Quark Core than Wd-e. Pauli's Exclusion Principle Forbids 2 d-Electrons in Any Neutron to Stay in Same State. Many Stable Chemical Elements Contain

More Neutrons than Protons Since Neutrons Do Not Decay Simultaneously. To Be Stable with Long Enough Lifetime, Nuclei on Stability Island Don't Demand Equal Number of Protons to Capture Relocating d-Leptons Emitted from Bound Neutrons.

Note: Two d-Electrons in Neutron Identical and Indistinguishable. But They Cannot Be in Same State. Thus, Sd-e and Wd-e are Two Distinguishable States for the Two Indistinguishable Identical Electrons.

- u-Quark Can ***MF-Stick*** both Electron/Positron for Short Period of Time as long as They Have Right ***Spin-Correlation*** and θ ***-Zone Locations***.
- To Be MF-Attractive, Two Spinning Charges DO NOT Have to Have Opposite Charges. As MF-Attraction Occurs at Shorter-than-MD Distances, It Overcomes Coulomb-Type Repulsion btw Two Charges of Same Type and Same Sign.
- c-Quark Formed by $ue^-\overline{\nu}_e e^+\nu_e$ in Excited State. Details of MB Excited State Not Very Clear. Sample Description Given in Section "MF (Magneton-Force)". Higher Excitation Turns c-Quark into t-Quark.
- DS-Excitation's LE and/or **QE** (**Q**uark-**E**xcitation), or Both.
- Because Isolated Single Quark Does Not Exist, QE is Excitation of **BC** uuu, **AntiBC** $\overline{u}\overline{u}\overline{u}$, or HMC $u\overline{u}$. Since c-Quark Never Appears alone (for instance, in Meson $c\overline{q}$ c-Quark Appears together with $\overline{q}$), Heavy Mass of c-Quark Due to Higher LE and/or QE, or Both.
- Every Quark and Every Anti-Quark Contains One u and One $\overline{u}$ Respectively. QE is Actually **UE**: Either u-Excitation OR $\overline{u}$ **-Excitation**. Because Neither u nor $\overline{u}$ Can Appear along, Every **UE** is Excitation of BC uuu, AntiBC $\overline{u}\overline{u}\overline{u}$, or HMC $u\overline{u}$. Higher MB DS-Excitations May Include Both UE and LE.
- Non-Spherosymmetric, θ -Depending, & Spin-Correlation-Depending Features of MT-Forces Tell: Radial Components of MT-Forces Can Be Attractive or Repulsive in Full/Near-Full Scale, Suppressed, or Turned off as Spins Change Directions or Interacting Spinning Charges Change Their θ -Zone Locations.
- More Examples and Details of How DS and MFs Explain Experimental Facts Given in Last Section "DS-Explanations of Experimental Facts…", Starting from Page 54 after Section " MFs (Magneton Forces)" and Ending at Page 112.

DS of Unstable Leptons (Muons and Taus)

LMC e^+e^- in Tau in Higher Excitation than LMC e^+e^- in Muon. Large Mass Difference btw Tau and Muon Provides with Very Large Available Energy to Open Decay Channels for Hadroproduction. Let $(e^+e^-)_1$ & $(e^+e^-)_2$ Be LMC of Muon & Tau respectively. DS of Muon & Tau is Written: $\mu^- = (e^+e^-)_1 e^-\overline{\nu}_e \nu_\mu$ and $\tau^- = (e^+e^-)_2 e^-\overline{\nu}_e \nu_\tau$. Sometimes Subscript Omitted and Context Tells LMC e^+e^- is for Muon or Tau.

Muons and Taus Must Meet Criterion S½abc to Be Called Elementary Fermions just Like Electrons. S½abc is the Criterion that Made Physicists to Conclude that Electron's Spin is ½. That S½abc is Impossible to Execute for Tau since Tau Lives too Short. It is Difficult to Execute for Muon as Muonic Atoms Live No Longer than $2.19\cdot10^{-6}$ Seconds. As Long as MTH Has Not Been Proved, We May Consider Their DS. DS of Muon together with DS of Quarks Immediately Demonstrate Unusual Power to Explain All Kinds of Experimental Facts.

- Experimental Errors in Measuring Total 4-Momentum of Final State Particles May Be Larger than Rest Energies of ν_e , ν_μ, and Their Antiparticles. Low-Energy NFMs ($\nu_e\nu_\mu\nu_\tau$ and their antiparticles) May Exist among Final State Particles but Have Eluded Detections. Decay Modes with < are Not Established Facts such as

$$\mu^+ \to$$

$$e^+\bar{\nu}_e\nu_\mu\,,<1.2\%\,/\,e^+\gamma\,,<1.2\cdot10^{-11}/\,e^+2\gamma\,,<7.2\cdot10^{-11}/\,e^+e^-e^+\,,<1.0\cdot10^{-12}\,.$$

- Precious Clue Can Be Seen in Following Confirmed Muon Decay Modes

$$\mu^+ \to$$

$$e^+\nu_e\bar{\nu}_\mu\ (\sim100\%),\ e^+\nu_e\bar{\nu}_\mu\gamma\ \ ((1.4\pm0.4)\%),\ e^+\nu_e\bar{\nu}_\mu e^+e^-\ \ ((3.4\pm0.4)\cdot10^{-5}).$$

- Soon in Last Section "DS-Explanations of Experimental Facts in SA Physics" It will be Seen How Decay Modes Put Lights on DS and How DS Explains Decay Modes.
- DS of Muons Immediately Predicts **Non-Existence** of Such Decays as

$$\mu^- \to e^- + \gamma, \qquad \mu^- \to e^- + \gamma + \gamma, \qquad \mu^- \to e^- + e^+ + e^-.$$

$\bar{\nu}_e\nu_\mu$ Not Particle-Antiparticle Pair and Never Annihilate into $\gamma\,/\,\gamma\gamma\,/\,e^+e^-$. Experiment's Proved Non-Existence of Such Events Since Very Beginning of Particle Physics. Muon's DS Reveals: μ^+/μ^- Not Excitation States of e^+/e^-.

- After Physicists Stopped Calling Muon 'Meson', Term 'Meson' was Reserved for **Hadronic Mesons**. Now We Still Call Hadronic Mesons **Mesons** but We May Call Muon and Tau **Leptonic Mesons**. Each Meson Contains HMC $u\bar{u}$. Leptonic Meson Muon Contains *Heavy* LMC e^+e^-, Not **Positronium** e^+e^-.
- Positronium e^+e^- Composite Particle. Its Size Same as Hydrogen Atom. MF Negligible at Atomic Distances. Low Speed in CMF Makes Magnetic Force btw Moving Charges Negligible too. Their Small KEs in CMF & Negative Coulomb PE Make Mass of Positronium Close to but Less than $2m_e$.
- LMC e^+e^- Produced from Deep SA Vacuum with Distance btw e^+ and e^- Much Shorter than Size of Positronium e^+e^-. Energy Needed to Produce Heavy LMC e^+e^- in Muon μ^+/μ^- Comes from What's Released in Annihilation of HMC $u\bar{u}$ in Pion π^+/π^- or Other Meson when It Decays. As Heavier HMC $u\bar{u}$ in Heavier Meson with UE and J=1 Label Annihilates, More Energies Available to Produce More than One LMC e^+e^- so that 2 or Even More Muons May Be Produced.

Later Proved TMET Shows Large ***Positive*** TME of MF-***Bound*** LMC e^+e^- of Small Size Less than 0TMED as IC when Produced in deep SA Space with Needed Energy Coming from HMC $u\bar{u}$ Annihilation which Triggers Decay of Charged Pions and Other Mesons. Dominating Pion-Decay Mode with DS

$$\pi^+ \rightarrow \mu^+ + \nu_\mu \text{ as } u\bar{d} = u\bar{u}e^+\nu_e \rightarrow e^+e^-e^+\nu_e\bar{\nu}_\mu + \nu_\mu \ (\sim 100\%)$$

Due to Nearly Complete Domination (~100%) of $u\bar{u}$ Annihilating into Two PAPs e^+e^- and $\nu_\mu\bar{\nu}_\mu$. Similarly, Dominating Muon-Decay Mode with DS

$$\mu^+ \rightarrow e^+\nu_e\bar{\nu}_\mu \text{ as } e^+e^-e^+\nu_e\bar{\nu}_\mu \rightarrow e^+\nu_e\bar{\nu}_\mu \ (\sim 100\%)$$

is Due to Domination (~100%) of e^+e^- Annihilating into Virtual Photon and/or Z-Boson Absorbed by Leptons $e^+\nu_e\bar{\nu}_\mu$ to Gain Large Kinetic Energies in CMF. Decay Mode $\mu^+ \rightarrow e^+\nu_e\bar{\nu}_\mu e^+e^-$ ($(3.4 \pm 0.4) \cdot 10^{-5}$) Tells: Chances're very small for LMC e^+e^- Pair Not to Annihilate but to Flee MF-Bound State with Large Positive TME into Free Region with Same Large Positive TME via QT. As LMC e^+e^- Annihilate into Virtual Photon and/or Z-Boson Absorbed by $e^+\nu_e\bar{\nu}_\mu$ to Gain KE, There is Small Chance (~1.4%) to Simultaneously Emit Observable Photon. Another Possibility for Decay Mode $\mu^+ \rightarrow e^+\nu_e\bar{\nu}_\mu\gamma$ is LMC e^+e^- Annihilating into ***Two*** Real Photons Followed by DS-Photoelectric Encounter with One Photon Fleeing out without DS-Photoelectric Encounter.

Inside π^+, HMC $u\bar{u}$ has very small chance (~0.013%) to Annihilate into virtual Photon and/or Z-Boson Absorbed by e^+ and/or ν_e to Yield Rare Decay Mode $\pi^+ \rightarrow e^+ + \nu_e$. Pion Meson Core Annihilation Channel with PAP Production $u\bar{u} \rightarrow e^+e^-\nu_\mu\bar{\nu}_\mu$ Dominates All Other Channels including One with Virtual γ and/or Virtual Z-Boson. This Domination due to VI, Giving PPC Priority to Consume Energy Released in $u\bar{u}$-Annihilation. Conservation of Energy Allows PPC for $u\bar{u}$-Annihilation but Forbids PPC $e^+e^- \rightarrow u\bar{u}$ for e^+e^--Annihilation in Muon Decay.

- Muon Mass Inequality below

$$m_\mu = 105.66 MeV / c^2 >> 3m_e + m_{\nu_e} + m_{\nu_\mu}$$

Seems to Be Major Obstacle for Compositeness of Muon. Surprisingly, TMET's Proved MFBF: TME of EqM 2B-System with MF Doing Binding is Positive at All Distances Shorter than 0TMED and the Shorter the Distance, the Larger the Positive TME. Thus, LMC e^+e^- Much Heavier than 2 Times Electron's Mass, Contributing Most to Muon-Mass. (SDL in last section "DS-Explanations of Experimental Facts in SA Physics")

- In Following Section of MF, Formula of MF Shows that MF is Proportional to r^{-4}, Almost Vanishes at Atomic or Longer Distances, while Dominates Coulomb Force at All Distances Shorter than MD. In CI, Furious MF Attraction btw e^+e^- at Short SA Distance where They're Born Causes Near-c Speed Circular Motion with Large γ–Value while Centripetal Acceleration Very Close to Upper Bound $a = c^2 / r$ for Circular Motion of Radius *r*. Rigorous Calculation in Circular Motion CI Shows a Sample Circular Motion with Radius $r = 5.03 \cdot 10^{-13} cm$ Requiring $\gamma = 154.425$ in order to Obey Relativistic Law of Motion with Near-0 LV Omitted in 3rd GS, Giving LMC e^+e^- Mass $105.12 MeV / c^2$, Very Close to Muon-Mass $105.7 MeV / c^2$. Other Leptons in Muon Originally in Pion Have Much Less KEs and Only Make Small Contributions to Muon-Mass.

Samples of How DS with 4TDSP Easily Explain Tau Decay Modes below:

- In Decay $\tau^- \to \mu^- \bar{\nu}_\mu \nu_\tau$ (17.39%), De-Excitation of $(e^+e^-)_2$ Turns it into $(e^+e^-)_1$, Releasing Energy in Form of $\nu_\mu \bar{\nu}_\mu$ and with ν_τ Flying away for Some Reason ($(e^+e^-)_1$ Unable to Bind ν_τ OR ν_τ Absorbs Virtual Z-Boson Released together with $\nu_\mu \bar{\nu}_\mu$ Gaining Enough Energy to Flee to Free Region). At Same Time, ν_μ Relocates into $(e^+e^-)_1$ so that They together with $e^- \bar{\nu}_e$ Form μ^- and Let $\bar{\nu}_\mu$ Go.
- In Decay Mode $\tau^- \to e^- \bar{\nu}_e \nu_\tau$ (17.82%), $(e^+e^-)_2$ Annihilates into Virtual Force Carriers Absorbed by Relic DSLs in τ^- and Sending Them Free.
- In Decay Mode $\tau^- \to \pi^- \pi^0 \nu_\tau$ (25.51%), $(e^+e^-)_2$ Annihilates into 2 $u\bar{u}$ -Pairs. One Captures DSLs $e^- \bar{\nu}_e$ Forming π^- while Another $u\bar{u}$ Pair becomes π^0. At Same Time, ν_τ, Losing Attraction from LMC $(e^+e^-)_2$, Goes away.
- In Decay Mode $\tau^- \to \pi^- \pi^+ \pi^- \nu_\tau$ (9.31%), $(e^+e^-)_2$ annihilates into 3 $u\bar{u}$ -Pairs, One e^+e^- Pair, and One $\nu_e \bar{\nu}_e$ Pair. Produced DSLs $e^- \bar{\nu}_e$ and $e^+ \nu_e$ Relocate in 2 $u\bar{u}$ -Pairs Turning Them into $\pi^- \pi^+$ respectively while Relic DSLs $e^- \bar{\nu}_e$ Left after $(e^+e^-)_2$ -Annihilation Relocate in 3rd $u\bar{u}$ -Pair Turning it into Another π^-. Once again, ν_τ, Losing Attraction from LMC $(e^+e^-)_2$, Goes away.
- In Decay Mode $\tau^- \to \pi^- \nu_\tau$ (10.91%), $(e^+e^-)_2$ Annihilates into a $u\bar{u}$ -Pair that Captures $e^- \bar{\nu}_e$ to Form π^- and Let ν_τ Go.

MFs (Magneton Forces)

- MF btw Magnetons with *Spin* MMs Being $\vec{\mu}_1$ and $\vec{\mu}_2$ Can Be Easily Deduced by Using 2 Well-Known Formulas in CED: Formula of Magnetic Field $\vec{B}_1$ Produced by *Spin* MM $\vec{\mu}_1$, Called Source MM, and Formula of Magnetic Force $\vec{f}_{12}$ that's Exerted on Spin MM $\vec{\mu}_2$ by Magnetic Field $\vec{B}_1$:

(H11) $$\vec{B}_1 = \frac{3\vec{\rho}(\vec{\rho} \cdot \vec{\mu}_1) - \vec{\mu}_1}{r^3}, \quad \vec{f}_{12} = \vec{\nabla}(\vec{B}_1 \cdot \vec{\mu}_2) \quad (\vec{\rho} \equiv \frac{\vec{r}}{|\vec{r}|} \equiv \frac{\vec{r}}{r})$$

Note: For $\vec{f}_{21}$ Exerted on $\vec{\mu}_1$ by $\vec{B}_2$ Produced by $\vec{\mu}_2$, We Call $\vec{\mu}_2$ Source MM.

Remark: MF in (H11) Contains No Another Component, Called **Orbital Force**, of *Mag Force* Exerted on 2nd Charged Particle by Mag Field $\vec{B}_1$. Orbital Component of Mag Force Due to Electric Current Created by Translational Motion of 2nd Charged Particle, Not by its Spin. Orbital Force is Negligible Only if Speed of 2nd Charged Particle's Translational Motion isn't Sufficiently Near c. Moreover, if Source Charged Particle is Orbiting and 2nd Charged Particle is Far Outside such Local Orbit, then Source MM $\vec{\mu}_1$ Should Be **Net MM** as Vector Sum of Source Charged Particle's Spin & Orbital MMs. SDL

- Choose Spherical Coordinate System (r,θ,ϕ) with Source MM $\vec{\mu}_1 = |\vec{\mu}_1|\vec{k}$ in Positive Zenith-Direction and at Origin. Consider Parallel and Anti-Parallel Cases Only ($\vec{\mu}_1 \times \vec{\mu}_2 = \vec{0}$). We Have

(H12) $$\vec{f}_{12} = \vec{\nabla}(\vec{B}_1 \cdot \vec{\mu}_2) = \vec{\mu}_1 \cdot \vec{\mu}_2 (\frac{3-9\cos^2\theta}{r^4}\vec{\rho} - \frac{6\sin\theta\cos\theta}{r^4}\vec{\theta})$$

- MF Force Field **Non-Spherosymmetric**, θ **-Depending**, Playing Vital Role in SA Small Spaces so that MF-Bound d-Electrons in Nucleons Have **Orbital Quantum Number** Even in Ground State and S-Wave States (Ground or Excited) of d-Electron Do Not Exist at All in SA World. S-Wave State Exists Only if its Size is Much Larger than MD so that MF Almost Vanishes with its Non-Spherosymmetry Ceasing to Play a Noticeable Role.
- Quantum Eqs in Non-Spherosymmetric and θ -Depending Force Fields Have Never Been Solved and May Be Unsolvable.
- For Parallel and Anti-Parallel Spin MMs of Two Elementary Fermions, Well-Known Quantum Connection $\mu \equiv |\vec{\mu}| = |q|\hbar/2mc$ and (H4) Lead to

(H13) $$\mu \equiv |\vec{\mu}| = |q|\hbar/2mc = 1.930664 \cdot 10^{-11} cm \cdot \frac{m_e}{m}|q|$$

$$\vec{\mu}_1 \cdot \vec{\mu}_2 = 1.930664 \cdot 10^{-11} cm \cdot sign(\vec{\mu}_1 \cdot \vec{\mu}_2)\mu_1 |q_2| \frac{m_e}{m_2}$$

$$= 3.727463 \cdot 10^{-22} cm^2 \cdot sign(\vec{\mu}_1 \cdot \vec{\mu}_2)|q_1 q_2| \frac{m_e m_e}{m_1 m_2}$$

$$\vec{\mu}_u \cdot \vec{\mu}_{e^-} = 1.930664 \cdot 10^{-11} cm \cdot sign(\vec{\mu}_u \cdot \vec{\mu}_{e^-})\mu_u e$$

$$= 2.484976 \cdot 10^{-22} cm^2 \cdot sign(\vec{\mu}_u \cdot \vec{\mu}_{e^-}) e^2 \frac{m_e}{m_u}$$

$$\vec{\mu}_u \cdot \vec{\mu}_u = 1.930664 \cdot 10^{-11} cm \cdot sign(\vec{\mu}_u \cdot \vec{\mu}_u)\mu_u 2e/3$$

$$= 2.484976 \cdot 10^{-22} cm^2 \cdot sign(\vec{\mu}_u \cdot \vec{\mu}_u) \frac{m_e m_e}{m_u m_u} 4e^2 / 9$$

- (H12) Tells: at $\theta = \theta_{zero1} \equiv \cos^{-1} 1/\sqrt{3}$ / $\theta = \theta_{zero2} \equiv \pi - \cos^{-1} 1/\sqrt{3}$, about 54.7^0 / 125.3^0 , *Radial* Component of MF Vanishes. We Call Them **<u>0R-MF Angles</u>**. <u>Region 0</u> (**R0**) Contains all Points where θ is Equal or Close to 54.7^0 or 125.3^0 . For any 2 Magnetons of Same Kind, MF btw them Turned off or Greatly Suppressed if They are in R0 at Each Other with Parallel or Antiparallel MMs.
- Non-Radial Component of MF Vanishes at $\theta = 0, \pi/2, \pi$. Within Any Sphere Centered at Source MM, One May Define **A-Region** and **R-Region** where MF is <u>A</u>ttractive and <u>R</u>epulsive respectively if Radial Domination Condition $|\vec{f} \cdot \vec{\rho}| \geq 2.5 |\vec{f} \cdot \vec{\theta}|$ Satisfied. Factor 2.5 is Just to Show Domination. Important Thing is Not Such Factor's Value but How to Use e.m., Wk, and St MFs to Explain SA Experimental Facts. Radial Domination Occurs in 2 3-Dim Regions.
- <u>Region 1</u> (**R1**) <u>at Source Magneton</u> with $-10.5^0 \leq \theta' \leq 10.5^0$ and <u>Region 2</u> (**R2**) <u>at Source Magneton</u> with $0 \leq \theta \leq 20.4^0$, $159.6^0 \leq \theta \leq 180^0$. We May Call $\theta' \equiv \pi/2 - \theta$ *Elevation* (Angle). By '<u>at Source Magneton</u>' We Mean: it's Center of Centrosymmetry of R1 or R2. We Focus On R1 or H-Region and R2 or V-Region ('H' and 'V' Mean 'Horizontal' and 'Vertical' Respectively) at Each Source Magneton. R1 Looks Like Non-Uniform Disc with thickness Increasing with Radius. We Call It **Bat-Disc**, Formed as Batman Sweeps Bat-like Object over Space for 360^0. R2 are Two Circular Cones (Up and Down) whose Vertices Coincide at Origin with Zenith Axis as Their Axis of Symmetry.
- For **Parallel** MMs, **R1**/R2 are **Repulsive**/Attractive Regions Respectively. For **Anti-Parallel** MMs, **R1** and R2 are **Attractive** and Repulsive Respectively.
- Each Plane Containing Zenith Axis Intersects R1 and R2 with Section Looking Like a **Fan** that Has 4 **Vanes**. When Rotating about Zenith Axis, They Sweep Over Space to Form R1 and R2. 4-Vane Figure is Useful for Us to See How d-Electrons in Ground and Excited States Move in SA Spaces and How DS-Symmetry and DS-Asymmetry Determine Mesons' Lifetimes Qualitatively and Half-Quantitatively. We May Call R1 and R2 **H-Vane** and **V-Vane** at Source Magneton Respectively. (H/V Means Horizontal/Vertical)
- R1 and R2 are 3-dim Space Regions while Their New Names H-Vane and V-Vane Remind Us of Shapes and Positions of Their Sections and of Themselves. Note: Their Shapes May <u>Deform</u> as More Magnetons Crowd together inside a Sub-Atomic Composite Particle. Example: H-Vane at u-Quark in π^- where d-Electron Stays is Deformed due to Repulsion from $\bar{u}$ Nearby.
- Before Dirac Eqn with MF-Binding is Solved, We Don't Have Many Clues about Ground State Region. But We Know that It Must Be Net-Attraction Region, where Resultant Force is Attractive whether or not Repulsive Force Component Exists. If Repulsive Force Exists, then Attractive Force Must Dominate in that Region. MF btw u and e^- Becomes Attractive in R1/R2 when Spin Vectors of u and e^- Parallel/Anti-Parallel. For $u\bar{u}$ Meson Core Binding, Spin Directions Depend on which is Ground State Region (R1 or R2) in Same Manner as Seen in ue^- Binding inside d-Quark. For uuu Baryon Core 3-Body Binding, Due to

Insolvability of both Classical and Quantum MB-Problems, We Do Not Know How 3 u-Quarks Moving. Even Classical 3B or Any MB-System Not Solvable. Although a Special Solution of Classical EqM 3B-Problem with Same Sort of Attractive Forces btw Each 2 Bodies Can Be Figured Out under Special IC: 3 Bodies Move on a Plane Forming Equilateral Triangle, Each One Circling about their Mass Center with Exactly Same Speed and Obeying Either Relativistic or NR Classical Law of Motion, Such State May Be very Unstable and any Small Perturbation Quickly Destroys Such Circular Motion State Leading to 3B-System with Motion Completely Unknown and Unsolvable.

- We Can Propose Sample Spin Directions and Compare Theoretical Outcomes with Empirical Data. Sample Spin Direction Arrangement Must Simultaneously Explain Observed Preferred Spin Direction of beta Particles in EMF, Unstable Neutrons with Very Slow Decay and Continuous Spectrum of beta-Particles, Stable Free Protons, Nucleons' MMs, Lifetimes of Non-Nucleon Hadrons and of Unstable Leptons Much Shorter than Neutron's, Lifetime Details such as Much Longer Lifetime of Charged Pions than that of Neutral Pion, Longer Lifetime of Charged Pions than Charged Kaons though Pion and Kaon with Same Charge Have Exactly Same Elementary Constituents in View of DS and They May Decay into Same Set of Final State Particles, …. SDL
- Spin Directions of d-Leptons Determine Their θ **-Locations** inside Hadrons, which Determine Their Decay Speed (SDL). Here 'θ-Location' Means MPR in Certain θ **-Zone**.
- For Ground State of MF-Binding We Give <u>Sample Arrangement</u>, Called **R1G**, with R1 as Ground State Region so that u-Quark and e^- Have Parallel Spin Vectors while OAM of e^- is Anti-Parallel to its Spin. OSOD Feature of Orbital and Spin Motions Produces Double Attractions as Shown Before. OSOD Can Be Easily Seen as We Apply Hand-Rules without Computations.
- For 1st & 2nd Excitation States of MF-Binding We Give a <u>Sample Arrangement</u>, Called **R2R1E**, with R2 as 1st Excitation State Region and R1 as 2nd Excitation State Region. Quantum Spin Flip Occurs as Excitation from R1G to **R2E** Takes Place. R2E is Non-Radial Excitation. **R1E** is 2nd Excitation, Either 1st Radial or 1st Orbital Excitation with Orbit in R1 Farther away from u-Quark or u-Quark Core than It is in R1G. Unstable Neutrons with Decay Much Slower than Slowest Hadron and Lepton Decay Have Been Explained before with ENM where Quasi-Bound d-Electron Wd-e Also Stays in R1 Farther Away from u-Quark or u-Quark Core. Such <u>Arrangement</u> of Quasi-Bound d-Electron's Spin and Location is Called **R1QB**.
- Above Sample Arrangement Can Also Be Labeled as R1E_0, R2E_1, & R1E_2, where E_0, E_1, and E_2 Refer to Ground, 1st Excited, and 2nd Excited States Respectively. See How to Use Sample Arrangement in DS to Explain Many Details of Hadrons' Lifetimes Later. We Call Them ***<u>Sample</u>*** Arrangement for We Actually Don't Know Solutions and Details of Excited States Even for 2B-System if Resultant Binding Force Includes MF-Component. Only Thing That is Sure Right Now is (r,θ)-Depending Wave Functions for Ground as well as for Excited States Caused by (r,θ)-Dependence of MF Field. Any Descriptions of Details Can Only Be Sample Descriptions.

- LE Has Negligible Impact on u-Quarks' Motions due to Heaviness of u-Quarks.

  Note: Creation or Annihilation of, For Instance, $d\bar{d}$, $s\bar{s}$, $c\bar{c}$, $\mu^+\mu^-$, $\tau^+\tau^-$, ... is Just Outcome of PAP Creation or Annihilation, in View of DS with Any One of Them Containing 3 or 5 Constituent Elementary Particles.

- Without EMF, Ground State Nucleons in Nucleus Tend to Have such Directions of Spin that 2 Neighboring Nucleons' MMs are in Opposite Directions (Anti-Parallel) Similar to *Shell Model* where 2 Neighboring Neutrons Tend to Have Anti-Parallel MMs. EMF Changes Directions of Spins of Neutrons in a Nucleus (such as ${}^{60}_{27}Co$) such that Their MMs Point Preferably at Same Direction as EMF when Irregular Thermal Motion's almost Stopped in Condition of Super-Low Temperature. Preferred Direction of beta Particles' Spin Vectors, Observed by Wu and Many Experimenters, Could Only Be Preferred Direction before Decay Occurred for So Many Electrons Emitted in beta Decay Moved Very Fast and Did Not Have Enough Time to Change Spin Directions during Very Short Period of Flying Time. Note: Even in Very Slow Motion, Quantum Electrons' MMs Do Have TWO Possible Directions (Parallel and Anti-Parallel to EMF), as Seen in Stern-Gerlach Experiment. DS Naturally Explains Observed Preferred Spin Direction of beta Particles in EMF. Extension of Electric Charge/MM to Wk Charge/MM Explains Observed Preferred Spin Direction of $\bar{\nu}_e$. See Details in Next Section "DS-Explanations of Experimental Facts in SA Physics".
- Geometric Shapes of R1 and R2 Indicate Possible Directions of OAM of d-Electron Moving around u-Quark that Binds it. As MB-Problem is Simplified to One u-Quark Binding d-Electron, We May Do Qualitative and Half-Quantitative Study. Magnetic Field Produced by Source u-Magneton's Spin Motion (Spin MM) Shown in (H11) Exerts Magnetic Force on Electric Current Produced by d-Electron's *Orbital* Motion, Called **Orbital Force**, Denoted by $\vec{f}^{orb}$. In Spherical Coordinate System, $\vec{v} = v\vec{\phi}$ Orbital Velocity Component of d-electron as It Moves on zenith Plane about zenith Axis (MM Line of Source u-Quark). Using Lorentz Force Formula and (H11) We Get

(H14) $$\vec{f}^{orb} = e\mu_u v(\vec{\theta}_0 - 3\cos\theta\vec{\theta})/cr^3 \quad (\vec{\theta}_0 = \vec{\theta}\,|_{\theta=0} = \vec{\phi}\times\vec{k}\,)$$

- With (H12), (H13), and (H14), Total Magnetic Force Exerted on Spinning and Orbiting d-Electron by Magnetic Field Produced by Spinning Electric Charge of u-Quark is

(H15) $$\vec{f}_{ue^-} = \vec{f}_r + \vec{f}_\theta + \vec{f}_{z0}$$

$$\vec{f}_r = \frac{3e\mu_u}{r^3}\frac{1.93\cdot 10^{-11}cm}{r}\cdot sign(\vec{\mu}_u\cdot\vec{\mu}_{e^-})(1-3\cos^2\theta)\vec{\rho}$$

$$\vec{f}_\theta = -\frac{3e\mu_u}{r^3}\left[\frac{1.93\cdot 10^{-11}cm}{r}\cdot sign(\vec{\mu}_u\cdot\vec{\mu}_{e^-})\sin 2\theta + \frac{v}{c}\cos\theta\right]\vec{\theta}$$

$$\vec{f}_{z0} = \frac{e\mu_u}{r^3}\frac{v}{c}\vec{\theta}_0 \quad (\vec{\theta}_0 = \vec{\theta}\,|_{\theta=0} = \vec{\phi}\times\vec{k}\,)$$

- In **R1**, $1-3\cos^2\theta > 0$. Radial Component in Direction of $sign(\vec{\mu}_u \cdot \vec{\mu}_{e^-})\vec{\rho}$. It is **Attractive** if They Have **Parallel Spin Vectors**. 3rd Component toward Z-Axis if $v<0$, i.e., if Orbital Velocity of d-Electron in Direction of - $\vec{\phi}$ with its **OAM** in Direction **Opposite** its **Spin** Vector. This is Quantitative Proof of **OSOD** Feature for d-Electrons without Using Hand-Rules.
- In **R1**, θ is btw 79.5^0 & 100.5^0, $-1.64 < 3\cos\theta < 1.64$, and $\vec{\theta}$ is Close to $-\vec{k}$. **Negative** v Simultaneously Produces Central-Like Orbital Force, with Force Component $\vec{f}_{z0} = e\mu_u v\vec{\theta}_0 / cr^3$ Turned toward u-Quark & its MM Line (Zenith Axis). Another Force Component $\vec{f}_\theta$ Shown in (H15) Turned toward xy-Plane. R1 is Attraction Region ($\vec{f}_r$ Opposite to $\vec{\rho}$) if d-Electron and u-Quark Have Spin Vectors Parallel to Each Other. Surely, $3-9\cos^2\theta > 0$ in R1. Hence, Attraction $\vec{f}_r$ in R1 Requires Negative $sign(\vec{\mu}_u \cdot \vec{\mu}_{e^-})$. Such Sign together with Negative v also Make $\vec{f}_\theta$ toward xy-Plane since both $\cos 2\theta$ and $\sin 2\theta$ are Positive/Negative Above/Below xy-Plane.
- **R2** is Attraction Region if d-Electron and u-Quark Have Anti-Parallel Spins. Attractive $\vec{f}_r$ in R2 Requires $sign(\vec{\mu}_u \cdot \vec{\mu}_{e^-}) > 0$ since $3-9\cos^2\theta < 0$ in R2.
- In **R2**, $0 \le \theta \le 20.4^0$ (Up Cone), $159.6^0 \le \theta \le 180^0$ (Down Cone), and $\vec{\theta}$ Close to $\pm\vec{\theta}_0$ in Up and Down Cone Respectively. Since both $\cos\theta$ and $\sin 2\theta$ are Positive/Negative in Up/Down Cone, $\vec{f}_\theta$ Close to $-\vec{\theta}_0$ in Both Up and Down Cones of R2 if v is **Positive**. Positive v Makes $\vec{f}_{z0}$ Tend to Push d-Electron away from u-Quark's MM-Line. But $\vec{f}_\theta$ Dominates $\vec{f}_{z0}$ in R2. One Can See This by Knowing that $-\frac{3e\mu_u}{r^3}\frac{v}{c}\cos\theta\vec{\theta}$ is Part of $\vec{f}_\theta$ that is Nearly Opposite to $\vec{f}_{z0} = e\mu_u v\vec{\theta}_0 / cr^3$ with Larger Norm in R2 for $\cos\theta\vec{\theta}$ Nearly in Same Direction as $\vec{\theta}_0$ in both Up and Down Cone of R2 while in R2 $|\cos\theta| > 0.937 > 1/3$.
- R2 (Up/Down Cones) is Attractive Region for 2 Parallel MMs such as ue^- with Anti-Parallel Spin Vectors.
- In R2 at u-Quark, d-Electron's Spin Anti-Parallel to u-Quark's and Anti-Parallel to Zenith Component of Its Orbital Angular Momentum.
- Once d-Electron's Spin Flips Over with its Spin Vector Becoming Anti-Parallel to u-Quark's Spin, It Becomes Excited, Jumping from R1 (H-Vane) to R2 (V-Vane). Then d-Quark Becomes s-Quark. An s-Quark is just a d-Quark in 1st **LE** (**L**eptonic **E**xcitation).
- c-Quark Contains All Constituents of d-Quark plus an $e^+\nu_e$ Pair, Called **c-Leptons** (Exactly Same as $\bar{d}$ -Leptons or Anti-d-Leptons), with d-Leptons $e^-\bar{\nu}_e$ inside c-Quark being in 2nd LE or in 1st **LUE** (**L**eptonic-**u**-Quark **E**xcitation), if Excitation Energy Lower or Higher than **LKoE** (**L**epton **K**n**o**ckout **E**nergy). Due

to Heavy u-Quark Mass, LE Puts Insignificant Impact on u-Quark, But Not Vice Versa in a Sub-Model Called **RQC** (**R**egular **Q**uark **C**ore) Sub-Model. Any u-Quark Always Stays together with $\overline{u}$ in Meson Core or with Other Two u-Quarks in Baryon Core. **UE** (**u**-Quark **E**xcitation) always Means Meson-Core or Baryon-Core Excitation. Here Meson-Core Means HMC. We May Let UE Include AntiBC Excitation. Due to Influence of UE on Light DS-Leptons in RQC Sub-Model, UE May Have Large Impact on SA Leptons' States Causing Complexity of Excitation States of Hadrons.

- Inside c-Quark, d-Electron/c-Positron May Stay in H-Vane/V-Vane respectively with Their Spin Vectors Parallel to u-Quark's. In c-Quark, d-Electron in Radial Excitation Staying and Bound in R1 while c-Positron Stays in R2 Temporarily (Attractive Coulomb Force from d-Electron Does Not Allow c-Positron Staying in R2 for Long).
- **b**-Quark, Having Same Constituents as d, s-Quarks, is 3rd LE of d-Quark if *below* LKoE *otherwise* it's 1st or 2nd LUE Depending on whether c-Quark is 2nd LE or 1st LUE.
- ***t***-Quark Higher Excitation State than ***c***-Quark with its Constituents Same as in c-Quark.
- If R2 is Ground State Region, ENM is Hard to Realize.
- We Take *Approximations* $\cos\theta = 0$ and **1** in R1 and R2 Respectively. Using (H12) We Write Two *Approximate* MF-Formulas in R1 and R2 as

(H16)
$$\vec{f}_1 = \frac{3\vec{\mu}_1 \cdot \vec{\mu}_2}{r^4}\vec{n}, \qquad \vec{f}_2 = -\frac{6\vec{\mu}_1 \cdot \vec{\mu}_2}{r^4}\vec{n}$$

- For Each e.m. Spin MM $\vec{\mu}$ of an ***Elementary*** Fermion of Electric Charge q /Mass m, We Have Quantum Connection (Radiation Correction Omitted)

(H17)
$$\vec{\mu} = q\vec{S}/mc, \quad \mu = |\vec{\mu}| = q\hbar/2mc = q\frac{m_e}{m}\cdot 1.93\cdot 10^{-11}cm$$

$$\vec{\mu}_1 \cdot \vec{\mu}_2 = \mu_1\mu_2 sign(\vec{\mu}_1 \cdot \vec{\mu}_2)$$

$$= sign(\vec{\mu}_1 \cdot \vec{\mu}_2)q_1q_2\frac{m_e m_e}{m_1 m_2}\frac{1.24}{3}(3\cdot 10^{-11}cm)^2 \quad (\frac{\hbar}{m_e c} = 3.86\cdot 10^{-11}cm)$$

- Extension of **Electric Charge** to **St** and **Wk Charges** Leads to St and Wk MMs. Force btw 2 MMs of Same Kind is e.m., St, or Wk **MF**. For Parallel or Anti-Parallel MMs $\vec{\mu}_1$ and $\vec{\mu}_2$ of Same Kind, Formula of **MF** with Position Vector from $\vec{\mu}_1$ to $\vec{\mu}_2$ ($\vec{r} \equiv |\vec{r}|\vec{n} \equiv r\vec{n}$) Perpendicular to or Collinear with ($\theta = \pi/2$ or $\theta = 0, \pi$) MM-Line is Obtained from (H12)

(H18)
$$\vec{f}_1 = \frac{3\vec{\mu}_1 \cdot \vec{\mu}_2}{r^4}\vec{n} = sign(\vec{\mu}_1 \cdot \vec{\mu}_2)\cdot 1.24\frac{q_1q_2}{r^2}(\frac{3\cdot 10^{-11}cm}{r})^2\frac{m_e}{m_1}\frac{m_e}{m_2}\vec{n}$$

$$\vec{f}_2 = -\frac{6\vec{\mu}_1 \cdot \vec{\mu}_2}{r^4}\vec{n} = -sign(\vec{\mu}_1 \cdot \vec{\mu}_2) \cdot 2.48\frac{q_1 q_2}{r^2}(\frac{3 \cdot 10^{-11} cm}{r})^2 \frac{m_e}{m_1}\frac{m_e}{m_2}\vec{n}$$

- Condition for Domination of e.m., St, & Wk MFs over Coulomb-Counterparts at Certain Distances against the **Magneton Distance (MD)** r_m, Can Be Seen in **MT-C (Magneton-Coulomb) Ratio**

(H19) $$k_{MT-C} \equiv \frac{|\vec{f}_i|}{|q_1 q_2 / r^2|} = \frac{r_{mi}^2}{r^2},$$

$$r_{mi} \equiv 3.341\sqrt{1+\delta_{i2}} \cdot 10^{-11}\sqrt{\frac{m_e}{m_1}\frac{m_e}{m_2}}cm, \quad (i = 1,2.)$$

$$r_m \sim 3 \cdot 10^{-11}\sqrt{\frac{m_e}{m_1}\frac{m_e}{m_2}}cm$$

DS-Explanations of Experimental **Facts** in SA Physics

We've Seen DS-Explanations of Many Experimental Facts in Previous Sections of "III. SA Physics with DS". Now We Continue to Explain More Facts.

- **Original Idea** of Explaining All Chemical Elements with Larger *Mass Numbers* than *Charge Numbers* quite Similar to DS: One or More Protons in Each of Such Element Contain Electron(s) to Neutralize Part of Positive Charge of Nucleus and to Yield Smaller Charge Number than Mass Number. That Original Idea was Later Abandoned to Create a New Term '**Neutron**'. Physicists who Discovered it Viewed it as *Elementary* like Proton. Since then, Even after Quark Model Showed Nucleons' Compositeness, Dominating Idea that Nucleons Contain No Electrons Had Been Established. Reason to Abandon that Original Idea was because Crucial Effects of MF at Short SA Distances and its very Important Non-Spherosymmetric Feature were Ignored and SA Electron was Incorrectly Assumed to Have No OAM if Ever inside Nucleon. Without Electron's OAM and its Corresponding Orbital MM in Opposite Direction of its Spin MM, Spin of Electron, if Contained in Nucleon, and its Corresponding Spin MM indeed Failed to Explain Observed MM of *Nucleus* No Matter How One Combined Spin MMs of Electrons and Protons' Spin MMs. Just Like Without d-Electron's OAM and its Corresponding Orbital MM in Opposite Direction of its Spin MM, d-Electron's Spin and Spin MM Have Failed to Explain Observed MM of *Nucleon* No Matter How One Combined d-Electron(s)' Spin MM(s) and u-Quarks' Spin MMs. Now by Just Recognizing MF and its Simple Effect, One Can Easily Turn Failures into Stunning Success.
- Unknown Mass of u-Quark Determines Its Spin MM via Quantum Connection. Its Orbital MM Unknown and Depends on Its Orbital Quantum Number that's Close to zero in Ground State iff Asymmetric St and e.m. MFs btw u-Quarks are Negligible. It is Closely Related to Size of Quark Core and Mass of u-Quark. See

(H16, 22, 23) for Details.

- Estimation of u-Mass Subject to ***Toughest*** Constraint: It Must ***Simultaneously*** Explain Nucleon's MMs, Nucleons' Masses, and Mass of π^0. MB-Problems and MF-Binding Systems Unsolvable. Here to Explain Means to Qualitatively and Half-Quantitatively Explain. Now that Quantitative Solutions of Quantum Wave Eqs Not Available for MF-Bound Systems and MB-Systems, Even Toughest Constraint Unable to Determine Value of u-Mass. SDL
- Non-Existence of qq Meson Suggests that Anti-Parallel St Currents Do Repel Each Other just like Anti-Parallel Electric Currents. Their Velocities Must Be in Opposite Directions if Viewed in Center-of-Mass Frame of qq 2B-System.
- Mesons (Each Has $u\bar{u}$ HMC) Exist so that Parallel St Currents Attract Each Other Just like Parallel Electric Currents. Then Parallel St MMs in R1 Should Repel Each Other Just like Parallel (e.m.) MMs in R1 while Anti-Parallel St MMs in R1 Should Attract Each Other Just like Anti-Parallel e.m. MMs in R1. In Every Meson without Q-Excitation, $u\bar{u}$ May Have Parallel or Antiparallel Spin Vectors. Their Orbital Quantum Number Unknown. ***'Spin 0'***, as ***Statistic Label***, a ***Label*** Indicating Lack of Excitation against Spin-Spin/Spin-Orbit Couplings, Cannot Justify Opposite Spin Directions of $u\bar{u}$ in 'Spin 0' Mesons. But OSTP that Brings us to DS-Explanation of Pion-Masses Does Show Opposite Spins of $u\bar{u}$ in Ground State of HMC as It is in Both Neutral & Charged Pions. **OSTP** at Near **0R-MF Angle** Suppresses MF No Matter 2 Interacting Spin MMs Parallel or Antiparallel. But Possibility of Antiparallel Spins Can Not Prove J=0 Label for Pion due to OAM of u and $\bar{u}$. Though OSTP Can Suppress MF, MF Fields Produced by Each Spin MM of u or $\bar{u}$ with Electric and St Charges Remain Non-Spherosymmetric to DSLs Bound by that $u\bar{u}$ HMC. Their Influences on Quantum Orbital Number of any such DSL Remain Quantitatively Unknown. In Spite of Complicated OAM/TAM Issue of SA Composite Particles, DS with MFs Has Explained Masses of Pions.
- Later We'll See **Small Sizes of HMC** $u\bar{u}$ Shown in < **HMC** $u\bar{u}$, **u-Mass**, and π^0 **-Mass Eqn** > Required by Simultaneous and Coherent Explanations of Proton-Mass, π^0-Mass, Heavy u-Mass, and Nucleons' MMs. Orbital Size of d-Electron in π^- Must Be Much Larger than Size of HMC. Only **Antiparallel Spins** of HMC $u\bar{u}$ in OSTP and in Tiny Space Can Yield a **Net MM** to Bind Light d-Electron in SA Small Space that's Much Larger than Size of HMC. Moreover, Only **Antiparallel Spins** of HMC $u\bar{u}$ in OSTP Can Automatically Obey Conservation Law of Total Angular Momentum During Precession. **OSTP** Suppresses Furious Repulsion Leading to **Small Radial Attraction** at Angles Nearby and Slightly Smaller than 0R-MF Angle. Due to OSTP, Orbital Plane of $u\bar{u}$ and the Plane that is Perpendicular to Spins and Spin MMs of $u\bar{u}$ Form an Angle. The Latter Must Be Orbital Plane of d-Electron Bound by MF via Magnetic Field Produced by Net Spin MM of $u\bar{u}$ 2B-System. That is: Orbital Plane of d-Electron in π^- Must Be Perpendicular to Spins of $u\bar{u}$ in HMC, but Orbital Plane of $u\bar{u}$ in π^- Not Perpendicular to Their Spin Vectors due to OSTP. It's Similar to Sun-Earth-Moon System: Moon's Orbital Plane Not Same as Orbital Plane of Earth.

- If $u\bar{u}$ -Pair alone Ever Forms a π^0 then Experimenters May Want to Measure its MM and Tell Whether or Not u and $\bar{u}$ Have Parallel or Antiparallel Spin Vectors. But π^0 Produces No Trace of Visible Length for its Lifetime too Short ($\sim 8.4 \cdot 10^{-17} s$). Experimenters Can Not Measure its MM to Suggest whether or not We Have Parallel Spin Vectors of u and $\bar{u}$ inside. It's MF Dynamics that Proves their Parallel or Antiparallel Spins. But as Explaining Small Mass of π^0 under Heavy u-Mass Setting that Simultaneously Explains Nucleons' MMs, We Found that Parallel or Antiparallel Spins of $u\bar{u}$ in π^0 Tilted against Orbit, **Suppressing** MFs & Positive TME Feature of MF-Binding to Yield **Negative** TME by Severely Reducing 0TMED and Critical Distance so that Mass of π^0 Can Be Much **Lighter** Even than Mass of Single u-Quark. SDL

We are Familiar with Empirical Lifetime Spectrum of Hadrons and Leptons. Now We Focus on More Details of Lifetimes of Mesons.

< **Slowly** and **Fast** Decaying Mesons >

- **SDM**s (Slowly **D**ecaying **M**esons) with Lifetime Ranging from $10^{-8}s$ to $10^{-13}s$ & **FDM**s (**F**ast **D**ecaying **M**esons) with Lifetime Ranging from $10^{-17}s$ to $10^{-24}s$ Can Be Explained via DS.
- Without Exception, FDMs Either Mesons with J=1 Label Showing Orbital Excitation of Their HMC $u\bar{u}$ always Returning to Lower Level Very Quickly like Baryons of J=1 Label (see discussion of long lived Ω later).
- FDMs with J=1 Label $\rho^{\pm}, \rho^0, \omega, \phi, J/\psi, Y, K^{*\pm}, K^{*0}$, and All D*-Mesons. In View of DS, 'Vector Mesons' FDMs for De-Excitation of Orbital Excitation of HMC $u\bar{u}$ Goes Very Quickly.

  Note: Asterisk D-Mesons & Asterisk B-Mesons Not Determined for J-Label while Asterisk B-Mesons' Lifetimes Not Yet Measured. (see PDG on Web)
- FDMs with J=0 Label $\pi^0, \eta, \eta', \eta_c$. In View of DS, They All DS-Symmetric, Lacking Significant e^--$\bar{u}$ or e^+-u Electric Repulsion to Slow Down Annihilation of $u\bar{u}$. Such Repulsions Seen Clearly in *DS-Asymmetric* Mesons Such as $\pi^{\pm}$.
- Some FDMs with J=1 Label Also DS-Asymmetric Similar to $\pi^{\pm}$, such as $\rho^{\pm}$. Pion and Rho with Same Non-zero Charge Even Have Same Quark Content and Same DS-Content. DS-Asymmetric Mesons Decay Fast if Having J=1 Label and Slowly if Having J=0 Label. This is because Excited HMC of J=1 Label Quickly Return to Ground State to Finish Decay and to Produce DS-Asymmetric Mesons of J=0 Label. Further Decay of DS-Asymmetric Mesons of J=0 Label as Middle Stage Particles Not Counted as Part of Their Decay Process. e^--$\bar{u}$ or e^+-u Repulsion That Exists in All DS-Asymmetric Mesons is Much Weaker in Time Period of De-Excitation of HMC Since Excitation of HMC from J=0 to J=1 State and De-Excitation of HMC from J=1 to J=0 State is Sort of Spin-Flipping Forth and Backwards or Some Unknown De-Excitation. It Happens and Finishes Very Quickly. Said Repulsions Have No Way to Slow down Motion of De-Excitation but They Do Slow down Motion of Annihilating $u\bar{u}$.

- MF-**BM** (**B**inding **M**echanism) in $d(ue^-\bar{\nu}_e)/\bar{d}(\bar{u}e^+\nu_e)$ Composite Systems with

R1G Requires Parallel Spin Directions of ue^- & $\bar{u}e^+$ Pairs. Charged Pions're **DS-Asymmetric**. Single e^- / e^+ in π^- / π^+ Repels $\bar{u} / u$ with Coulombian Force. Such Repulsion Slows Down $u\bar{u}$ Annihilation to Yield Much Longer Lifetime of $\pi^\pm$ than that of π^0. How Repulsion Slows down $u\bar{u}$-Annihilation in Decay of DS-Asymmetric Mesons with Label J=0 will Be Discussed on Page 84. Before We See and Explain Lifetimes of All Other DS-Asymmetric Mesons of J=0 Label Always Much Longer than that of DS-Symmetric Mesons, We Now Explain Lifetimes of Muon and Positronium Ps 2B-System e^+e^-.

< **Lifetimes** of **Muon** and **Positronium** 2s-Ps/*o*-Ps/*p*-Ps >

- Longer Lifetime of Muon than Positronium e^+e^- Also Due to Repulsion from DS-Asymmetric Content e^+ or e^- in Muon. It Slows down Annihilation of LMC e^+e^- in Muon, just like e^+ or e^- in Charged Pion Slowing down Annihilation of HMC $u\bar{u}$ (see page 84). Pion-Life Shorter than Muon Simply because HMC Has Much Smaller Size than LMC (see pages 79, 83, 85).
- In Sharp Contrast, Life-Inequality for Ps 2B-System e^+e^- in 2s-Ps/*o*-Ps/*p*-Ps States

$$1.1\mu s > 1.42 \cdot 10^{-7} s > 1.244 \cdot 10^{-10} s$$

Partially Due to Distance Inequality:

$$\text{d(2s-Ps)} > \text{d}(o\text{-Ps}) > \text{d}(p\text{-Ps})$$

Ignition Time of e^+e^--Annihilation Sensitive to Distance. Large Distance in 2s State Yields Microscopically Long Life of 2s-Ps (1.1μs). Strangely, Distance btw e^+e^- in LMC Much Shorter than what's btw e^+e^- in *p*-Ps & *o*-Ps while Muon-Decay Even Slower ($2.19 \cdot 10^{-6} s > 1.42 \cdot 10^{-7} s$). It Happens because DS-Asymmetric Content (e^- in μ^- and e^+ in μ^+) Slows Down Annihilation of LMC e^+e^-, just like DS-Electron/DS-Positron in π^- / π^+ Slows down $u\bar{u}$-Annihilation of HMC Making Charged Pions Live Longer than 0-Pion.

- (H19) Tells: At Distance Nearby $10^{-8} cm$, MF btw e^- & e^+ about 10^{-5} Times Weaker than Coulomb Force btw Them, Playing Small Role in Spectrum Issue and Issue of Geometric Shape of MPR. Ground State of Positronium Close to S-State. At SA Distances, MF and its Non-Spherosymmetric and θ-Depending Feature Play Important Roles in Binding and Annihilation. In R1, MF btw e^- & e^+ Attractive and Repulsive for *o*-Ps (Anti-Parallel MMs/Parallel Spins) and *p*-Ps (Parallel MMs/Anti-Parallel Spins) respectively. In R2, MF btw e^- and e^+ Repulsive and Attractive for *o*-Ps and *p*-Ps respectively. e^+e^--Annihilation in *o*-Ps Goes through R1 while It in *p*-Ps Goes through R2. Shorter Life & Faster Annihilation of *p*-Ps than *o*-Ps due to Larger Attractive MF in R2. See (H16) for MF in R1 and R2.
- For Positronium, MF Brings about Measurable Effect Such as Difference btw its Two States: *p*-Ps and *o*-Ps. For Hydrogen Atom, MF Seems to Have Caused Negligible Effects or Hyperfine Effects Only. Reason is Obvious: Magnitude of MM of Proton Much Less than That of Electron or Positron.

- Quantum States of MF-Bound Particle Unknown/Complicated. For Instance, We Don't Know Details of Non-Radial Excitations. We Adopt Sample Spin Flip.
- **Spin Flip** of $u\bar{u}$ 2B-System Leads to **Non-Radial** Excitation of HMC $u\bar{u}$ with Opposite Spin Directions of $u/\bar{u}$ (without OSTP) and with $\bar{u}$ in V-Vane at u, or Equivalently, u in V-Vane at $\bar{u}$. Such $u\bar{u}$ Spin-Flip Takes Place as Pion Meson Turns itself into Rho Meson after Receiving Certain Amount of Energy and Gets Excited. It Also Occurs as $\eta/\eta'/\eta_c/\eta_b$ Turn Themselves into $\omega/\phi/J/\psi/\Upsilon$ Respectively after Receiving Some Energies and Getting Excited.
- Spin Flip May Be Quantized, Somehow θ-Depending now that MF Field is θ-Depending. We Do Not Know Details.
- **Spin Flip** of d-Electron Turns d-Quark into s-Quark by Making It Relocating from R1 to R2 at u-Quark.
- Common Feature of **Spin-Flip UE** is Extremely Fast De-Excitation with Such Short Lifetimes of Vector Mesons of $J=1$ and Baryons of $J=3/2$, Ranging from 10^{-20} to 10^{-24} sec. **Only Exception:** $8.21\cdot 10^{-11}$ sec is Ω's Lifetime. It May Be an Evidence that Ω Has No *u*-Core Excitation While Delta and Σ^* Baryons with 3/2 J-Label Do Have *u*-Core Excitation. In Addition, Evidence of J=3/2 Label for Ω Doesn't Exist for Composite Particles Don't Spin and TAM of Their Constituent Particles Not Measurable, as Carefully Shown Before in Section "Unknowns, Spin Misconception, TAM Disability, …". Symmetry, Represented by Group Theory in Mathematics, Was Thought to Be Powerful Enough to Predict and Explain Existence of Particles. A Natural Extension of Such Idea is to Propose Some Symmetries, like *Supersymmetry*, that Has Been Used to Predict Existence of New Particles Never Discovered. Among All Photos, Taken in Past 70 Years, No Single One was Found to Support Existence of Such Hypothetic Particles. Symmetry in Group Theory Not Mathematical Principle of Natural Philosophy for Composite and Elementary Particles. Dynamics and Structures that Have Played Backbone Role in Physics Since Newton, Einstein, and Founders of QM and QFT w/o RC Cannot Be Replaced with Symmetries. Quark-Model Most Important Milestone on Way to Reveal SA Secretes because of its Quark Content Scenario So Right Everywhere in SA World for All Hadrons (Even DS Cannot Change Quark Content of Any Hadron though DS Further Considers Deeper Structure of d, s, c, b, and t Quarks), Not because of Hypothetic Relation btw Group-Symmetries & Existence of Hadrons. Truth is: Experimenters Had Easily Found many new Hadrons far beyond Original Meson Octet/Baryon Octet/Baryon Decuplet since Long Time Ago.
- *Atomic De-Excitations* Belong to Atomic Electrons. By *SA De-Excitations* We Mean De-Excitations of Excited Nuclei, Excited DS-Leptons in s, c, b, t Quarks, Excited Q-Cores (HMC/BC/Anti-BC) in Meson/Baryon/Anti-Baryon, Possibly in t, b, c Quarks also, and Excited LMC in Tau. Higher J-Labels are for Hadrons with Non-Radially Excited Q-Cores. LMC in Tau Radially Excited. Non-Radial Excitations would Make e^+e^- in LMC Depart R1 at Each Other so that MF Attraction Greatly Reduced Unable to Bind Light e^+e^- within SA Small Spaces. If Experiment Shows Non-Existence of Non-Radially Excited State of Muon, It May Imply that Non-Radially Excited State of LMC e^+e^- Cannot Live Long

Enough to Serve as LMC to Stick Some Other Leptons to Form Muon or Tau. Non-Radial Excitation in R2 would Force e^+e^- to Oscillate and End up with Annihilation Immediately so that It Could Not Be Treated as Particle. Some States of Hadrons May Be Due to Radially Excited Q-Cores.

- Crucial Difference btw Atomic and SA De-Excitations Seen in ERM: Energy Release in Atomic De-Excitation Does Not Go via Measurable $l\bar{l}$ / $\nu\bar{\nu}$ / $q\bar{q}$ Pair Production Channels due to Conservation of Energy. Atomic De-Excitations, at Least Apparently, Release Photons Only. SA De-Excitations Originally Release $u\bar{u}$, e^+e^-, and $\nu\bar{\nu}$ Pairs as well as Photons. In Many SA De-Excitation Cases PAPP Channels Main Channels while PR Channel Suppressed. If Unification Philosophy Right for ERM then Unified ERM Requires Existence of Radiation of LE (Low Energy) $\nu\bar{\nu}$ Pairs in Atomic De-Excitations. So far LE NFMs Not Detectable. Future Experiment May Finally Verify if ERM also Works for Atom as Discussed below:
- Experimenters Seem to Never Try to Create Atomic Excitation of Isolated Single Atom or Atom in Near 0-K Environment, Free from Thermal Collision and from Getting Excited in Thermal Collision and from Being Hit by Any Uncontrolled Intruding Photon, Hit Only by Single Photon in Control. Then Single Photon Detection Technology would Reveal If Conservation of Energy Permits Low Energy $\nu\bar{\nu}$ -Pair Production in Atomic De-Excitation. We Do Not Know if There's Such Technology to Allow Experimenters to Prove or Disprove Any Missing Photon Radiation in Atomic De-Excitation if Detection of Low Energy $\nu\bar{\nu}$ Particles Much More Difficult than Above Said Single Atom and Single Photon Technique.
- PAPP Now is Uncertain in Atomic Physics. But Experimenters Proved PAPP in SA Collisions and SA De-Excitations almost Every Day in Past 100 Years, in Recorded Cosmic Shower Events and HE Collision Events in Accelerators, as long as We View These Events in Terms of DS. Surely, Energy Conservation Allows ERM to Go through PAPP Channels in SA Physics.
- K/L Shell Electrons Atomic Electrons. **Electron-Capture** Not Atomic Nor SA De-Excitation: Shell-Electrons Step Down to Lower Potential Regions around u-Quark Cores of Protons. It's Called **ASA** (**A**tomic-**S**ub-**A**tomic) De-Excitation.
- Continuous Energy Spectrum of Emitted beta-Particles Proves Non-Discrete Feature of Quasi-Bound d-Electron Wd-e in Neutron. Solution of 2B-Problem Proves that *Ground State Energy*, as Sum of KEs of All Constituent Particles and PE of Entire System in Ground State, is Constant (Time-Independent). But Each Individual Particle in Unknown and Unsolvable Potential Field Produced by its Partner in Unknown Motion. Solved Quantum 2B-System Never Makes Any Individual Particle's Motion Solved.
- Quantum State, Such as Quantum State of Wd-e in Neutron, Bound or Quasi-Bound, Has Never Been Solved and May Be Unsolvable. But We'll soon See:

1BEx Proves **MBT** (**MB-Theorem**) about *Time-Varying* Feature of Individual Particle's ME. MBT and **MBP** (**MB-Postulate**) Naturally Explain *Continuous* Energy Spectrum of β -Particles. Also, MBT, MBP, DS, ENM, and MF-Binding Feature Shown in TMET Explain Simultaneously Stable Free Proton with its d-Quark Never Decaying, Never Having Positive or 0 or Near-0

ME and Unstable Neutron with 1 of its d-Electrons (Wd-e) Having Positive, 0, or Near-0 ME in about 15 Minutes.

< **Bizarre Uniqueness** in SA World >

All SA Composite Particles're Unstable Except Free Protons.
Such Bizarre Uniqueness Can Only Be Explained in Terms of DS with MB-Theorem and MB-Postulate. To Explain beta Decay, SM without DS Has to Postulate W-Particles and Decay Modes

$$d \rightarrow u + W^- , \quad W^- \rightarrow e^- + \bar{\nu}_e$$

Then Such Idea was Immediately Disproved by Observed Stability of Protons which Contain Stable **dQ**s (**d**-**Q**uarks).

Among All Unstable SA Composite Particles, Neutrons Only Ones with Average Lifetime Much Longer than All Other Unstable SA Particles. Reasons: Neutron Not Excited State, β -Decay Not Fast SA De-Excitation, Electric Repulsion Co-Exists with Attractive Forces, and SA Magnetic Force Constantly Changes Direction of Velocity. (See more reasons in Appendix)

Note: Charged Particle Makes Spiral Track Moving in Uniform Magnetic Field as Seen in Bubble Chambers. As Uniform Magnetic Field Bends, Twists, and Becomes Complicated just like SA Magnetic Fields inside Neutron Produced by u-Quarks and d-Electrons with Rapid and Complicated Motions, We do Expect: Spiral Tracks of Charged Particles Also Bend and Twist, Forming '***Knitting-Ball***'. For Single Bound d-Electron (SBd-e) in Proton & Sd-e in Neutron, 'Knitting-Ball' is Closed while for Wd-e in Neutron, It's Open but Greatly Postpones Final Leaving Time of Wd-e. DS, Including This Explanation of Microscopically Long Lifetime of Neutron Similar to Bohr Model, Needs to Be Proved Theoretically and Quantitatively. In Sharp Contrast with Bohr's Model, Insolvability's Proved by DS Revealing Reality of MB-Systems and Other Difficulties in SA Physics while Bohr's Model was Soon Proved Theoretically and Quantitatively by Great Schrödinger and Dirac. If this Feature of SA Physics Can Be Recognized, Many Physicists would Stop Searching for Quantitative Theoretical Results. In History, Mathematicians Finally Abandoned Their Hard Works to Try to Get and Prove Formulas of Solving Polynomial Eqs of Higher Degrees after Évariste Galois Proved Impossibility.

- **PNMI** (**P**roton-**N**eutron **M**ass **I**nequality) $m_n > m_p + m_e + m_\nu$ due to Change in TME of Proton's 5-Body System after One More Pair of d-Leptons Step in. Denote by $p \,/\, p'$ Proton's 5-Body Composite System before/after Extra Pair of d-Leptons $e^-\bar{\nu}_e$ Step in to Form Neutron n . Due to Large Perturbation from New Comers ($e^-\bar{\nu}_e$), 5-Body System p Changes to p' with its 5 Constituent Particles Same as in p but in Different State. Neutron n 100% Sure Not p plus $e^-\bar{\nu}_e$ but p' plus $e^-\bar{\nu}_e$. No Empirical/Theoretical Evidences of such Inequality as $m_n > m_{p'} + m_e + m_\nu$. $m_{p'}$ Empirically Immeasurable and Theoretically Unknown. PNMI can't Be Treated as Evidence Supporting SM without DS. DS is So Natural and Powerful. It Explains too Many Tough Experimental Facts SM without DS Has Failed to Explain Since the Beginning of SM without DS.

- *Stable* Free Proton, *Unstable* Neutron, *Continuous* Energy Spectrum of β - Particles, and other Related Facts, such as PNMI, Can Be Further Explained via MB-Theorem, MB-Postulate, and **1BEx** Seen below:

Every n-Body (n=2, 3, 4, ...) System Can Be Divided into 2 Sub-Systems: 1-Body System plus (n-1)-Body System. By 1BEx of n-Body Problem We Mean **1-Body** in **Resultant Force Field** Produced by (n-1)-Body System. If That 1-Body Far away from Constituent Particles of (n-1)-Body System Bound in Small Space, (n-1)-Body Can Be Treated as Single Particle in 1BEx and Can Be Represented by Force Field It Produces, such as Atomic Electron in Hydrogen where Constituent Particles Bound inside Proton are Treated as Single Charged Particle Represented by Coulomb Central Field it Produces. Then 1BEx of Hydrogen Becomes Solvable. In SA World, Insolvability of n-Body System with n>2 Means Impossibility to Know Resultant Force Field Produced by (n-1)-Body System. Even if Such Complicated Resultant Force Field is Revealed, It Must Be ***Time-Dependent*** (see details below) and Very Complicated so that 1-Body in Such Force Field Leads to Unsolvable 1BEx. Anyone Who Solves 1BEx Solves n-Body Problem. Even for 2B-System Bound by Coulomb Type Force, which Has Been Solved, Only TME of Entire System is Proved to Equal Eigen-Energy of Fictitious Particle of Reduced Mass in Central Coulomb Type Field as Energy Solution of such 1BEq (**1-B**ody **Eq**uivalence). Energy and Quantum State of Each Individual Particle Remain Unknown. A 2B-System is Solvable if its 1BEq is Solvable. 1BEx of SA Composite Systems Unsolvable. But 1BEx Can Easily Prove MB-Theorem (See Page 35). 1BEx Shown below.

< **1BEx** of n-Body Schrödinger Eqn >

SA Interactions Depend on Position Vectors, Velocities, and Spins. Operators of NR KEs, PE of n-Body System, and n-Body Schrödinger Eqn are Written

(H20a) $$\hat{E}_{Ki} \equiv -\hbar^2\nabla_i^2/2m_i = -\frac{\hbar^2}{2m_i}\left(\frac{\partial^2}{\partial x_i^2}+\frac{\partial^2}{\partial y_i^2}+\frac{\partial^2}{\partial z_i^2}\right) \quad (i = 1,2,...n),$$

$$U = U(\vec{r}_1, d\vec{r}_1/dt, \vec{S}_1, ..., \vec{r}_n, d\vec{r}_n/dt, \vec{S}_n)$$

$$i\hbar\partial_t\psi = (\Sigma\hat{E}_{Ki} + U)\psi$$

Even if Solution $\psi = e^{-iEt/\hbar}\phi(\vec{r}_1,...,\vec{r}_n)$ were Stationary, Insolvability still Visible. In QM, E Means ME. Picking jth Constituent Particle for 1BEx, We Write Schrödinger Eqn for nB System as

(H20b) $$(\hat{E}_{Kj}+U)\phi = \widetilde{E}\phi, \quad \widetilde{E} \equiv E - (\sum_{i\neq j}\hat{E}_{Ki}\phi)/\phi$$

PE U Unknown before MB Problem Solved to Tell what're Position Vectors and

Velocities which Determine *U*. (H20b) Not Solvable. However, **MB-Theorem** is Proved. Statement of MB-Theorem and Proof Given on Page 35.

< Empirical **Evidences** of **MB-Postulate** >

Statement of MB-Postulate Given on Page 36. For 2B-System with PE of Form $U = U(x_2 - x_1, y_2 - y_1, z_2 - z_1)$, SSE (Solution of Schrödinger Eqn) Has Been Studied. It's Proved that TME Same as 1BEq of Fictitious Particle with Reduced Mass in Central Potential Field of PE $U = U(x, y, z)$. TME of Such 2B-System Has Constant Value Determined by IC. If PE $U = U(x, y, z)$ Leads to Bound States in 1BEq, Such 2B-System Also Bound with TME Having Discrete Constant Values Depending on ICs. Even in This Example of Solvable 2B-system, ME and Wave Function of Each Individual Particle Remain Unknown as Seen in 1BEx. For MB-System, We Can Only Propose *MB-Postulate* and Make Sure It Agrees with *Experimental Facts* as Shown Below:

Constant TME Corresponds to Observed *Stable Mass* of SA Composite Particle such as Proton and Neutron. *Time-Varying* and *Non-Discrete* ME of Wd-e in Neutron Can Explain beta *Decay*, *Continuous* Spectrum of beta Particles, and Variety of Lifetimes of Neutrons Simultaneously. Lifetimes of Free Neutrons Vary for They Born to Have Different IC for Wd-e, Leading to its Different ME Function as Existing Solution of 1BEx, Meaning that It Takes Different Time to Turn its ME to be Near Zero or Positive and to Open QTC for Decay to Occur.

MEs of SBd-e in Proton and Sd-e in Neutron Also Non-Discrete, Time-Varying, and IC-Depending but Never Have Positive, Zero, or Near-Zero Negative Value so that It's Bound at All Times, Never Being Able to Relocate in Free Regions Outside Nucleons via QT, Agreeing with Fact of Stable Free Proton and Fact of Non-Existence of Double-Electron Emitting from One Neutron. ME of Wd-e in Neutron Time-Varying but May Have Positive, Zero, or Near-Zero Negative Value in a Variety of Lifetimes (Averagely about 15 Minutes). Positive, Zero, and Near-Zero Negative Values of ME Allow Wd-e in Neutron to Step Out and QT-Relocate in free Regions Outside Neutron or Nucleus. In Neutron, Sd-e Exerts Repulsive Coulomb Force on Wd-e. Repulsive Force Adds Positive Values to ME. Due to Lack of Repulsive Force from Another d-Electron, Single d-Electron in Proton Can Have Negative ME, Not Near Zero at All Times & Never Being Able to Relocate Outside. With Repulsive Force from Wd-e, Sd-e in Neutron Can Still Have Negative ME Not Near Zero at All Times for It's Closer to u-Quark and u-Core. Nucleons are MB-Systems in Ground States. β-Decay NOT SA De-Excitation.

In Some Special MB-Systems Time-Dependence of a Constituent Particle's ME Negligible. Planets in Solar System Attract Each Other with Gravitational Force Much Smaller than What is Exerted on Each of Them by Much Heavier Sun. This is Why Kepler's Laws with Stationary Elliptical Trajectory and Constant ME of Each Planet Could Match Observation and Newton's Calculation via 1-Body Treatment So Accurately. Perturbations from Other Planets Play very Tiny Effects Only. Repulsive Coulomb Forces btw Atomic Electrons in Atom with

Small Z (>1) Also Much Smaller than Attractive Coulomb Forces Exerted on Them by Nucleus. MB-Problems Unsolvable. But 1-Body Approximation for Atoms with Nuclear Charge *Ze* Proves to Be in Good Agreement with Empirical Spectra if Z Not Large. Most Probable Distance btw Innermost Electron and Nucleus is $r_z = r_b / Z$. Most Probable Distance btw Outermost Electron and Nucleus is Approximately Equal to Bohr Radius r_b (Effective Charge to Bind Such Outermost Electron is Roughly e Not Ze). Suppose Z Electrons Evenly Occupy Space of Volume $4\pi r_b^3 / 3$. Most Probable Distance btw Neighboring Atomic Electrons is Approximately ${}_e^e d = \sqrt[3]{4\pi r_b^3 / 3Z}$. Thus Repulsion btw any 2 Atomic Electrons is Much Weaker than Attraction from Nucleus Ze:

(H21a)
$$f_{\max}^{rep} \sim e^2 / {}_e^e d^2 = \sqrt[3]{9Z^2} e^2 / \sqrt[3]{16\pi^2} r_b^2 = 0.385 \sqrt[3]{Z^2} e^2 / r_b^2 ,$$

$$f_{\min}^{att} \sim Ze^2 / r_b^2 > f_{\max}^{rep} , \qquad f_{\min}^{att} / f_{\max}^{rep} \sim 2.60 \cdot \sqrt[3]{Z} .$$

This Insures Domination of Attraction from Nucleus over Repulsions from Other Atomic Electrons. Repulsions from Most Other Atomic Electrons Much Weaker than $f_{\max}^{rep} = e^2 / {}_e^e d^2$ for Distances are Much Longer than ${}_e^e d$. Said Repulsive Forces Not in Same Direction. Their Resultant Also Much Weaker than Attraction from Nucleus. Electrons in Atom Repel and Stay away from Each Other as far as They Can so that Perturbations from Other Atomic Electrons Cause Small Effects Only and 1-Body Approximation that Treats Many-Body Atom as One Atomic Electron in Coulomb's Time-Independent Central Field of Source Charge Ze as Single Charge Could Yield Good Results for Ground State Atomic Electron. Hydrogen Atom MB-System if Quarks in Proton Taken into Account. But It Can Be Treated as One Atomic Electron in Coulomb Central Field Produced by Effective Charge e for Charged Constituent Particles in Proton Stay in Small SA Spaces while Atomic Electron Stays far away within Atomic Space much Larger than SA Spaces. Heaviness of Proton Makes 2B Treatment of Hydrogen Atom Yield Nearly Same Results as 1-Body Approximation. Above Mentioned Exceptions Don't Apply to SA Composite Particles.

Stable Proton with SBd-e (Single Bound d-Electron) Tells: SBd-e 1B-System in Unknown Time-Dependent Force Field, Produced by Other Four Constituent Particles of Proton, Has Time-Dependent ME ***Negative*** Enough at All Times.

- Unstable Neutron/beta Decay Tell: Wd-e in Neutron is 1-Body System in Unknown Time-Dependent Force Field, Produced by Other Six Constituents of Neutron. It Has Time-Dependent ME ***Negative*** Enough at All Times before Decay Occurs. Time-Depending ME of Wd-e in Neutron Dooms to Change its Value from Negative to Non-Negative or Positive in a Variety of Lifetimes Averagely about 15 Minutes.
- We're Able to See Such Amazing Physical Picture Not Because Any One Can Solve MB-System and Prove It but Because Basic Empirical Facts and DS Join Forces to Enable Us to Penetrate Seemly Impenetrable Dense 'Fog' that Covers SA World.

< Three **Non-De-Excitation** DS-Channels of Decay: **KC**, **PPC**, **QTC** >

- $\pi^0\pi^\pm$ are in Ground States and Must Decay via Non-De-Excitation Channels.
- Somehow One May Treat LMC e^+e^- in Muon as in Deep SA MF-Excitation State while Positronium e^+e^- as in Ground State with MF Playing Small Hyperfine Effect Only at Atomic Distances. LMC e^+e^- in Muon Produced in Deep SA Space with e^+e^- So Close to Each Other that Annihilation Channel Dominates. Please Remember, at Distances Shorter than 0TMED, the Higher the Excitation, the Larger the Positive TME and the Smaller the Binding Size.
- Three Types of Non-De-Excitation DS-Channels of Decay are

1. **KC** (**K**inetic **C**hannel) of PAPA: PAP May Annihilate into ***Real*** Carrier(s) of Force (so far only Real Photons Seen) and/or Timelike ***Virtual*** Force Carrier(s) for Some ***Bound*** Particle(s) to Absorb, Gain KE, Change ME to Positive, and Become Free.

   Remark: Energy Conservation Forbids Free Elementary Particle to Absorb or Emit any Timelike ***Real*** Particle (Force Carrier) and Only to Gain or Lose KE, as Proved in Chapter 1. SR and QFT without LV Proved Long Time ago: Free Electron Cannot Absorb nor Emit Massless Real Photon to Gain or Lose KE for Conservation of 4-Momentum Forbids Such Event. Theory with LV and Theory without LV Reach Same Conclusion in this Issue: Free Elementary Particle Can't Absorb Nor Emit Real Force Carrier. Experimenters Never See Violation of This Conclusion.

   Elastic Scattering, however, Allows Elementary Particle Other than Force Carrier to Absorb or Emit ***Virtual*** Force Carrier and Change its 4-Momentum when Another Elementary Particle is Present to Emit or Absorb the Said Virtual Force Carrier. But Transferred 4-Momentum then Must Be **Spacelike**. Each Individual Force Carrier is either Massless in Massless QFT or Massive in Massive QFT. Transferred p^μ That Serves as Label for Internal Line in Feynman Diagram is Forbidden to Be Carried by any Individual Force Carrier. It's Just Effect of Virtual Force Carrier Exchange. QFT itself Explicitly Shows: p^μ of each Individual Virtual Force Carrier Appear in 3-dim Propagator. 4-dim Version of Propagator Contains Mathematical Parameter Irrelevant to Energy Carried by Individual Force Carrier. Use of Such Mathematical Parameter is due to Application of Integration of Complex Function. (A Mathematician Said: Greatest Discovery in 20th Century is Complex Plane.) But It's Not a Physical Quantity though It Has Energy Dimension. This View of QFT Immediately Explains why We See Examples of Neutral Currents with Transferred Energy Much Smaller than Z-Boson's Rest Energy.

   Photoelectric Effect Proves that Bound Electron Can Absorb Real Photon to Change its ME from Negative to Positive and Become Free. Atomic Physics and QM Proves: Electron Bound in Atom Can Absorb a Real Photon to Jump to Higher Level while Excited Electron Bound in Atom Can Emit Real Photon to Return to Ground State or Lower-Level State.
   All Things above Justify KC of Decay in DS Model. Such Particular DS Decay

Channel is SA Version of Some Things We See in Non-SA World.

2. **PPC** (**P**air **P**roduction **C**hannel) of PAPA: PAP May Annihilate and Turn Itself into Another PAP or More than One New PAPs. Note: PPC of SA De-Excitation is Denoted as PAPP. PPC of PAPA Non-De-Excitation Cannel.

3. **QTC**: In Many Cases, MFs Create Bound or Quasi-Bound States for DSLs with Positive ME so that They Can Penetrate Deep PW through QT Fleeing into Free Regions with Same Positive ME. QTC May Be Dominating or Be Dominated by Other Channels of Decay.

PPC is ***Shut*** in Muon-Decay due to Energy Conservation. Energy Released in Muon-LMC e^+e^- Annihilation Not Enough to Produce Lightest $u\bar{u}$ 2B Bound System. It's Open in Tau-Decay for Tau-LMC Much Heavier.

PPC is ***Open*** and ***Dominating*** in $\pi^{\pm}$ -Decay (SDL)

KC ***Dominates*** in $\mu^{\pm}$ -Decay for **PPC** is ***Shot*** in $\mu^{\pm}$ -Decay (SDL)

KC ***Suppressed*** in $\pi^{\pm}$ -Decay for **PPC**'s ***Open*** & ***Dominating*** in $\pi^{\pm}$ -Decay

QTC ***Suppressed*** in $\mu^{\pm}$ -Decay: It Has Little Chance ($\sim 3.4\cdot 10^{-5}$) for LMC e^+e^- Bound with TME>0 to Penetrate Deep PW and Become Free with Released $e^+\nu_e\bar{\nu}_\mu$ Becoming Free too and Finishing Muon-Decay. This is a PQTR.

QTC ***Suppressed*** in $\pi^{\pm}$ -Decay: $\pi^+ \to e^+\nu_e\pi^0$ and $\pi^- \to e^-\bar{\nu}_e\pi^0$ ($1.036\cdot 10^{-6}$).

QTC ***Shut*** for HMC $u\bar{u}$ in Any Meson Decay for $u\bar{u}$ Bound with TME Negative Enough at All Times.

Please Note, Atomic Electrons Have No Chance to Penetrate PW through QT and Become Free if No Any Energy Received: Energy Conservation Forbids Them to Change Negative TME to Positive Automatically without Any Energy Received. QT Obeys Conservation Law of Energy. But LMC e^+e^-, Bound in Deep Negative PE Region, Have Positive TME According to TMET with MF-Binding. QT Allows Them to Penetrate Deep PW and Become Free with Same Positive TME. But in Muon Decay, QTC Severely Suppressed by KC in LMC-Annihilation while Another Annihilation Channel PPC is Forbidden in Muon Decay for Reaction

Muon-LMC $e^+e^- \to u\bar{u}$ Forbidden by Energy Conservation

Mass Inequality $m_\pi > m_\mu$ Implies Larger Mass of $u\bar{u}$ Composite Bound System than LMC e^+e^- Composite Bound System while Heavy u-Mass Required by Small Nucleons' MMs Gives Larger Value of $2m_u$ than Heavy Mass of 0-Pion that is $u\bar{u}$ Composite Bound System (HMC).

< Sharp **Contrast** btw **Muon**'s Decay and **Charged Pion**'s Decay >

- **QTC** for u and $\bar{u}$ in ***Pion*** Decays Leading to Free u and $\bar{u}$ among Final State Particles is Forbidden (***Shut***) Showing Negative Enough TME of HMC $u\bar{u}$ Bound System. $u\bar{u}$ Can Tilt Their Spins to Turn off MFs Completely or Nearly Completely so that Dominating IPE Binding Leads to Negative TME of $u\bar{u}$.

- **QTC** for DSLs in Charged-***Pion*** Decays Narrowly Open, Suppressed by PPC of $u\bar{u}$ -Annihilation.
- **QTC** in ***Muon*** Decays Leading to Free e^+e^- among Other **FSP**s (**F**inal **S**tate **P**articles) is ***Open*** but ***Narrow***, Suppressed by e^+e^- Annihilation Channel.
- **PPC** of $u\bar{u}$ Annihilation in ***Pion*** Decay is ***Allowed*** by Energy Conservation. Once PPC Open, It Immediately Dominates **KC** of $u\bar{u}$ -Annihilation Due to **SAVI** with **PAPP** Having Priority to Consume Energies Released in SA De-Excitations, Collisions, and Annihilations as Effects of Virtual Force Carrier Exchanges. Inside π^+ , HMC $u\bar{u}$ Has Very Small Chance (~0.013%) to Annihilate into Virtual Photon and/or Z-Boson Absorbed by e^+ and/or ν_e . Charged Pion's Decay Mode via KC $\pi^+ \to e^+ + \nu_e$ Rare. PPC $u\bar{u} \to e^+e^-\nu_\mu\bar{\nu}_\mu$ of HMC Annihilation in π^+ Decay Dominates Other Channels due to SAVI, which also simultaneously Explains FQA Fact no matter QKoE Higher, Lower, or Much Lower than Energy Scales of Accelerators.

Remark: Bound Atomic Electrons Can Only Relocate to Other Atoms Nearby through QT and Be Bound There. Without Receiving Any Energies, They Can Never Escape from Bound States and Become Free for Free Regions Don't Take any Particle with Negative TME as They Initially Have in Bound States inside Atoms. Positive TME of Final State Particles in Muon Decay Simple Fact seen every day in Labs. MF-Binding & TMET Theoretically Proves Positive TME in Some SA Bound States.

Now Let's Use Abbreviations PiD (Pion Decay), PiDKC, PiDPPC, PiDQTC, MuD (Muon Decay), MuDKC, MuDPPC, MuDQTC to Write π^+ / μ^+ Decay Modes (Channels) in Terms of DS:

$$\pi^+(u\bar{u}e^+\nu_e) \to$$

PiD**KC**, Rare 0.0123% :

$$e^+\nu_e,\ (u\bar{u} \to \text{Virtual } \gamma/Z \text{ Absorbed by } e^+\nu_e),$$

PiD**PPC**, Dominating 99.99%:

$$\mu^+(e^+e^-e^+\nu_e\bar{\nu}_\mu)\nu_\mu,\ (u\bar{u} \to \text{Virtual } \gamma/Z \to e^+e^-\nu_\mu\bar{\nu}_\mu),$$

PiD**QTC**, Rare $1.036 \cdot 10^{-6}$:

$$u\bar{u}e^+\nu_e,\ (\text{Bound } u\bar{u}e^+\nu_e \to \text{Bound } u\bar{u}\ (\pi^0) + \text{Free } e^+\nu_e$$

$$\mu^+(e^+e^-e^+\nu_e\bar{\nu}_\mu) \to$$

MuD**KC**, Dominating ~100%:

$$e^+\nu_e\bar{\nu}_\mu,\ (e^+e^- \to \text{Virtual } \gamma/Z \text{ Absorbed by } e^+\nu_e\bar{\nu}_\mu),$$

MuD**PPC,** Forbidden 0%:

$$\pi^0(u\bar{u})e^+\nu_e\bar{\nu}_\mu,\ (\text{LMC } e^+e^- \to \text{Virtual } \gamma/Z \to u\bar{u}),$$

MuD**QTC**, Rare $\sim 3.4 \cdot 10^{-3}$%:

$e^+e^-e^+\nu_e\bar{\nu}_\mu$, (Bound $e^+e^-e^+\nu_e\bar{\nu}_\mu$ →Free $e^+e^-e^+\nu_e\bar{\nu}_\mu$)

- Negative TME of $u\bar{u}$ in π^0 is a Key Reason for π^0-Mass to Be Lighter than Heavy u-Mass. Heavy u-Mass in HQMT and OSOD in DS with MF-Binding Explain Anomalous MMs of Nucleons. At Same Time, Heavy u-Mass Allows St Coulomb Force to Bind $u\bar{u}$ in π^0 within SA Short Distances as MF and St MF are Turned off Quite Completely due to OSTP.
- In Sharp Contrast Again, If e^+e^- Tilted Their Spins and Turned off MF Completely or Nearly Completely, They Could No Longer Form LMC e^+e^- of SA Small Size. In Positronium, e^+e^- Turn off MF So Completely Mainly due to Atomic Distances Much Longer than MD btw e^+e^-.
- Positronium e^+e^- Lives Shorter than LMC e^+e^- of Muons $\mu^+(e^+e^-e^+\nu_e\bar{\nu}_\mu)$ and $\mu^-(e^+e^-e^-\bar{\nu}_e\nu_\mu)$ Because DS-Electron/DS-Positron Slows Down LMC e^+e^- Annihilation in μ^-/μ^+, just like DS-Electron/DS-Positron in π^-/π^+ Slows down $u\bar{u}$-Annihilation of HMC Making Charged Pions Live Longer than 0-Pion. $d\bar{d}$ Quark-Content in 0-Pion, if any, Has DS-Symmetry of Both DS-Electron and DS-Positron. They Annihilates Very Quickly due to Small Mass and Large MF Attraction btw e^+e^- in DS of PAP $d\bar{d}$. They Annihilates So Quickly that They Unable to Slow down $u\bar{u}$-Annihilation so that 0-Pion Has Very Short Lifetime even if Mix $d\bar{d}$ Exists in 0-Pion. (SDL about lifetimes)

< DSL-Knockout >

- Ironic FQA Fact Refers to Ironic Fact that No Single Unbound Quark Has Ever Been Found among MSPs and FSPs in All Recorded Events. But Free Electrons and Free NFMs Found among MSPs and FSPs Every day.
- We Have Explained FQA Facts with Two Different Possible Interpretations, i.e., QNC with AVI and QC with NLVUC on Pages 42, 43.
- Quarks are Heavy and Bound within Tiny SA Spaces. It's Possible that QKoE is above NLVUC so that Quarks Can Never Be Knocked Out. But DSLs are Light and Bound within Larger SA Spaces. LKoE is Unlikely to Be above NLVUC. Just like QNC with AVI, DSLs Might Also Be Knocked Out Millions of Times Every day in Cosmic Showers and Collisions in Accelerators.
- It's Not Clear in Direct Observation whether Lepton-Knockout Event did Occur for $u\bar{u}$ Pairs Produced in Pair-Productions of DI Collisions May Capture DSLs via QT before Knockout Can Happen. Namely, Whether or Not LKoE is Far Below, Below, Close to, or Higher than Energy Scales of Large Accelerators, DSLs May Relocate via QT in **PW**s around $u\bar{u}$ Pairs Produced in Pair-Productions. Here by PW We Mean Potential Well or Negative Potential Region Usually Lower than its Surrounding.

< **SPS** (**S**un-**P**lanet-**S**atellite) System >

- In View of DS, Meson is Produced in Processes with Common $u\bar{u}$ -PAPP, in Most Cases Followed Immediately by $u\bar{u}$ -Capturing QT-Relocating DSLs, either Constituents of Struck Hadron and **Heavy Leptons (Muons/Taus) or Coming from Leptonic Pair Production such** as e^+e^-, $\nu_e\bar{\nu}_e$, and/or $\nu_\mu\bar{\nu}_\mu$. If there is any Neutral Meson whose Quark Content is $u\bar{u}$ without $d\bar{d}$ and $s\bar{s}$, then It is Produced in $u\bar{u}$ Pair Production that is by Chance Not Followed by Capturing any DS-Leptons. Say again that OSTP Shows: HMC $u\bar{u}$ in π^0 May Be Unable to Bind d-Electron and $\bar{d}$ -Positron Simultaneously to Form $d\bar{d}$.
- Single d-Quark ($ue^-\bar{\nu}_e$) in Proton is a **SPS** (**S**un-**P**lanet-**S**atellite) System. In **SQC** (**S**mall **Q**uark **C**ore) Sub-Model, Quark Core Serves as 'Sun'. At Present, We Do Not Know Exactly Largest Size of SQC and Smallest Size of RQC.
- From (H24d) on Page 77 and HMC $u\bar{u}$ Ground State Table on Page 79, Size of HMC $u\bar{u}$ as π^0 May Range Approximately from $2\cdot 10^{-15}$ to $3\cdot 10^{-19}cm$ as u-Mass May Range Approximately from 320 MeV/c^2 to 30 GeV/ c^2. Small Size of HMC Indicates that HMC Serves as 'Sun'. For HMC $u\bar{u}$ EqM 2B-System, We Have Hypothetic Quantum State OSTP to Suppress MF Greatly to Explain Small Masses of Pions under Heavy u-Mass Settings. But for BC uuu EqM 3B-System, Author Couldn't Figure Out its Possible Quantum State.
- Charged Pions're 4B-Systems in View of DS: $\pi^-(d\bar{u}) = ue^-\bar{\nu}_e\bar{u}$ for Instance. Tiny Heavy HMC $u\bar{u}$ Effectively Single Heavy Particle with Net MM 2 Times as Large as u-Quark's MM and without Net Charge. DSLs in π^- Effectively Bound by Net MM of $u\bar{u}$, which Serves as 'Sun'. It Allows Us to Ignore Contribution from $\bar{\nu}_e$ for its mass too Small to Make Meaningful Contribution to Mass of Charged Pion π^-. Once OSTP Leads to a Net MM of Tiny HMC $u\bar{u}$, its MF-Field without Any Suppression Plays Key Role to Bind d-e^- in π^- within SA Small Space much Large than Tiny HMC $u\bar{u}$. Difference btw Hydrogen Atom and π^- Clear: MF-Field Produced by Proton-MM Negligible at Atomic Distances while MF-Field Produced by Net MM of $u\bar{u}$ is Not at All Negligible to d-Electron Bound within Small SA Space. Consequently, π^- Not Solved and May Be Unsolvable in SA QM Even though its 4B-System Has Been Reduced to 1B-System while Hydrogen Atom Solved in QM long time ago. However, Physical Picture We See Allows Circular Motion CI to Solve eqn of Motion in Relativistic Mechanics. SDL in Last Section.
- Two Composite Particles with Same Set of Constituent Elementary Particles Not Identical if Identical Constituents inside Two Composite Particles in Different States. Two Composite d-Quarks in Neutron Not 2 Identical Particles.
- *Exclusion Principle* Never Says that 2 Different States Must Be Stable Bound States. Continuous Energy Spectrum of beta Particles Indicates non-Discrete and non-Stationary Features of Energies of Neutron's Constituent Particles before They're Emitted, though Isolated Free Composite Particles, as 2B/MB-Systems in Whole, always Seem to Have Lifetime Constant Stationary Rest Masses no matter They are Stable or Have Lifetimes Ranging from about 15 Minutes, 10^{-6} sec to 10^{-17} sec or Shorter such as 10^{-24} sec.

< C-Q Facts >

- Example of PCQM is Called **1st C-Q Fact**. It's So Natural Since Quantum Theory Adopts Same $E-\boldsymbol{p}$ Relation Drawn from Classical Mechanics for Heavy Bodies. PCQM Shown before is Relativistic Version of 1st C-Q Fact. Due to 1st C-Q Fact and PCQM, We Allowed to Use Classical Theory to Get Hint and Clue under Condition that Quantum Eqs Not Solvable as Certain Types of Departure from IPE are Significant. Classical Imitation Encouraged by 1st C-Q Fact and PCQM. MF Yields Large Departure at SA Distance since it's Inversely Proportional to r^4. Regardless of whether Quantum Wave Eqs in Various MT-Force Fields Solvable, 1st C-Q Fact and PCQM Allow Us to Get Clues for Quantum MF-Effects from Powerful Classical Imitations and Examples.
- **W.K.B.** Approximation Works for Quantum Particle of Small Mass such as an Electron if Interaction Force is of Very Small Magnitude AND its KE is Very Large. We Call This **2nd C-Q Fact**. If One of Two Conditions Does Not Hold, Electron Remains Non-Classical and W.K.B. Approximation Cannot Be Close to Reality. DS-Electron/Positron in Hadrons or Heavy Composite Leptons Do Have Very Large KE. But Interaction Forces of Very Large Magnitudes within SA Short Distances May Invalidate Classical Condition of Radiation. Severe Radiation was Predicted by CED for Electron Moving with Near-c Speed in Circular or Quasi-Circular Motion with Extremely Large Acceleration Written as

$$a = or \approx u^2 / r = (c-\varepsilon)^2 / r\,, \quad r < or \ll r_{sa} \equiv e^2 / m_e c^2 = 2.82 \cdot 10^{-13} cm$$

- 2nd C-Q Fact May Forbid Large Radiation from ANY Charged DS-Leptons with Positive TME, Even if Not Completely. Significant Radiation May Not Happen Even if TME of Some Charged DS-Leptons in Neutron and in Muon is Positive. Please Note, Constant TME Only for Entire Composite System, Not for Any Individual Constituent Particle in 2B or MB-System. If Solution of Quantum 2B System Does Not Tell whether or not Individual Charged Particle Can Radiate in Positive TME Condition, We May Get Hint from Well-Known Conditions to Validate W.K.B Approximation, as 2nd C-Q Fact.
- 2nd C-Q Fact is Supported by Lack of Severe Radiation in beta Decay and during Lifetimes of Unstable Composite Particles. It Allows Neglecting Radiation Fields of Bound and Quasi-Bound Charged DS-Leptons. Thus, We May Ignore 'Acceleration Fields' and Only Take Account of 'Velocity Fields' via Liénard-Wiechert Potentials for Point Charge. 'Velocity Fields' Produced by Point Charge in Translational Motion without Spin Written (J.D. Jackson, "Classical Electrodynamics", 2nd Edition, John Wiley & Sons, p 657, 1975):

$$\vec{E} = e[\gamma^{-2}(1-\vec{\beta}\cdot\vec{n})^{-3} R^{-2}(\vec{n}-\vec{\beta})]_{ret}\,, \ \vec{B} = [\vec{n}\times\vec{E}]_{ret}\,.$$

Effect of Retardation on MF Field Produced by Spinning Charge is Unknown. Since Constituent Particles as Sources of Force Fields inside SA Composite Particles Move Fast and Distance btw Any 2 Interacting SA Constituent Particles is Short, One Can Supposedly Expect Severe Retardation Effect in SA Physics. Even for Atomic Electron, Effect of Retardation is Supposed to Yield Departure from Coulomb Field: $\vec{n}_{ret}$ & R_{ret} almost Constants Due to Much Larger Atomic Sizes than Nucleons'. But Retardation Factor $[(1-\vec{\beta}\cdot\vec{n})^{-3}]_{ret}$ Always Changes its Value as Quark Moves around in Bound State. We Did Not Understand why such Departure from Static Charge's Coulomb Field Did Not Invalidate Solutions of Dirac Eqn in Static Central Coulomb Field. Dirac Complained that People Studied this Field and that Field, but Nobody Studied Velocity Field. May Be, Quantum Wave Eqs in Such Velocity Field Not Solvable. We Now Propose:

< **VFP** (**V**elocity **F**ield **P**ostulate) >

Retardation Factor $[(1-\vec{\beta}\cdot\vec{n})^{-3}]_{ret}$ in Velocity Field Produced by u-Q and d-Electron inside Proton Time-Varying, Making Velocity Field Produced by Proton Radically Different from Static Coulomb Field Produced by Fictitious Point-like Static Charge e that Represents Nucleus of Hydrogen. MPR of Atomic Electron Far Away from Nucleus so that It, with Slow Motion, Not Sensitive to Rapid Local Motions of u-Quarks and d-Electron. Unknown Solution of Quantum Eqn in such Time-Varying Velocity Fields Gives Nearly Same Energy Levels of Bound Electron as Seen in Solutions of Quantum Wave Eqs in Coulomb Field. **OR** Quantum Systems that Invalidate W.K.B. Approximation and Turn off or Severely Suppress Classical 'Acceleration Fields' May Also Turn off or Severely Suppress Retardation Shown in Classical 'Velocity Fields' in Unknown Manner.

- We Call VFP Issue **3Q-Retardation Puzzle** for Proton Contains 3 Quarks in SM w/o DS. If Proton Contained only One Quark, No Retardation would Occur in Proton-Rest Frame. Namely, before Quark Model Reveals Compositeness of Proton, There's No Retardation Issue. Composite Proton Contains More than 2 Constituent Particles in Rapid Motion. Such Reality Makes Us Think why Static Coulomb Field Produced by Fictitious Rest Charge without any Inner Structure Could Yield Good Theoretical Result in QM to Match Experimental Facts.

  Remark: Current Version of VFP Contains 2 Parts Joined by Key Word 'Or'. We've No Theoretical Way to Prove or to Disprove VFP for We Don't Know How to Solve Quantum Eqn in Time-Varying Velocity Fields with Motions of Source Particles.

- Classical and Quantum MB-Systems Not Solvable. In Classical Imitation, We Need Solve Classical 2B-Systems to Get Clues for HMC $u\bar{u}$ and LMC e^+e^-. A Solvable 2B-System is Classical EqM 2B-System in Uniform Circular Motion and Bound by Specific Type(s) of Forces with Retardation Omitted or without Retardation. Motion is Given before We Solve because Electrodynamics since Maxwell/Lorentz/Einstein Has Only Been Able to Find Out Fields Produced by Source Charge whose Motion is Given OR Motion of Charged Particle in Given Fields.
- Even in Classical Imitation, We Can Show Insolvability Brought by Retardation if Classical 'Velocity Fields' with Retardation are indeed Reality of Quantum Bound Systems at Same Time when 'Acceleration Fields' Entirely Dismissed.

< Three Parts of **TMET** with **Proof** >

- **TMET** (**TME-T**heorem):

Part I. Circular Motion of Classical 1-Body in Every *Coulomb*-Type Central Force Field with IPE Always Leads to ***Negative*** TME, True for All Circular Binding Sizes $0<r<\infty$. Equal-Mass 2B Circular Motion with ***Coulomb***-Type Binding Force alone Also Have ***Negative*** TME, True for All Circular Binding Sizes $0<r<\infty$.

Pf : Omitted. 1B-Case was Proved in Newton's Time. Equal-Mass 2B-Case is Special Case of Part III.

Part II. 1-Body Circular Motion in Any Central Force Field **alone**, where Force $\vec{f}=-a_{MF}\vec{r}/r^5$ is of ***MF***-Type and Attractive ($a_{MF}>0$), Has ***Positive*** TME for All Circular Binding Sizes. Its Equal-Mass 2B Version Also Has ***Positive*** TME, True for All Circular Binding Sizes.

Pf: Omitted. Part II is Special Case of Part III.

Part III. **1-Body** Circular Motion in Central Resultant Force Field of Any *Coulomb*-Type **plus** Any *MF*-Type $\vec{f}=-(k_C/r^2+a_{MF}/r^4)\vec{r}/r$ Has ***Positive*** TME for All Circular Binding Sizes Smaller than 0TMED r_{0TME} where TME Equals 0 and ***Negative*** TME for All Circular Binding Sizes Larger than 0TMED r_{0TME}. Equal-Mass **2B Circular Motion** with above Resultant Interaction Force Has ***Positive*** TME Rigorously Proved to Be **True** at **All** Circular Binding Sizes Smaller than r_{0TME} while ***Negative*** TME True at **All** Circular Binding Sizes Larger than r_{0TME}. At Unique **Critical Distance** r_c $(r_c>r_{0TME})$, TME Has Minimum Negative Value. Such Bizarre Positive TME for Such Special Bound State Has Been Rigorously Proved to be True for Any Positive Values of k_C and a_{MF} and at ANY Radius (Circular Binding Size) r Smaller than 0TMED.
Note: MT-Forces in (H12) and (H15) Depend on Both r and θ. So far TMET is Proved for Special Circular Motion on Zenith Plane $\theta=\pi/2$. Non-Radial Component of MF Outside Zenith Plane Has No Influence on Classical Motion if Such Classical Motion Takes Place on Zenith Plane. While It Must Have Influence on Quantum Motion. Such Influence on Quantum Motion Unknown since Quantum Eqs in MF Field Unsolved and May Be Unsolvable.

Pf: Since LV Omitted in 1st, 2nd, and 3rd GS, γ is Constant in Every Uniform Circular Motion and Eqn of Motion in **1B Uniform Circular Motion** is

(H22) $$f=k_C/r^2+a_{MF}/r^4=m\gamma u^2/r, \qquad \beta^2\gamma=fr/mc^2,$$

$$\beta^2=k\gamma^{-1}, \qquad \gamma=\frac{k+\sqrt{k^2+4}}{2}, \qquad k\equiv fr/mc^2=\frac{1}{mc^2}(\frac{k_C}{r}+\frac{a_{MF}}{r^3}).$$

$$U=\int fdr=-k_C/r-a_{MF}/3r^3=-kmc^2+2a_{MF}/3r,$$

$$E_k = m\gamma c^2 - mc^2, \qquad T_{ME} = E_k + U = mc^2 \frac{\sqrt{k^2+4} - k - 2}{2} + \frac{2a_{MF}}{3r^3}$$

Immediately, $a_{MF} = 0$ Leads to TME T_{ME} Negative for any $k > 0$ and any Size of Binding $r < \infty$. This is 1B-Case in Part I. While $k_C = 0$ Yields $a_{MF} / r^3 = kmc^2$ and Positive TME T_{ME} True for any $k > 0$ and any Binding Size $r < \infty$. It's 1B-Case in Part II. In General, Both k_C and a_{MF} are Positive. One Can Check

(H22a) $T_{ME} = 0 \Leftrightarrow 3mc^2kr^3 = a_{MF}(\sqrt{k^2+4} + k + 2) \Leftrightarrow \sqrt{k^2+4} + k + 2 = 3 + \dfrac{3k_C}{a_{MF}} r^2$

$$\Leftrightarrow \sqrt{k^2+4} + k = 1 + \frac{3k_C}{a_{MF}} r^2$$

Above Eqn Not Solvable. But Last Eqn Proves: There is 1 and Only 1 Solution $r = r_{0TME}$. To Reach Such Conclusion under Condition of Insolvability One Needs to Look at LHS and RHS of Last Eqn in (H22a): LHS Monotone Decreasing from Super Large Positive Value to Almost 2 as r Increasing from near-0 to Long Distances while RHS Increasing from 1 to Super Large Positive Values as r Increasing from near-0 to Long Distances. Two Graphs Must Intersect at 1 and Only 1 Point. Its r-Coordinate is Just 0TMED r_{0TME}. Moreover, TME is Positive at any r Shorter than r_{0TME} and Negative at any r Longer than r_{0TME}.

Asymptotic Behavior of TME-r Curve Can Be Seen in (H22): As r Gets Shorter and Shorter Positive TME Gets Larger and Larger While as $r \to \infty$, TME Approaches 0. Existence of Critical Distance r_c where TME Takes Minimum Value Seen below

(H22b) $dT_{ME} / dr = 0 \leftrightarrow 4a_{MF} = [1 - k/(k^2+4)^{1/2}](k_C r^2 + 3a_{MF}) \leftrightarrow$

$$1 - (1 + 4/k^2)^{-1/2} = 4(k_C r^2 / a_{MF} + 3)^{-1}$$

Again, Last Eqn in (H22b) Not Solvable but Proves: There is 1 and Only 1 Solution $r = r_c$. Obviously, $r_{0TME} < r_c$.

More Rigorous Proof of 1B-Case Omitted. Rigorous Proof for Equal-Mass 2B-Case is Shown below. One Can Easily Repeat Rigorous Proof for 1B-Case. **Equal-Mass 2B-Systems** Important for HMC and LMC are such Systems.

Consider Special Classical Solutions: Uniform Circular Motion of Radius r. Distance btw two Equal-Mass Particles Bound by Attractive Force is $2r$. Write

(H22c) $f = k_C / 4r^2 + a_{MF} / 16r^4 = m\gamma u^2 / r$,

$$\beta^2 = k\gamma^{-1}, \quad \gamma = \frac{k+\sqrt{k^2+4}}{2},$$

$$k \equiv fr/mc^2 = \frac{1}{mc^2}\left(\frac{k_C}{4r} + \frac{a_{MF}}{16r^3}\right), \qquad E_k = 2m\gamma c^2 - 2mc^2,$$

$$U = \int 2fdr = -k_C/2r - a_{MF}/24r^3 = -2kmc^2 + a_{MF}/12r^3$$

$$T_{ME} = E_k + U = mc^2(\sqrt{k^2+4} - k - 2) + \frac{a_{MF}}{12r^3}$$

Again, $T_{ME} = 0$ is Unsolvable Eqn. Value of r_{0TME} is Algebraically Unknown. Existence and Uniqueness of Solution, However, Can Be Rigorously Proved:

(H22d) $T_{ME}/mc^2 = \sqrt{k^2+4} - k - 2 + \frac{a_{MF}}{12mc^2r^3} = \frac{a_{MF}}{12mc^2r^3} - \frac{4k}{k+2+\sqrt{k^2+4}}$

$$= \frac{a_{MF}(k+2+\sqrt{k^2+4}) - 4k\cdot 12mc^2r^3}{12mc^2r^3\cdot(k+2+\sqrt{k^2+4})} = \frac{A-B}{D}$$

$$A = k+2+(k^2+4)^{1/2}, \quad B = \frac{12k_C}{a_{MF}}r^2 + 3, \quad D = \frac{12mc^2r^3\cdot(k+2+\sqrt{k^2+4})}{a_{MF}}.$$

Note: $\frac{4mc^2kr^3}{a_{MF}} = \frac{k_C}{a_{MF}}r^2 + \frac{1}{4}$ True for k -Function Defined in (H22c).

As r Increases, A is Monotone Decreasing from Super Large Positive Value near $r=0$ to Nearly 4 at Very Long Distances while B is Monotone Increasing from 3 at $r=0$ to Super Large Values at Very Long Distances. Hence Curves of Functions A and B Must Have 1 and Only 1 Intersection Point, Simultaneously Proving that TME Vanishes at r_{0TME}, Positive for All $r < r_{0TME}$, and Negative for All $r > r_{0TME}$.

1st Special Case is Pure Coulombian or Coulomb-type Force with $a_{MF} = 0$ and $T_{ME} = mc^2(\sqrt{k^2+4} - k - 2)$ ($k = \frac{k_C}{4mc^2r}$), **Negative** at All Distances. One Can See that as $a_{MF} \to 0$, $r_{0TME} \to 0$ too.

2nd Special Case is Pure MF or MF-type Force with $k_C = 0$ and

$$T_{ME}/mc^2 = A/D, \quad A = k-1+(k^2+4)^{1/2},$$

$$D = \frac{12mc^2r^3 \cdot (k+2+\sqrt{k^2+4})}{a_{MF}}, \quad k = \frac{a_{MF}}{16mc^2r^3}.$$

A, D, and T_{ME} **Positive** at All Non-Zero Distances. One Can See: $r_{0TME} \to \infty$ as $k_C \to 0$.

In General, both a_{MF} and k_C Not Zero. Existence and Uniqueness of Critical Distance r_c Can Also Be Proved Rigorously:

(H22e) $$4r^4 dT_{ME}/dr = [1-k/(k^2+4)^{1/2}](k_C r^2 + 3a_{MF}/4) - a_{MF} =$$

$$(1-k/\sqrt{k^2+4})(a_{MF}/4)[4k_C r^2/a_{MF} - (k^2+1+k\sqrt{k^2+4})] = D(A-B),$$

$$A \equiv 4k_C r^2/a_{MF}, \quad B \equiv k^2+1+k\sqrt{k^2+4}, \quad D \equiv (1-k/\sqrt{k^2+4})(a_{MF}/4).$$

As r Increases, B is Monotone Decreasing from Super Large Positive Value near $r=0$ to Near 1 at Very Long Distances while A is Monotone Increasing from 0 at $r=0$ to Super Large Values at Very Long Distances. Two Graphs of Functions A and B Have Exactly 1 Intersection Point. Its r-Coordinate Denoted by r_c. It Simultaneously Proves: TME Monotone Decreasing in Interval $(0, r_c)$, Monotone Increasing in Interval (r_c, ∞), and Takes Minimum Value at r_c. Such Minimum Value Obviously Negative.

Eqs $T_{ME}=0$ & $dT_{ME}/dt=0$ Unsolvable. Values of r_{0TME} & r_c Algebraically Unknown. But Once k_C and a_{MF} Given One Can Easily Get Numerical Values of r_{0TME} and r_c as Accurately as One Wants.

Note: ***V-Shape*** of TME-r Curve Nearby r_c May Create Bunch of Bound States Nearby r_c, Called **NCB-States** (**N**ear-**C**ritical **B**ound **States**) with Small Level-Differences. Mass Spectrum of Hadrons Contains States with Small Differences in Mass.

- Using Classical Imitation and **TMET** We'll Explain Heavy Mass of ***Muon*** in Terms of DS under Condition of ***Positive*** TME and Small Mass of Stable Leptons that Form Muon as well as ***0-Pion*** Mass under Condition of ***Negative*** TME and Heavy Mass of u-Quark (Heavier than 0-Pion Mass).
- Classical TMET Proves Unexpected MFBF of TME that is of Vital Importance as We Explain Mass Spectrum of SA Composite Particles and Their Empirical MMs with Masses and MMs of Their Constituent Elementary Particles.
- MT-Forces in (H12) and (H15) Depend on Both r and θ. So far TMET Proved for Special Circular Motion on **Zenith Plane** $\theta = \pi/2$. Non-Radial Component of MF Outside Zenith Plane Has No Influence on Classical Motion if Such Classical Motion Takes Place on Zenith Plane. While It Must Have Influence on Quantum Motion. Such Influence on Quantum Motion Unknown since Quantum Eqs in

MF Field Unsolved and May Be Unsolvable.

- Heavy u-Mass and Relatively *Small Pion Mass* Further Indicate ***Negative*** TME and Require Dismissal/Suppression of MF and St MF for HMC $u\bar{u}$ Produced in any Event of HE Collision, Cosmic Shower, or DS De-Excitation in Some Decay Modes. Once MF Suppressed Greatly, 0TME Distance Shifts to Much Shorter One so that $u\bar{u}$ Binding Size Becomes Larger than 0TMED & TME Becomes Negative. It Can Be Naturally Done with OSTP.
- 1st Application of TMET Leads to **Sample Interpretation** of π^0-**Mass** under Condition of Heavy u-Quark, Serving as Example of Mass of Composite Particle Smaller than Mass-Sum of its Constituent Particles.

  Large u-Mass Explains MMs of Nucleons. D.H. Perkins Did so. DS with OSOD Proves d-Electrons' Contributions to MMs of Nucleons Small. 1st Application of TMET is Powerful Enough to Explain Masses of π^0, p, and n and Nucleons' MMs Simultaneously/Half-Quantitatively.

< **HMC** $u\bar{u}$, **u-Mass**, and π^0**-Mass Eqn** >

Rigorous Classical Circular Motion Solution of Equal-Mass 2B-System Bound by Coulomb-Type Force alone Has Been Obtained in TMET. Now We Repeat Similar Calculations Particularly for Pure St Coulomb Force Case:

Let $f_C = k_C / r'^2$ with $k_C > 0$ and $r' = 2r$, where r Radius of Circular Motion of Equal-Mass 2B-System. Let Origin of Chosen Coordinate System Coincide with Center of Circular Motion. For $\gamma < 10^3$ We May Set $d\lambda/dt = 0$, $\lambda = 1$. For Uniform Circular Motion, Speed's Constant ($d\gamma/dt = 0$ with $\vec{u}\cdot d\vec{u}/dt = 0$). Centripetal Acceleration is $d\vec{u}/dt = -r^{-2}u^2\vec{r}$. We Have

(H23a) $$\vec{f}_C = -k_C r^{-3}\vec{r}/4 = d(m\lambda\gamma\vec{u})/dt = m\gamma d\vec{u}/dt = -m\gamma r^{-2}u^2\vec{r},$$

$$2f_C dr = dU_C \quad \text{(Work-PE Eqn for Equal-Mass } \mathbf{2B}\text{-System)}$$

$$U_C = -\frac{k_C}{2r} = -\frac{\eta_C e^2}{2r} \qquad (\eta_C \equiv k_C / e^2)$$

$$\beta^2 = \frac{k_C / mc^2}{4r}\gamma^{-1} = \eta_C \frac{m_e}{m}\frac{r_{sa}}{4r}\gamma^{-1} = k\gamma^{-1}, \quad \beta^4 + k^2\beta^2 - k^2 = 0$$

$$k \equiv \eta_C \frac{m_e}{m}\frac{r_{sa}}{4r}, \quad r_{sa} \equiv \frac{e^2}{m_e c^2} = 2.8177471 \cdot 10^{-13} cm$$

Now We Write $\gamma - r$ Relation and TME T_{ME} as

(H23b) $$\beta^2 = \frac{2}{1+\sqrt{1+4/k^2}}, \quad \gamma = \frac{k+\sqrt{k^2+4}}{2}, \quad (k \equiv \eta_C \frac{m_e}{m}\frac{r_{sa}}{4r}),$$

$$T_{ME} = 2mc^2(\gamma - 1) - \frac{\eta_C e^2}{2r} = mc^2(\sqrt{k^2+4} - k - 2) < 0, \ (\forall r > 0).$$

Equal-Mass 2B Composite System with Small Enough Binding Size and with 'long' Enough Lifetime is Always Treated as Composite Particle with Rest Mass

(H23c) $$m' = 2m + T_{ME}/c^2 = m(\sqrt{k^2+4} - k), \quad (k \equiv \eta_C \frac{m_e}{m}\frac{r_{sa}}{4r})$$

We Now Make Following <u>Sample Setting</u> for $u\bar{u}$ St-Coulombian Bound System.

$$m = m_u = 320\, MeV/c^2, \quad \eta_C = \eta_C^{st} = 100\,.$$

We Ignore Small Contribution of DS-Leptons inside 0-Pion, if any, and Write

(H23d) $$m_{\pi^0} = m_u(\sqrt{k^2+4} - k), \qquad k = 25\frac{0.511}{320}\frac{r_{sa}}{r}\,.$$

If $k = 4.55$, then (H23d) Leads to 134.45 MeV/ c^2 as Mass of Zero-Pion. Classical Imitation of Ground State of HMC $u\bar{u}$ with Their Spin Vectors Tilted at 0R-MF Angle and Hence Turning MFs Off Completely Features IPE Binding Size Same as what One Can Get from (H23d) with Required Value $k = 4.55$:

(H23e) $$r = \frac{25}{k}\frac{0.511}{320} r_{sa} = 2.47 \cdot 10^{-15} cm$$

At Such Short SA Distance with $k = 4.55$, (H23b) Gives:

(H23f) $$\beta = 0.97768, \qquad \gamma = 4.76008.$$

Remark: Asymmetric MF Field is Not Sphero-Symmetric, Depending on r and θ. We May Have Discrete Values of Most Probable Angle θ While Ground State May Not Correspond to 0R-MF Angle Exactly.

Now We Consider **<u>Suppressed MF</u>** nearby **0R-MF** Angle for MF btw $u\bar{u}$ plus St Coulombian Force btw Them. We Modify MF $\vec{f}_1$ in (H18) and r_{m1} in (H19) with a Suppressing Factor k_{MF} When OSTP Occurs and Suppresses MF btw $u\bar{u}$. Write

(H24) $$f_{MF}^{Sup} = 1.24k'e^2\frac{(3\cdot 10^{-11}cm)^2}{(2r)^4}\frac{1}{k_u}\frac{1}{k_u} \qquad (k_u \equiv m_u/m_e,\ k' \equiv k_S/k_{MF})$$

$$k_S = \begin{cases} 1 \ \ if \ \nexists \ St \ \ MF \\ \eta_C^{st} \ \ if \ \exists \ St \ \ MF \end{cases}, \qquad k_{MT-C}^{\text{sup}} \equiv \frac{f_{MF}^{\text{sup}}}{\eta_C^{st} e^2/4r^2} = (\frac{r_m^{\text{sup}}}{r})^2$$

$$r_m^{\sup} = 1.67\sqrt{\frac{k'}{\eta_C^{st}}} \cdot \frac{1}{k_u} \cdot 10^{-11} cm$$

(H24a) $$f = f_C^{st} + f_{MF}^{\sup} = \frac{\eta_C^{st} e^2}{4r^2} + \frac{\eta_C^{st} e^2}{4r^2}\frac{(r_m^{\sup})^2}{r^2} = m_u \gamma a = m_u \gamma u^2 / r$$

$$\beta^2 = k\gamma^{-1}, \quad k = \frac{\xi_u}{4}(\eta_C^{st} + 3513.98724 k' \xi_u^2)$$

$$\xi_u \equiv \frac{r_{sa}}{k_u r}, \quad r_{sa} \equiv \frac{e^2}{m_e c^2} = 2.8177471 \cdot 10^{-13} cm, \quad k' \equiv \frac{k_S}{k_{MF}}$$

$$\beta^2 = \frac{2}{1 + \sqrt{1 + 4/k^2}}, \quad \gamma = \frac{k + \sqrt{k^2 + 4}}{2}.$$

where Suppressing Factor $k_{MF} >> 1$ Due to Small Deviation from 0R-MF Angle in OSTP. If St MF Never Exists, $k_S = 1$ due to Existence of MF. Otherwise, $k_S = \eta_C^{st}$ is Coefficient of St-Coulombian Force. Then

(H24b) $$2(f_C^{st} + f_{MF}^{\sup})dr = dU, \quad U = -\eta_C^{st} e^2 / 2r - \eta_C^{st} e^2 (r_m^{\sup})^2 / 6r^3$$

One Can Check Following PE and TME Eqs:

(H24c) $$U = -m_u c^2 (2k - 1171.32908\, k' \xi_u^3)$$

$$T_{ME} = 2m_u c^2(\gamma - 1) + U = m_u c^2(\sqrt{k^2 + 4} - k - 2 + 1171.32908 k' \xi_u^3)$$

At Large Enough r, ξ, k, and T_{ME} Almost Vanish. As r Decreases, ξ and k Increase. While T_{ME} Deceases to Minimum $T_{ME}^{\min} < 0$ and then Increases. At r_{0TME}, T_{ME} Equals Zero. At Shorter Distances, T_{ME} Becomes Positive.

Above Expression of T_{ME} and CME (H2) Lead to π^0-**Mass-Eqn** with $u\bar{u}$ Content Only without DSLs or with Small Contribution from DSLs Omitted:

(H24d) $$m_{\pi^0} = 2m_u + T_{ME} / c^2 = m_u(\sqrt{k^2 + 4} - k + 1171.32908 k' \xi_u^3)$$

$$k = \frac{\xi_u}{4}(\eta_C^{st} + 3513.9872 k' \xi_u^2)$$

$$\xi_u \equiv \frac{\xi}{k_u} \equiv \frac{r_{sa}}{k_u r}, \quad r_{sa} \equiv \frac{e^2}{m_e c^2} = 2.8177471 \cdot 10^{-13} cm, \quad k' \equiv \frac{k_S}{k_{MF}}$$

$$k_u \equiv \frac{m_u}{m_e} = \frac{m_{\pi^0}}{m_e(\sqrt{k^2+4}-k+1171.32908k'\xi_u^3)}$$

$$m_u = \frac{m_{\pi^0}}{\sqrt{k^2+4}-k+1171.32908k'\xi_u^3}$$

$$m_{\pi^0} = m_u(\sqrt{k^2+4}-k+1171.32908k'\xi_u^3)$$

Classical Circular Motion Imitation Could Get Verified Perfect Match with QM for Atomic Electron in Hydrogen Since r-Setting Could Use Solution in QM.

Unknown Value of u-Quark Mass and Value of Proton's Mass Produce Almost Zero Effect on Solution of QM in Atomic Physics. For Example, Fictitious Hydrogen with Fictitious Proton's Mass 5, 10, or More Times Heavier that Real Proton would Have Almost Same Solution as Real Hydrogen in QM and Atomic Physics. For $u\bar{u}$ 2B-System, Quantum Solution's Not Available for its 1BEq with Non-Spherosymmetric MF Field Involved Unsolved and May Be Unsolvable. Unknown Solution Must Heavily Depend on u-Mass, which is NOT Known and Cannot Be Determined by Solely Looking at Nucleons' MMs. Since Solution Unknown, We Don't Have Hint to Figure Out MF-Suppressing Factor k_{MF} and Parameter k'. We Can Use Some Ground State Sample Settings of k', ξ_u, and η_C^{st} to Put Lights on Mysterious Inner SA World. From Pion-Mass Eqn We See: Once 3 Parameters k', ξ_u (Equivalently, Product $k_u r$ or $m_u r$), and η_C^{st} are Determined, k, γ, k_u, and m_u are Also Determined, while r_0, as Radius of Circular Motion in Ground State, Can Also Be Determined:

(H24e) $$r_0 = r_{sa} / k_u \xi_u = e^2 / m_u c^2 \xi_u$$

Sample Settings of k', ξ_u, and η_C^{st} with **Results** (k, γ, k_u, m_u, r_0) for *Ground* State are Put into Table below, Called **HMC $u\bar{u}$ Ground State Table**. All Parameter-Settings are for Ground State Only. $r_{b,u\bar{u}}^{Dirac}$ is DBR of $u\bar{u}$ 2B-System Bound Solely by St Coulombian Force, with η_C^{st} and m_u Given in Same Row. $r_{m,u\bar{u}}^{\sup}$ MD btw $u\bar{u}$ with Suppressed MF and Corresponding Setting of k' and Consequent Values of k_u, m_u, r_0, OAM, IPE-Size $r_{b,u\bar{u}}^{Dirac}$, and MD $r_{m,u\bar{u}}^{\sup}$ with MFs Suppressed Greatly in Same Row.

Table Below Obtained by Sample Setting η_C^{st}=130. Unit for Distances/Sizes is *cm*. (Same table is on p.262)

HMC $u\bar{u}$ Ground State Table

k'	ξ_u	k	k_u	$m_u c^2$	r_0	OAM ($\hbar/2$)	$r_{b,u\bar{u}}^{Dirac}$	$r_{m,u\bar{u}}^{\sup}$
0.013	0.176	5.74	626.9	320.4 MeV	2.55 10^{-15}	0.48	2.05 10^{-14}	2.66 10^{-16}
0.01	0.2	6.57	675.1	345.0 MeV	2.09 10^{-15}	0.49	1.90 10^{-14}	2.17 10^{-16}
10^{-6}	1.0	32.50	4217	2155 MeV	6.68 10^{-17}	0.47	3.04 10^{-15}	3.47 10^{-19}
10^{-8}	5	162.5	19184	9803 MeV	2.94 10^{-18}	0.47	6.69 10^{-16}	7.64 10^{-20}
10^{-10}	15	487.5	58736	30.01 GeV	3.20 10^{-19}	0.47	2.19 10^{-16}	1.49 10^{-22}

QKoE $E(uKo) = 2m_u c^2 - m_{\pi^0} c^2 = 505.8,\ 555,\ 4174.6,\ 19470.8,\ 59893\ MeV$ Respectively. Even Smallest QKoE above Larger than Total Energy Needed to Produce Three Pions. AVI with Absolute Priority for PAPP to Consume Impact Energy and PQTR Explain FQA Fact in Terms of QNC. Now that at Least One Bound $u\bar{u}$ Pair is Produced Right before $u\bar{u}$ in Pion are Knocked Out and Each One in New Born Bound $u\bar{u}$ Pair Very Sensitive to Empty PW Wrapping Each Naked One Knocked Out from $u\bar{u}$ in Pion. Naked Ones and New Born Ones Emerge at Nearly Same Place and Same Time so that PQTR Easily Makes Hole Left behind by Knocked Out One Stuffed and No Naked Single Unbound Quark Could Appear among FSPs. In Cases of Very Heavy u-Mass and Very Large QKoE, QC Can Be True if QKoE is Larger than NLVUC. Besides, V-Shape of TME-r Curve May Create a Series of Bound States with Binding Sizes Smaller than r_c including All Bound States with Binding Sizes Less than 0TME Distance r_{0TME} and with Positive TME up to Super-Large Value. Distribution of Excited States in Inward Direction May Be another Reason of FQA Fact: Impact in Collision May Push Hadron to Inward Excited States with less and less Sizes of Binding instead of Knocking Out Quarks in Outward Directions.

OAM is for Reference Only. OAM Obtained in Circular Motion CI May Be Meaningless. For Example, OAM Vanishes in S-States while It's Nonzero in any Classical Circular Motion. We Don't Know if there is Example of PCQM that Includes Perfect Match in OAM.

Remark: If Sample Settings and Outcomes in 1st Two Rows Good, u-Mass is Larger but Not Much Larger than 1/3 of Proton's Mass. Pure St Coulombian Force without Any MF and Mag-Forces Could Only Bind Heavy $u\bar{u}$ with above u-Mass within SA Small Space with Size $r_{b,u\bar{u}}^{Dirac}$ (DBR) $\sim 2\cdot 10^{-14} cm$. Suppressed MF Equals and is Stronger than St Coulombian Force at MD $r_{m,u\bar{u}}^{\sup}$ about $2\cdot 10^{-16} cm$ and Shorter Distances respectively. Its Large Influence Reaches Longer Distances so that Binding Size Reduces from DBR $r_{b,u\bar{u}}^{Dirac}$ to r_0 about $2\cdot 10^{-15} cm$, which is btw IPE-Binding Size $r_{b,u\bar{u}}^{Dirac}$ and MD $r_{m,u\bar{u}}^{\sup}$. If Sample Settings/Outcomes in Last Two Rows are Better, We Have Much Smaller Binding Sizes/Heavier *u*-Quark. If Heaviest Meson is Found to Be about 9 *GeV*, for Example, We May Assume u-Mass to Be about 5 *GeV*.

- For Sample Settings with u-Mass Much Heavier than 1/3 of Proton's Mass, Tiny u-MM Makes Negligible Contribution to Nucleons' MMs. Then In Such Very Heavy u-Mass Models, OSOD Must Yield Net MM of Each d-Electron that Can Explain Nucleons' MMs. At Present, We Do Not Know if LV Space Force Plays a Role in QD in 3rd GS. Once LV Space Force Enters Stage of QD, Pion-Mass Eqn Must Be Modified.
- Unsolvable MB-Systems Do Not Allow Us to Get OAM Quantitatively. OSOD, at Present, Unable to Provide with Quantitative Result. So Mysterious *u*-Mass and Many Consequences Remain Unknown.
- It's Difficult to Measure Quantities in Sample Settings and Results. Free Quarks Not Available to Allow Experimenters to Measure η_C^{st} Accurately. Collection of **All Experimental Data Contain Unknown Effects of Unknown Solutions** of SA MB-Systems. As One Tries to Explain a Fact, Assumptions Must Be Made before Explanation. We Have Radical Uncertainty of Values of Many Important Quantities, No Matter How Many More Experiments will Be Performed in Coming 500 Years. DS Provides Physical Picture so Close to Reality that It Has Shown its Power to Explain Huge Number of SA Facts and Events.
Note: If Energy Released in SA De-Excitation or HE Collision Large Enough, Excited HMC $u\bar{u}$ with Larger OAM & Higher Spin Label $J=1$ Can Be Born to Become HMC for Rho ρ and Other Mesons with Label $J=1$. CI of Such Quantum State Might Be Classical **Elliptical** Motion. Such Classical Motion for Equal-Mass 2B-System May Be Unsolvable Even in Classical Theory. Absence of Classical and Quantum Solutions, However, Does Not Cover Physical Picture of Excited States of HMC $u\bar{u}$ & More Complicated Excited States of Meson-Systems with DS-Leptons Involved.

DRGGD (***De Rujula-Georgi-Glashow*** **Description**) of Hadrons:

They Consider Excitation States. Quark Model with Broken Symmetry without Excitation States Equivalent to Abandoning Basic Dynamics Corner Stones in All Branches of Modern Classical and Quantum Physics. Broken Symmetry was Proposed when Dense 'Fog' Covering Dynamics in SA World Didn't Seem to Be Penetrable. We Expect Original Thinker to Put aside His Idea that's Stood for so long because of Unsuccessful Potential Models and Other Dynamical Models as All of Them Missed Crucial Effects of DS with MFs.

DS and MFs Support DRGGD of Hadrons. We've Seen that Pion Mass is Never Accidentally Low but is Explained as Ground States with DS and Heavy u-Mass that Simultaneously Explain Nucleons' MMs as well as Heavier Mesons with Spin Label $J=1$ as Excited States just as Indicated in DRGGD.

< $\pi^{\pm}$-Mass Eqn >

$$m_{\pi^{\pm}} - m_{\pi^{0}} = (139.5702 - 134.9766) MeV/c^2 = 4.5936 MeV/c^2$$

Such a Small Difference Indicates Small Contribution DS-Leptons Make to $\pi^{\pm}$

that Contain Them. Since that Difference is Larger than Sum of Rest Masses of DS-Leptons in Charged Pion, ***ME*** of DS-Leptons Must Be ***Positive*** but Small. **PME** (**P**ositive **ME**) is Due to MFBF according to TMET. PME Opens QTC for Decay Allowing Bound DSLs to Penetrate PW Fleeing to Free Region through QT. DS Predicts Decay Modes $\pi^+ \to e^+ \nu_e \pi^0$ and $\pi^- \to e^- \overline{\nu}_e \pi^0$ Proved by Experimenters with Small Branch Ratio ($1.036 \cdot 10^{-6}$). In this Example, again, HMC $u\overline{u}$ Annihilation Channels Dominate QTC in Sense that $u\overline{u}$ Annihilation Occurs before Bound DSLs with Positive MEs would Escape via QT.

DS further Explains Difference btw Pions' Masses as Shown below with Very Small Contribution from NFM Omitted:

- HMC $u\overline{u}$ Ground State Table Tells: **HMC Size** ($<2.6 \cdot 10^{-15} cm$) **Much Less** than **MD** of MF btw ue^- or $\overline{u}e^+$ Deduced from (H19). We May Expect Much Larger Orbital Sizes of DS-Leptons than Orbital Size of HMC $u\overline{u}$. Thus, to DS-Electron/Positron in $\pi^{\pm}$, HMC Looks like Tiny Particle with 0-Net Charge So that Coulombian Force Negligible at all Places far away from HMC. MF exerted on d-Electron or $\overline{d}$ - e^+ Dominates Coulombian Force almost Completely with MFBF of **Positive ME**. Here We Use Term ME instead of TME for Tiny Heavy HMC $u\overline{u}$ Can Be Treated as a Fictitious One Particle that is at Rest while TME of a Charged Pion Equals ME of d - e^- in π^- or $\overline{d}$ - e^+ in π^+, that Moves around Rest Fictitious Particle. Such 1BEq Can Produce Good Result as We Study Unsolvable MB-System because HMC $u\overline{u}$ Not Only Very Small in Comparison with DS-Lepton's Orbit but also Very Heavy if We Compare it with Light DS-Lepton. Surely, NFM in Charged Pion too Light to Make Significant Contribution to Mass of Charged Pion. We Omit it in 1BEq.
- Antiparallel Spins of $u\overline{u}$ in Ground State HMC Lead to Parallel Spin MMs Yielding Single Net Spin MM Twice of u-Quark's Spin MM Effective for any Spin MM Far away from $u\overline{u}$ 2B-System. It Can MF-Bind One DS-Electron or One DS-Positron to Form π^- or π^+. To Be Bound by Such Net Spin MM of HMC in Ground State to Form π^- also in Ground State, d-Electron Must Have its Spin in Direction Parallel to that of Net MM of HMC and Stays in R1 at the Net MM such that It together with u-Quark in HMC Can Form d-Quark in π^- though it is Actually Bound by Entire HMC 2B-System $u\overline{u}$. Similarly, To Be Bound by Such Net Spin MM of HMC in Ground State to Form π^+ also in Ground State, Directions of DS-Positron's Spin and HMC's Net MM Must Be Opposite and DS-Positron Stays in R1 at that Net MM such that It together with $\overline{u}$ in HMC Can Form Anti-d-Quark $\overline{d}$ though it's Actually Bound by Entire HMC 2B-System $u\overline{u}$. If This Spin Correlation and MPR-Location is Correct for HMC $u\overline{u}$, DS-Electron, and DS-Positron <u>All in Ground State</u>, then $d\overline{d}$ Can Not Exist as Particle/Meson because d - e^- and $\overline{d}$ - e^+ would Have Parallel Spin MMs and They would Severely Repel Each Other if They were Simultaneously Bound by HMC. Later We will Study Spin-Correlation of HMC $u\overline{u}$ in Kaons.
- Using u-Mass in 1st Row of HMC $u\overline{u}$ Ground State Table, We Get Net Spin MM of HMC $u\overline{u}$ in Ground State Effective to Far Away Spin MM such as that of *d*-

Electron or that of $\bar{d}$ -Positron, by Using Quantum Connection for MM:

(H24f) $$\mu(\pi - HMC) = 2\mu_u = 2\mu_{\bar{u}} = \frac{2e\hbar}{3m_u c}$$

MF Formula (H12) Gives Approximation for Charged DS-Lepton (e^- or e^+) with $\theta \approx 90^0$ Bound in Charged Pion by Net Spin MM of Pion's HMC:

(H24g) $$f(e\ \ in\ \ \pi) = \frac{3\mu(\pi - HMC)\mu_e}{r^4} = \frac{m_e c^2 \xi^3}{k_u \alpha^2 r}, \quad \xi \equiv \frac{r_{sa}}{r}$$

Here Zenith Axis (z-axis) is in Same Direction of Net MM of Pion's HMC that's at Rest at Origin. This New Coordinate System is for Observation of Charged DSLs, Not for Observation of $u\bar{u}$ in HMC. Orbit-Spin Tilting in OSTP Tells Us that Orbit-Plane of $u\bar{u}$ Motion in CI Now is Not xy-Plane on which Charged DSL Moves about Net MM of $u\bar{u}$ in CI. A Big Problem is that Precession in OSTP Means: Spins and MM-Vectors of $u\bar{u}$ Must Rotate about Normal Vector of $u\bar{u}$ Orbital Plane. **Even Classical Motion of Charged DSL about such Net MM of $u\bar{u}$ in Rotation May Not Be Solvable.** To Penetrate Dense Fog and See Something We Fix Z-Axis on Rotating Net MM Vector and Use This Partially Co-Moving Coordinate System to Describe Charged Pion Approximately.

Circular Motion CI Leads to Eqn of Motion in Relativistic Mechanics as LV is Omitted in 1st Three γ-Sectors ($\gamma < 10^3$):

(H24h) $$f(e\ \ in\ \ \pi) = \frac{m_e c^2 \xi^3}{k_u \alpha^2 r} = m_e \gamma u^2 / r, \quad k\gamma^{-1} = \beta^2,$$

$$k \equiv \frac{f(e\ \ in\ \ \pi) r}{m_e c^2} = \frac{\xi^3}{k_u \alpha^2}, \quad \beta^2 = \frac{\sqrt{k^4 + 4k^2} - k^2}{2}$$

PE of $e^{\pm}$ in $\pi^{\pm}$ and Composition Mass of ($\pi^{\pm}$-Mass Eqn) Written

(H24i) $$U = -\frac{\xi^3}{3k_u \alpha^2} m_e c^2 = -\frac{1}{3} k m_e c^2$$

$$m_{\pi^{\pm}} = m_{\pi^0} + m_e \gamma c^2 - k m_e c^2 / 3$$

With Possible Values of $k_u \equiv m_u / m_e$ Shown in HMC $u\bar{u}$ Ground State Table We Now Put Corresponding Results of Proper Sample Setting of Parameter ξ of Orbital Size of Charged DS-Lepton in $\pi^{\pm}$ into $\pi^{\pm}$-Mass Table below:

< $\pi^{\pm}$-Mass Table >

k_u	626.9	675.1	4216.8	**19183.9**	58736.34
ξ	0.764	0.783	1.4422	**2.39**	3.471
k	13.3583	13.3532	13.3587	**13.3637**	13.3698
β^2	0.994458	0.994454	0.9944583	**0.99446238**	0.98586298
γ	13.4328	13.42796	13.433169	**13.43812**	13.44421
r	3.688 10^{-13} cm	3.599 10^{-13} cm	1.954 10^{-13} cm	**1.179** 10^{-13} **cm**	0.8118 10^{-13} cm

Contribution from NFM Ignored. Orbital Size r of Charged DS-Lepton in Charged Pion Proves it to Be SA Size. γ-Value in 2nd GS so that We Can indeed Use SR's RM (Relativistic Mechanics) with LV Omitted in CI.

Remark: From γ-Values & Orbital Sizes r in above Table We See that OAM of d-Electron Smaller than Expected: It's Less than 26% of Electron's Spin AM. This May Be due to Deviation from Circular Motion. Even in Spherosymmetric Coulomb Field, Deviation of Atomic Electron's Motion from S-State Could Get Large OAM such as $L=\sqrt{l(l+1)}\hbar$ with $l=1$. Now that Ground State of d-Electron in Non-Spherosymmetric MF Field Not S-State, its Classical Motion in CI Should Not Be Circular. Once in More General Closed Classical Motion, its OAM as an Integration Constant of Eqn of Motion in SR with Tiny LV Ignored for Classical Planar Motion on a Plane where Force Field is Central Can Have a Series of Values Larger than All OAMs of Circular Motions of Radius Noncircular Closed Classical Motion's Orbit Covers. Again, CI itself Unable to Determine Exact Value of Larger OAM. OSOD of d-Electron with a Small Net MM is a Must for DS to Agree with Observed Small Values of Nucleons' MMs. Large Net MM of d-Electron, if any, would Require Large Spin MM of u-Quark which would Demand Small u-Mass. Then Heavy Proton and its Stability Could Not Be Explained. In π^-, d-Electron is Moving in Pure MF-Field for Net Charge of HMC $u\bar{u}$ is Zero. But d-Electrons in Nucleons Bound by both MF and Coulomb Force. This Should Be Leading to Larger OAM. Another Possible Reason is that We Get OAM of d-Electron in π^- Smaller than Expected Since Precession of Net MM of HMC $u\bar{u}$ is Ignored to Simplify the Problem as We Try to Know the Motion of d-Electron in π^-. OSTP and Fast Orbital Motion of $u\bar{u}$ Circling Each Other Makes Net MM Tilted Forming Angle with Normal Vector of $u\bar{u}$ Orbital Plane but Rapidly Rotating about It. Although Motion of OSTP May Make Motion of d-Electron in π^- Bound by Net MM of $u\bar{u}$ Unsolvable, We May Expect Larger OAM of d-Electron due to such Motion of Net MM of $u\bar{u}$. More Radically, *OAM Obtained in Circular Motion CI Cannot Be Trusted.* Example of PCQM Shows Perfect Match btw Average Value of γ in Quantum Ground S-State and γ-Value Obtained in Circular Motion CI Using Classical Mechanics in SR. No Match for OAM: OAM Vanishes in S-States and is Nonzero in Classical Circular Motions. At Present, Author Does Not Know if there is Example of PCQM that Also Includes Perfect Match btw OAMs.

< **d-Electron** in <u>DS-Asymmetric</u> π^- **Slows down** $u\bar{u}$ **-Annihilation** >

Due to OSTP, d-Electron in π^- Mainly Stays in R1 at Net MM of HMC $u\bar{u}$. But MPR of d-Electron in π^- almost Centrosymmetric (Not Spherosymmetric) in Places Quite far away (Not Too Far) from $u\bar{u}$ while Not Centrosymmetric at Places Outside HMC $u\bar{u}$ and Close to it: At Such Places Electron Cloud **Denser on** u-Side Producing Coulomb Repulsive/Attractive Forces on $\bar{u}/u$ respectively to Slow down $u\bar{u}$ Annihilation. We Call such Forces **A**nnihilation **R**esistance (**AR**). (d-Electron Seldom Visits Place inside Tiny HMC $u\bar{u}$ so that Electron Cloud there Negligible.) Asymmetric Distribution of Electron Cloud outside and Near $u\bar{u}$ due to Simple Fact that u Coulomb-Attracts Electron Cloud while $\bar{u}$ Coulomb-Repels it and Norms of 2 Forces Differ Significantly at Places On or Near $u\bar{u}$ -Line outside $u\bar{u}$ but Not too Far from $u\bar{u}$. AR Exists in π^+ too.

Remark: Due to OSTP that Suppresses MFs Greatly, HMC $u\bar{u}$ Ground State Table Shows Small Possible Sizes of HMC $u\bar{u}$ such as $10^{-15}cm$ to $10^{-19}cm$.

Asymmetric Distribution of Electron/Positron Clouds Remains the Reason to Slow down $u\bar{u}$ -Annihilation in Charged Pions as Carefully Shown in above.

- 2nd Application of TMET Yields **Sample Interpretation** of *Heavy* **LMC** e^+e^- and **Muon-Mass** in Terms of DS $\mu^+/\mu^- = e^+e^-e^+\nu_e\bar{\nu}_\mu / e^-e^+e^-\bar{\nu}_e\nu_\mu$ and with MF-Binding Turned on **without OSTP**, Serving as Example of RCMI, where Composite Particle Heavier than Mass-Sum of its Constituent Particles.

< **LMC** e^+e^- and **Muon-Mass Eqn** >

Rigorous Classical Circular Motion Solution of Equal-Mass 2B-System e^+e^- as LMC Bound by Coulombian Force plus MF is Special Example of What is Shown in Part III of TMET. When Pion Decays via PPC, e.g., $\pi^+ \to \mu^+ + \nu_\mu$, HMC $u\bar{u}$ in π^+ Annihilate into 2 PAPs: LMC e^+e^- and $\nu_\mu\bar{\nu}_\mu$. LMC e^+e^- Captures $\bar{\nu}_\mu$ in $\nu_\mu\bar{\nu}_\mu$ Pair and DS-Leptons $e^+\nu_e$ Originally Contained in $\pi^+ = u\bar{d} = u\bar{u}e^+\nu_e$ to Form μ^+ ($\mu^+ = e^+e^-e^+\nu_e\bar{\nu}_\mu$) & Lets ν_μ Go. DSL e^+, if in Ground State, Make Very Small Contribution to μ^+ -Mass and π^+ -Mass. Then $m(LMC\ e^+e^-) = m_\mu$ is Accurate Enough in Our Half-Quantitative Treatment. Heaviness of μ^+ Mainly Due to Heavy LMC e^+e^-, Produced in Deep SA Vacuum with Very Large Positive TME. Heavy HMC $u\bar{u}$ Provides Enough Energy Released after Annihilation. Such Large TME Can Be Reality for Composite Muon that Lives Long Enough to Leave Visible Long Track because of MF-Binding Feature.

Settings, Eqs, and Results for LMC e^+e^- 2B-System Can Be Obtained from Their Counterparts in Pion 2B System Dynamics with Following Replacements

(H25) $\quad u\bar{u} \to e^+e^-,\ m_u \to m_e,\ k_S,\ k_u,\ k_{MF} \to 1,\ \eta_C^{st} \to \eta_C^{em} = 1,$

$$k' = k_S / k_{MF} \to 1,\ f_C^{st} \to f_C^{em},\ m_{\pi^0} \to m_\mu \approx m(LMC\ \ e^+e^-).$$

We Obtain TME of LMC e^+e^- 2B-System without OSTP and Muon-Mass Eqn:

(H26a)
$$T_{ME} = m_e c^2(\sqrt{k^2+4} - k - 2 + 1171.32908\xi^3),$$

$$k = \frac{\xi}{4}(1 + 3513.9872\xi^2),\quad \xi \equiv \frac{r_{sa}}{r},\quad r_{sa} \equiv \frac{e^2}{m_e c^2}$$

$$m_\mu \approx m(LMC\ \ e^+e^-) = m_e(\sqrt{k^2+4} - k + 1171.32908\xi^3)$$

Muon-Mass Eqn above Unsolvable Non-Linear Eqn of One Independent Variable r. Solution Exists but Unknown Even if We Know Mass of LMC e^+e^- Exactly. Sample Setting and Numerical Solution are

(H26b)
$$m(LMC\ \ e^+e^-) = 105.12\,MeV/c^2,\quad \xi = 0.56,\quad k = 154.418,$$

$$\gamma = 154.425,\quad r = r_{sa}/\xi = 5.03 \cdot 10^{-13} cm$$

(H19) Indicates: MD btw e^+e^- about $3 \cdot 10^{-11} cm$, Much Longer than Binding Size of LMC e^+e^- Shown above. IPE-Binding Size of e^+e^- 2B-System with **MF Turned Off** Equals Size of **Positronium** about $10^{-8} cm$. At Such Distance much Longer than MD btw e^+e^-, MF is Almost Zero. It Seems Impossible for Such weak MF to Cause a Bound State of e^+e^- 2B-System with Size btw IPE-Size and MD. On Contrary, Much Heavier u-Quark's IPE-Binding Size Much Smaller. Besides, DBR Shorter than SBR (Schrödinger Bohr Radius) Due to Small Factor $\sqrt{1-\alpha_s^2}$ in St Force Case. At Distances btw MD with Suppressed MF and IPE-Binding Distance of $u\bar{u}$ 2B-System, MF Still Has Large Magnitude though Suppressed. It's Able to Cause Bound State with Binding Size btw IPE-Binding Distance and Very Short MD btw $u\bar{u}$ with OSTP.

Please Note, (H26a) Can Be Used to Calculate Positronium's Mass as Ground State Mass of 2B-System e^+e^- as well as Mass of LMC e^+e^- in Tau, which is in Higher Excitation than LMC e^+e^- in Muon.

< **Positronium** e^+e^- and **Positronium-Mass Eqn** >

Although at Atomic Distances MF btw e^+e^- Not Exactly Zero, We Still Have Very Accurate Result after We Ignore MF to Get Binding Size: 2B-System e^+e^- Bound by Coulombian Force with MF Ignored is of Binding Size $2r_b$ Proved by Rigorous Solution of EqM 2B-System's Schrödinger Eqn. Then We Set Radius r of Classical Circular Motion in Classical Imitation to Be Equal to Bohr Radius ($r = r_b$). From (H26a) We Immediately Obtain Positronium Mass Eqn:

(H26c)
$$\xi \equiv \frac{r_{sa}}{r} = \frac{r_{sa}}{r_b} = 5.325 \cdot 10^{-5} \ \ (r_{sa} \equiv \frac{e^2}{m_e c^2})$$

$$k = \frac{\xi}{4}(1 + 3513.9872\xi^2) \sim 10^{-5},$$

$$m(Positronium \ \ e^+e^-) = m_e(\sqrt{k^2+4} - k + 1171.32908\xi^3) = 2m_e$$

Empirical Measurement of Energies of 2 Photons Produced in Positronium's Decay Proves above Mass Value: Each Photon's Energy Equals 0.511 MeV.

< **LMC** e^+e^- in **Higher** Excitation and **Tau-Mass Eqn** >

At Atomic Distances, e^+e^- in Positronium Bound by Coulombian Force. MF Only Causes Hyperfine Effects. Such IPE-Binding Yields Larger Binding Sizes for Higher Excitations. In Small SA Spaces, MF Plays Crucial Roles, Even Dominates Coulomb Force. MF-Binding Features Smaller Binding Size for Higher Excitation, Which Can Be Seen Clearly in Proof of TMET, together with Amazing Positive TME at binding sizes Less than 0TMED. From (26a) We Make Following Sample r-Setting and get Mass Eqs for Tau's LMC e^+e^- and for Tau.

(H26d)
$$\xi \equiv \frac{r_{sa}}{r} = 1.42, \ \ (r_{sa} \equiv \frac{e^2}{m_e c^2}, \ r = \frac{r_{sa}}{1.42} = 1.984 \cdot 10^{-13} cm)$$

$$k = \frac{\xi}{4}(1 + 3513.9872\xi^2) = 2515.744,$$

$$m_\tau \approx m(Tau's \ \ LMC \ \ e^+e^-) = m_e(\sqrt{k^2+4} - k + 1171.32908\xi^3)$$

$$= 1713.82 \ \ MeV/c^2$$

Note: Masses of Positronium, Muon, and Tau Explained via Classical Circular Motion Imitation Once Dynamical Effect of Resultant Force of Coulombian Force & MF is Calculated and DS of Muons and Taus Puts Light on Their Physical Pictures. Classical Imitation Brings about Perfect Match btw QM and CI Only if Quantum Wave Eqn Solved to Provide with Binding Size for Us to Determine r-Setting or r-Inputting in CI. Since Quantum Eqs in MF Field Not Solved and May Be Unsolvable, There is No Guarantee for Sample r-Settings in Muon-Mass Eqn and Tau-Mass Eqn to Be Alright.

Remark: We Can Get Accurate Agreement with Empirical Values of Masses of Pions & Muons by More Carefully Inputting Values of Some Parameters. But That Does Not Mean We Can Get Accurate Quantitative Results. By its Very Nature, MB-Systems in SA World Not Solvable and 2B-System with MFs Not

Solved and May also Be Unsolvable. Accurate/Quantitative Theoretical Results of Many Things in SA World will Remain Unknown/Uncertain in Future. Only Thing We and Future Generations Can Do: Get Right Pictures of Reality.

In Decay Mode $\pi^+ \rightarrow \mu^+ + \nu_\mu$, HMC $u\bar{u}$ in π^+ Annihilates to Produce LMC e^+e^- and $\nu_\mu\bar{\nu}_\mu$. Immediately, LMC e^+e^- Captures Relics $e^+\nu_e$ and $\bar{\nu}_\mu$ to Form μ^+ $=e^+e^-\ e^+\nu_e\ \bar{\nu}_\mu$. Even if we Ignore NFMs to Simplify Problems, How does LMC e^+e^- Captures and Binds DS-Positron during Microscopically Long Lifetime of Muon? A Sample **ZL** (**Z**one **L**ocation) and **SC** (**S**pin **C**orrelation) of $e^+e^-\ e^+$ 3B-System is Like This: Before Annihilation, $u\bar{u}$ Had Antiparallel Spins and Parallel Spin MMs to Join Forces to Bind Charged d-Lepton with its Spin MM Anti-parallel to Net Spin MM of $u\bar{u}$. Once $u\bar{u}$ Have Annihilated and LMC e^+e^- is Born, e^+, Originally in π^+, is Captured by LMC, Relocating via QT in R2 Region at e^+ in LMC. Both 2 Positrons Have Parallel Spins and Spin MMs, which Antiparallel to that of Electron in LMC. Then Electron in LMC MF-Binds 2 Positrons which are in R1 at Electron while 2 Positrons MF-Bind Each Other and are in R2 at Each Other. Roughly, Attractive MF in R1 Causes Circular Motion while Attractive MF in R2 Causes Oscillation along MM-Line. Furious Coulombian Repulsion btw 2 Positrons within SA Short Distances Suppressed by Attractive MF at All Distances Shorter than MD and They Never Annihilate during Oscillation, Making Muon Live Long Time. Such 3B-System (e^+ plus e^+e^- in Composite μ^+ with NFMs Omitted) Cannot Be Reduced to 2B-System or 1B-System as Approximation. But for 3B-System $u\bar{u}$ plus e^+ in π^+ with ν_e Omitted, We Have 2B-Approximation for HMC $u\bar{u}$ with Perturbation from Light e^+ Omitted and 1B-Approximation for e^+ in π^+ where Heavy HMC $u\bar{u}$ is Treated as Single Tiny Neutral Particle with Net Spin MM and at Rest. How to Calculate TME of above Said $e^+e^-\ e^+$ 3B-System Unknown. But We've just Calculated TME of e^+e^- 2B-System Produced in SA Space before e^+ Steps in.

Note: Before $u\bar{u}$-Annihilation, e^+ Made Small Contribution to TME/Mass of π^+ due to its Small ME which is Reserved as e^+ Relocates into LMC e^+e^- via QT. QT Preserves ME. Once LMC Captures e^+ to Form 3B-System $e^+e^-\ e^+$, its Mass and TME are Close to What We Have Calculated for 2B-System LMC e^+e^-.

- **PAPPs** that Consume Energies Either Released in SA De-Excitation in Decay Events, Except for Neutron Decay that is Triggered by ENM, Not by SA De-Excitation, Or Released in HE Collision Events are **Creation Events** in Deep SA Spaces. TME of Produced Pair is Either Negative, such as for $u\bar{u}$ HMC, or Positive, such as for e^+e^- LMC.
- As e^+e^- Circling Each Other, Local Electric Currents, Associated with Opposite Orbital Velocities of e^+e^-, are Always in Parallel Directions Causing Attractive Magnetic Forces if Retardation Omitted. We Have Ignored such Mag-Forces so Far. It Has Same Magnitude as Coulomb Force due to near-c Speed of Large γ-Value. MF-Domination at Short Distance Permits such Neglect.

- In **<u>Positronium</u>** e^+e^- Case, IC of e^+e^- **<u>Validates</u>** Solution of Dirac Eqn for 2B-System in Coulomb Field of Interaction with Mag-Forces (Large at Short Distances) Ignored at Atomic Distances. Such Solution Matches Empirical Facts and Data So Well since Wave Function as Part of Such Solution Reveals Very Small Probabilities for Distance btw e^+e^- to Be Short Enough like SA Short Distances.

  In **<u>LMC</u>** e^+e^- Case IC of e^+e^- **<u>Invalidates</u>** Solution of Dirac Eqn with Mag-Forces Totally Ignored. LMC e^+e^- Produced in Deep SA Spaces. Such IC Enforces Mag-Domination, Not Mag-Neglecting. LMC e^+e^- is Born to Be within Short SA Distances, Have Very Large KEs with Near-c Speed and with Large γ -Value and Very Large Positive TME. e^+e^- in LMC are Born to Have KEs Much Larger than Very Large Absolute Value of Very Negative PE at Short SA Distance. Exact Binding Sizes in SA World Unknown, but We May Plug Series of Small SA Radii into Classical Eqn for Circular Motion of Electron in Coulomb Field and See Clearly Series of Circular Orbital Motions (Simplest Classical Examples of Bound Systems) with Near-c Speeds & Large γ Values.
- Three u-Quarks in Every Baryon Forms 3B u-Quark Core. Directions of Their Spins and Motions Unknown/Unclear due to **MBD** (**M**any **B**ody **D**ifficulty). For Strong Force Does Not Act on Electrons and Wk Forces Exerted on d-Electron by any Neutrinos/Antineutrinos and u-Quarks are Negligible, Motions and Spin Directions of d-Electrons are Relatively Easy to Figure Out. In SQC Sub-Model, We Can Even Treat 3 u-Quarks in Nucleon or Baryon as Point-like u-Quark Core Represented by a Point-like Electric Charge with a Net MM.
- Since Neutrinos/Antineutrinos Only Exert Wk Forces on Other Particles, Their Existence inside SA Composite Particles, Motions, Spins, and Wk MMs Have Little Influence on d-Electrons and u-Quarks. To Figure Out Motions and Spin Directions of u-Quarks and d-Electrons, We Can Ignore DS-Neutrinos and DS-Antineutrinos.
- Bohr's Model Soon Proved by Schrödinger & Dirac via Theoretical/Quantitative Results, in Good Agreement with Experimental Data. However One Can Neither Prove Nor Disprove DS Quantitatively for **MMD Unsolvable**. Qualitatively and Half-Quantitatively Speaking, There're No Phenomena DS with MFs Has Failed to Explain.

  Under Condition of **Insurmountable Obstacle** to Get Quantitative Solutions and Explanations, DS Has Demonstrated its **Power** to Explain Experimental **Facts**, Including Many 'No-Way' to Explain Facts So Naturally and Coherently, and So Overwhelmingly.
- Baryon Number Conservation Never Requires Nor Leads to Stable Hadrons just because of DS that Includes d-Leptons and Their Never Stable Excited States.
- All Free Protons/Anti-Protons are Stable and All Other Hadrons, Free or Bound, are Unstable. But Bound Protons Can Be Changed to Neutrons in Some Decay Modes. DS also Explains Such Bizarre Phenomenon. SDL
- Identical d-Leptons in Free as well as in Bound Neutrons, Repel Each Other, Causing beta Decay. QT Relocates d-Leptons Released from Bound Neutron to Proton Nearby Bound in Same Nucleus and Changes Proton/Neutron Bound inside Same Nucleus into Each Other. Bound Neutrons Also Unstable. Chemical

Elements on Stability Island Have Long enough Lifetimes to Be Called Stable, though Their Nuclei Contain Neutrons (Sometimes Even More than Protons). All Such ***Ironic*** and ***Bizarre Facts*** Must Be Explained. Using DS, We Have Explained All of Them and More to Come….

- If Both Target and Incident Charged Particles with MMs are Composite or Just Target MM is Carried by Composite Charged Particle, Departure from Inverse PE and Coulomb Scattering Occurs at Distance Approximately Equal Sum of 2 Size-Radii. If Incident Particle is Electron, its Size-Radius is Negligible. For Any Many-Body Composite System or Electron Colliding with Target Nucleus, We Expect that Departure from Inverse Potentials/Coulomb Scattering Takes Place at Distance Larger than or Nearly Equal to Size of Target Composite System: Once Distance btw Mass Center of α -Particle or Electron and Gold Nucleus Reaches Distance Smaller than 27fm or 1.2 fm, a Constituent Nucleon of α -Particle or Incident Electron May Hit a Nucleon in Target Nucleus So That Such Close Contact Occurs within MD btw 2 Constituent Particles Involved in Close Contact, Triggering Large MF and Departure from Inverse Potential and Coulomb Scattering.
- Deeper Probing Triggers Not Only MF-Departure from IPE Scattering but Also Inelasticity of Vacuum with PAPP, In Most Cases Followed Immediately by QT-Relocating/Capturing to Create Seemly Complicated Middle Stage Particles.
- MF & Vacuum Inelasticity Make Rutherford/Mott/Hofstadter Methods Unable to Obtain Finer Pieces of Empirical Information about Hadrons' Sizes. Sizes of Baryon Quark Cores Remain Unknown.

< **MDs** without **OSTP** >

- It Seems that MD for d-Electron to Bind d-$\bar{\nu}_e$ Can Be Obtained from (H19) by Setting $m_1 = m_e$, $m_2 = m_\nu = k_\nu 10^{-6} m_e$: $r_m = 3\cdot 10^{-8} cm/\sqrt{k_\nu}$. This is **Wrong** for (H19) Omits Orbital MM of d-Electron. Above MD is Just MD btw Electron and $\bar{\nu}_e$ where No Local Orbital Motion of Electron Involved such as Atomic Electron in S-State (but Not d-Electron):

(H27a) $$r_m(e^-\bar{\nu}_e) \sim 3\cdot 10^{-8} cm/\sqrt{k_\nu}\,, \qquad k_\nu = 10^6 m_\nu / m_e$$

- Above MD Not Shorter than Atomic Distance for $k_\nu < 1$. Later We'll Explain Why d-Electron and u-Quark Core Can Join Forces to Bind $\bar{\nu}_e$ while Atomic Electron Only Has Small Perturbation Effect on d-$\bar{\nu}_e$ Even within Distance Smaller than Their MD. Here We Simply Point Out that d-Electron's Spin and Orbital Motions are Highly Oriented Relative to u-Quark's MM Vector and to u-Quark Core's Net MM Vector, Required by MF-Binding Condition, while Light Atomic Electron's Spin MM Has No Any Relatively Fixed Direction Due to Negligible MF-Binding in Atomic Electron Case. If in EMF, Atomic Electron's Spin Still Has 2 Possible Directions: Same as or Opposite to EMF, Causing Two Discrete Spectrum Lines in Stern-Gerlach Experiment after Hydrogen Atoms Pass through Non-Uniform Magnetic Field.

- MF-Binding in R1 is Characterized by d-Electron's Spin Parallel to u-Quark's Spin or u-Quark Core's Net/Effective MM. MF-Binding with **Asymmetry** Must Lead to **Orbital** Motion with Positive Orbital Quantum Number Even in Ground State. Quasi-Bound d-Electron Wd-e in Neutron Also Stays in R1 with its Spin Parallel to u-Quark's Spin or u-Quark Core's Net/Effective MM.
- Inside Neutron, Bound d-Electron's Small Net MM Has Little Effect on Quasi-Bound d-Electron Staying in R1 with Spin Parallel to Bound d-Electron's Spin. OSOD Make MD r_m for d-Electron to Bind d-Anti-e-Neutrino Much Less than Above Value Seen in (H27a). Spin MM of Electron about 657.411 Times Larger than Proton's Empirical MM. If $|\vec{f}_i|$ of **Net MF** in (H19) Reduced 625 or More Times, Corresponding MD r_m will Be Reduced $\sqrt{625}=25$ or More Times. Due to OSOD, d-Electron's Net Wk MM is Reduced at Same Rate as Its Net MM.
- Let μ_e^{swk} and μ_{d-e}^{nwk} Be Magnitude of Electron's Wk Spin MM and d-Electron's Net Wk MM Respectively with $\mu_e^{swk}=625k_{d-e}\mu_{d-e}^{nwk}$. For d-Electron Binding $\bar{\nu}_e$ via Wk MF, We Write

(H27b) $$r_m(d-e^-\bar{\nu}_e)<1.2\cdot 10^{-9}cm/\sqrt{k_\nu k_{d-e}}\ ,$$

$$k_\nu \equiv m_\nu 10^6/m_e\ ,\quad k_{d-e}=1.6\cdot 10^{-3}\mu_e^s/\mu_{d-e}^n=1.6\cdot 10^{-3}\mu_e^{swk}/\mu_{d-e}^{nwk}$$

This May Imply Atomic Size for MPR of Bound d-$\bar{\nu}_e$ like Atomic Electron.
- d-Electron's Orbital Quantum Number and Orbital MM, Its **Net MM**, and Parameter k_{d-e} **<u>All Have Unknown Values</u>** Due to Absence of Solution for any MB Quantum System. Empirical MM-Values of Nucleons Unable to Determine Them for HQM Theorem Only Proves that $m_u > or >> M_p/3$ and Fundamental SA Quantity m_u Remains Unknown.
- OSTP of HMC $u\bar{u}$ in CI of its Ground State is 2-Dim. BC uuu is 3B-System. We Do Not Know How a 3-Dim OSTP Suppresses MF and St MF Greatly for BC uuu. Even if MFs and St MFs among uuu are Suppressed Greatly, We Still Know Nothing about OAM and Orbital MM of Each u-Quark in BC uuu. MMs of Nucleons/Baryons are Vector Sums of Spin MMs and Orbital MMs of Three u-Quarks and DS-Electrons and/or DS-Positrons. Rigorously Speaking all above MB-Systems **<u>Permanently</u>** Forbid Accurate Quantitative Explanations of Their Empirical MMs. Hypothetic Zero-OAM and Zero Orbital MM Baseless.
- As Long as d-Electron's Net MM is Small Enough, d-Electrons and DS Can Be Ignored in Nucleon-MM Issue and Theoretical Values of Nucleons' MMs Can Be Determined Mainly by Adjusting u-Mass with d-Mass Almost Same as u-Mass. Recently, Donald H. Perkins ("Introduction to High Energy Physics", 4th Edition, World Web, p. 130-132) Showed His Work on This Direction. He Obtained $\mu_n=-(2/3)\mu_p$. Empirical Ratio is -0.685. Relative Error is less than 3%. For Other Baryons, His Calculated MMs Agree with Observations within Discrepancies at 10%-20% Level. To Obtain Such Good Results, He Has Set $m_u=m_d=336$ MeV and $m_s=509$ MeV.

- It's Also Possible that u-Mass Much Heavier than Perkins's Mass-Setting. Then MD btw 2 *u*-Quarks Becomes Much Shorter than 1fm Because of (H19) and Empirical MMs of Nucleons Mainly due to Net MMs of d-Electrons. It Points to another Direction to Explain Empirical MMs of Nucleons, Requiring to Find out OAM of Single d-Electron in Proton and 2 OAMs of Wd-e and Sd-e in Neutron. Such Unknown Theoretical Values May Or May Not Be Obtained in Future.
- Because of Uncertainty Principle, Small Enough Binding Size Δr Yields Large Enough Δp, which Implies Near-c Speed with Large γ -Value. As Long as u-Mass is Large Enough, Quark Cores' Sizes Can Be Small Enough.
- In Presence of Empirical Values of Nucleons' MMs, Still, We are Unable to See Which One, SQC or RQC, is True. Even if Heavier u-Mass than Perkins' One and SQC are True, His Work Obviously Remains **Important Step** to **Seriously Face Conflict** btw **Empirical MMs** of **Nucleons** and **Theoretical Prediction Drawn from Light u-Mass Setting**. Perkins Settings Satisfy HQM Theorem.
- In MPR of Atomic Electron, MFs Exerted on It by u-Quarks/d-Electron(s) are Negligible. Magnetic Fields Produced by Local Currents Associated with orbital and Spin Motions of u-Quarks and d-Electron(s) too Weak in Atomic Electron's MPR to Enforce Preferred Direction of Its Spin MM. Namely, Atomic Electron Does Not Have Preferred Direction of Its Spin MM in Absence of EMF. Even in Presence of EMF, 2 Discrete Spectrum Lines Seen in Stern-Gerlach Experiment Tell that Both 2 Directions of Atomic Electron's Spin MM (Same as or Opposite to EMF's Direction) are Possible. None is Preferred. Without Preferred or Fixed Direction and Due to Much Larger than MD Distance btw Atomic Electron and a SA Particle (u-Quark or d-Electron), **Powerful Spin MM** of **Atomic Electron Only Has Perturbation Effect** on **u-Quarks/d-Electrons**.
- From (H27b) It Follows that d-$\bar{\nu}_e$ May Have Orbital Size as Large as Atomic Electron's. Average Distance btw Atomic Electron and d-$\bar{\nu}_e$ May Be of Same Order of Magnitude as Distance btw d-Electron and d-$\bar{\nu}_e$. Crucial Difference is that d-Electron is Bound by u-Quark and/or u-Quark Core with MF Involved to Cause Such SA Binding. MF-Binding Causes **Spin-Correlation** such that Their Spin/Orbital Vectors are Oriented with Fixed Correlated Directions. Atomic Electron Not MF-Bound. Its Spin Has No Fixed Correlated Direction. Without Any Fixed Direction, Spin Wk MM of Atomic Electron is Not Powerful at All Even to d-$\bar{\nu}_e$ Nearby within MD btw Atomic Electron and d-$\bar{\nu}_e$. Moreover, Atomic Binding Sizes of DS-NFMs Bound by Resultant Force of Wk MF and Wk-Coulombian Force Imply Small KoE and DS-NFMs Knocked out Every Minutes Millions of Times in Collisions Seen in Cosmic Shower Events and in Accelerators all Over the World. Just like Co-Events of Q-Knockout and DSL-Knockout, PAPP of NFMs Occurs Right before NFM-Knockout due to AVI and PQTR Helps to Fill up Holes Left behind by Knocked out NFMs. V-Shape of TME-*r* Curve of DS-NFM Makes a Series of Excited States in Inward Direction, just like V-Shape of TME-*r* Curves of Bound Quarks/d-Electrons Does, Helping to Consume/Absorb Impact Energy.
- MD for u-Quark to Bind $\bar{\nu}_e$ Obtained from (H19) by Setting $m_2 = m_\nu = k_\nu 10^{-6} m_e$ and $m_1 = m_u = k_u m_p / 3 = 612.05089 k_u m_e$ with $k_u >$ or $>> 1$ (HQMT). MD for u-Quark

to Bind d-Anti-e-Neutrino is

(H27c) $$r_m(u\bar{\nu}_e) = 1.212628\cdot 10^{-9}\,cm/\sqrt{k_\nu k_u}\ ,$$

$$k_u \equiv 3m_u/m_p = m_u/612.05089m_e > or >> 1$$

Approximation in (H19): $r_m \sim 3\cdot 10^{-11}\sqrt{m_e m_e/m_1 m_2}\,cm$. Then MD for u-Quark to Bind d-e^- is Obtained by Setting $m_2 = m_u = 612.05089\,k_u m_e$ and $m_1 = m_e$:

(H27d) $$r_m(ue^-) \sim 1.2\cdot 10^{-12}\,cm/\sqrt{k_u}\ ,\qquad k_u > or >> 1$$

- Also, MD btw 2 u-Quarks is Obtained from (H19) by Setting $m_1 = m_2 = m_u$

(H27e) $$r_m(uu) \sim 4.9\cdot 10^{-14}\,cm/k_u$$

- As u-Mass Increases, k_u Increases while Both $r_m(uu)$ and $r_m(ue^-)$ Decrease. (H27d, e) Tell Us: $r_m(uu)$ Decreases Faster than $r_m(ue^-)$.

Remark: (H27c, d, e) Give MDs btw 2 Elementary Fermions Carrying Same Type of Charges and at Least One is u-Quark. Orbital MM of u-Quark Omitted. This May Not Be Right. We Do Not Have Empirical Values of u-Quark's Mass, OAM, and its Binding Size. Three u-Quarks in any Baryon Have Severe 3-Body Problem. It's Very Difficult to Know How They Move. Still We Can Figure Out if St MF May Play Important Role to Yield Orbital Motion and Small Energies of Radial Excitations in SQC Sub-Model. If St MF Can Be Ignored, Coulomb Type St Force Must Dominate. HMC $u\bar{u}$ Ground State Table on Page 79 Serves as 2B-Example with MFs Suppressed by OSTP. OAM of u Equals OAM of $\bar{u}$ for They Move about Same Center. Since They Have Opposite Charges, Two Orbital MMs Antiparallel effectively Yielding 0-Net Orbital MM for Magnetons far away from HMC $u\bar{u}$. But for BC uuu as 3B-System, OAMs, Orbital MMs, and their Net Values Totally Unknown. Thus, Empirical MMs of Nucleons Can Not Be Explained Quantitatively in any Future. In Mathematics, Unsolvable Eqs are Much More than Solvable Ones. We Know that 3B-Systems Unsolvable and Accurate Theoretical Results Unknown in any Future. Still, Power of Scientific Reasoning Manifests itself When It Tells Us Crystal Clearly Why Quantitatively Accurate Explanations Cannot Be Made. For BC 3B-System, Future Study May Adopt Different Postulates about Basic Motions, Spin-Correlations, and u-Mass and Then Deduce Theoretical Consequences and Compare. We Can Always Improve Knowledge/Understanding by Comparing Different Ways to Explain Experimental Facts. No Comparison, No Advancement.

- Surely, we Needs Incoming Electron to Have Large Initial Momentum in Order to Probe Deeply into Quark Core if It's Really Smaller or Much Smaller than 1 Fermi. Such Large Momentum Requires Large Initial Energy. In Deep Inelastic Scatterings and Collisions, Particle Productions May Cover Needed Clues to

Sizes of Quark Cores. Consider Particle 1 Hitting Rest Target Nucleon (Particle 2). Large-Angle Scattering of HE Particle 1 Means Large $\Delta\vec{p}_1 \equiv \vec{p}'_1 - \vec{p}_1$ that Must Cause Recoil of Particle 2 with Large $\Delta\vec{p}_2 = -\Delta\vec{p}_1$ if Large LV Space Force Doesn't Appear to Violate Momentum Conservation. In Elastic Collision with $\Delta E_1 + \Delta E_2 = 0$, Once $|\Delta E_1|$ is Larger than Required Energy of e^+e^- or Pion-Pair Production, Particle Production Takes Place to Absorb and Consume Impact Energy. This Makes Deep Inelastic Scattering Unable to Help Us to Get Smaller Upper Bound of Hadrons' Sizes, No Matter How We Enlarge Energy Scale of Largest Accelerators. If a Collision Happens without Enough Recoil to Conserve Total Momentum, Like Observed Bizarre Rare Jets with Nothing to Balance in Opposite Directions, Such Events May Be Evidences of LV Space Force.

< **5BF** of Nucleons with **Preferred** Spin Direction of β-Particles >

- A Basic Theory is Required to Explain **5BF** (Five **B**asic **F**acts) about Nucleons:

1. **Stable** Free Protons (No Single One Found Decaying in Billions of Photos).
2. **Unstable** Free Neutrons with Lifetimes **Much Longer** than Other Unstable SA Particles, about 10^8 longer than slowest decaying ones among them.
3. **CES** (**C**ontinuous **E**nergy **S**pectrum) of β-Particles.
4. Observed **MMs** of **Nucleons** (Empirical Values of Nucleons' MMs).
5. **Preferred Spin Direction** of beta Particles in EMF, Bizarrely Parallel to EMF. We've Explained 1st Three Facts before.

Following **Spin** & **Orbital Directions** of d-Electrons Help Us to Explain above Five Facts Simultaneously.

- Spin Vector of Bound d-Electron in Proton (SBd-e) or Neutron (Sd-e) is Parallel to a *u*-Quark's Spin and u-Quark Core's Net MM as it Stays in R1 with Position Vector Relative to u-Quark and/or u-Quark Core Perpendicular to their Spins so that the said u-Quark and the u-Quark Core MF-Attract d-Electron and Bind it. Once One More d-Lepton Pair Enters Proton and Turns it into Neutron, Pauli's Exclusion Principle Makes 'New Comer' Be in Different Quantum State. It Also Stays in R1 but Further Away from u-Quark Core than Previous One. For New Comer Wd-e (Quasi-Bound One), Its **Spin** Vector Also **Parallel** to u-Core's Effective or Net MM. OSOD Yields Small Net MMs of d-Electrons in Neutron.
- Observed Preferred Spin Direction of beta Particles in EMF Reminds us of Two Fundamental Facts: Both CED and Experiment Prove that Classical Heavy Body always Turns its MM into ***Same*** Direction of EMF while for Quantum Electron, Free or Bound in Atom, Has Two Possible Directions of its Spin MM (Opposite to or Same as Direction of EMF), ***None is Preferred***. These 2 Facts Have Been Explained in CED and QM Respectively. How Can a Basic Theory that Covers Both Classical and Quantum Phenomena Explain Such **Bizarre Phenomenon**: Beta Particles Flying out from Cobalt with Spin MMs Preferably Pointing at direction ***Opposite*** to Direction of EMF?
- OSOD and 2 d-Electrons' Same Spin Direction Parallel to u-Quark Core's Net MM Explain Observed Preferred Spin Direction of β-Particles:

Neutrons in Nuclei of Source Matter Preferably Have Their Net MMs Pointing at

EMF's Direction before β-Decays Occur. Nucleus of Cobalt $^{60}C_0$ Used in Wu's Experiment was Very Heavy with Small Quantum Effect so that Its Net MM (Vector Sum of All Constituents' MM Vectors) Could Have Preferred Direction Parallel to EMF just like Spinning Heavy **Classical** Charged Metal Ball in EMF. CED Tells that Magnetic Field Always Exerts Force on **Classical** MM and Force It to Turn Its Direction to Same Direction as EMF. Although Magnetic Force Vanishes as Classical MM in Direction Anti-Parallel to EMF, such Equilibrium is Not Stable. Any Perturbation Would Cause Deviation from such Direction with Magnetic Force Appearing and Turning It Around until its Direction Points at EMF's Direction. Parallel Directions of u-Quark Core's Net MM and Quasi-Bound d-Electron's Spin Vector are also Parallel to Net MM of Neutron as long as d-Electrons in Neutron Have Orbital MMs with Magnitudes Very Close to that of Their Spin MM and/or with Orbital MM Larger than Spin MM so that OSOD Lead to Net or Effective MMs of d-Electrons with Small Enough Magnitudes and/or with Same Direction as u-Quark's or u-Quark Core's Net MM. Preferred Spin Direction of beta-Particles Has Been Proved Same as EMF. Quantum State of Spin Shows that Spin Vector of Electron is Either Parallel or Anti-Parallel to EMF. None is Preferred. But Observed Preferred Direction of Emitted Electrons' MM Vectors in Neutron Decay is in ***<u>Opposite</u>*** Direction of EMF. This Implies 2 Things: No Enough Times for Them to Adjust Directions of Their Spin and MM Vectors during Flying Time while Observed Preferred Direction of Their Spin Vectors was Also Preferred Spin Direction of d-Electrons in Neutrons in Same EMF before They Fly Out. Electron's Quantum Particle. Free Electrons and Electrons Bound by Electric Forces (such as Atomic Electrons Bound by Coulombian Forces) Don't Let Their MM Vectors Have Preferred Directions. Even in EMF, Unbound Electrons and Atomic Electrons Always Point Their Spin MM Vectors at EMF's Direction **OR** at its Opposite Direction, as Seen in Stern-Gerlach Experiment. No Preferred Direction for such Quantum Particle's MM. But d-Electrons're **MF-Bound**. MF-Binding Requires **Spin-Correlation** btw u-Quark or u-Quark Core and d-Electron. Each u-Quark is MF-Bound too so that three u-Quarks in Nucleon are Spin-Correlated with One Another and with Heavy u-Quark Core. MM of Heavy Classical Body Has One & Only One Preferred Direction in EMF, Same as EMF. Heavy u-Quark Core and Heavy Nucleon are btw Quantum and Classical Particles. Nucleons Inside Cobalt's Nucleus are MF-Bound with Spin-Correlation. Heavy Cobalt's Nucleus Contains 60 Nucleons. Definitely, It's More Classical than Quantum. Thereby, its Net MM indeed Has Preferred Direction Same as EMF. Even if quantum effect of Each Individual Neutron in Cobalt-60 Not Negligible, Directions of Their Net MMs, Statistically Speaking, Must Have One Preferred Direction, Same as EMF, which Net MMs of Classical Heavy Nuclei of $^{60}C_0$ Substance Must Point at Preferably. While d-Electrons' Spins, due to MF Spin-Correlation, Required by MF-Binding, Have Same Direction as Net MM of Neutron's BC, which in Same Direction as Net MM of Neutron because of Smallness of d-Electrons' Net MM. Finally, We Now Know why beta Particles Have Preferential Spin Direction Same as EMF.

- Bound d-Electron Inner One while Quasi-Bound d-Electron Outer One so that Orbital Motion of Bound d-Electron Produces <u>Local</u> Electric Current Relative to

Quasi-Bound d-Electron just like Spin Motion that Produces Local Currents. Such Local Orbital Current Produces Effective MM, Called **Orbital MM**. OSOD Yields Small Net MM of Bound d-Electron Causing Small Effect on Any Later Comer with Spin and Orbital Motions so that d-Electron that Steps in Later Can Stay in R1 with Spin Parallel to Bound d-Electron's Spin. Then Coulombian Repulsive Force btw 2 d-Electrons in Neutron is Only Significant Dynamical Reason to Cause ENM. Free Neutron never Seen to Emit 2 Electrons to Become Spin ½ *uuu*-Baryon because Quasi-Bound d-Electron Stays Farther away from u-Quark Core and is Always 1st One to Be Pushed Out from Neutron so that Simultaneous Pushing Out Can Never Happen. Once Quasi-Bound d-Electron is Pushed Out, Neutron Becomes Proton without Repulsive Force of ENM so that Bound d-Electron in Proton Can Be in Stable Bound State.

- Quasi-Bound d-Electron in Neutron Also Exerts Repulsive Force on Bound One. It Cannot Push away Bound One for It is Bound, Closer to u-Quark or u-Quark Core, and u-Quark and u-Quark Core Exert More Powerful Attractive Force on Bound d-Electron than on Quasi-Bound d-Electron.

- Unsolved Dirac Eqn in MF Field is Not Capable to Prove Nor Disprove Larger Magnitude of Orbital Angular Momentum than that of Spin. Neither it's Capable to Prove Nor Disprove Small Difference btw Magnitudes of Two Angular Momenta (Spin/Orbital) of Each d-Electron in Nucleons. Microscopically Very Long Life of Neutron and Nucleons' MMs (Much Less than Electron's Spin MM), however, Join Forces to Support Small Difference btw Magnitudes of Orbital and Spin Angular Momenta of Each d-Electron in Nucleons. Large Difference would Yield Large Net MM of d-Electron, which would Immediately Make MM of Proton Larger than Empirical Value and too Large Repulsive Force btw Large Net MMs of d-Electrons in Neutron Leading to beta-Decay Much Quicker than Observed Long Ones with Empirical Lifetimes of Free Neutrons about 10^{10} Times Longer than Longest Lifetimes of Unstable Mesons Such as Charged Pions and Kaons.
- Identical d-Electrons Do Repel Each Other. Large Coulomb Repulsive Force at Such Short SA Distances Can Only Cause Slow Decay of Neutron because MF-Attractive Forces from u-Core Greatly Reduce such Large Coulomb Repulsion. Parallel Net MMs and Parallel Spins of Two d-Electrons in Neutron Yield No Large MF Due to OSOD and Smallness of Their Net MM. Thus, DS and MF Explain Simultaneously above Said **5BF** (Stable Free Proton, Unstable Neutron Decaying very Slowly, Continuous Energy Spectrum of beta Particles, MMs of Nucleons, and Observed Preferred Spin Direction of beta Particles in EMF).
- As Pointed Out Before, If Electrons and Anti-e-Neutrinos Emitted from Cobalt Reversed Directions of their Velocity-Vectors & Spin-Vectors, Parity Violation and Handedness of Leptons in SM would Remain True. DS Explains why It Did Not Happen That Way. Parity Violation, Handedness in SM, and DS with MFs Join Forces to Fully Explain beta-Decay Experiment.

< **Stability** Island with **Decaying** Neutrons >

- Once One d-Lepton Pair $e^{-}\bar{\nu}_e$ are Pushed out from Neutron Bound in Nucleus, They May Leave Nucleus and Visible beta Decay Occurs or They May Go via

QT and Relocate in Nearby Proton Bound in Same Nucleus. QT Changes *n*/*p* in Nucleus into Each Other iff DS is Real.

- Chemical Elements on Stability Island with Large Z and More Neutrons than Protons Exist because Large Number of Neutrons in Each of That Kind of Elements Makes It Harder, Rare, or Very Rare for So Many Neutrons to Nearly Simultaneously Release $e^-\bar{\nu}_e$ Pairs. Thus, Less Protons Bound in Same Such Nucleus May Be Enough to Take and Relocate Them. The More the Neutrons, the Longer the Time Needed for Simultaneous or Nearly Simultaneous Decays to Happen. The Less the Neutrons, the Easier the Simultaneous Decays. Now We Understand Why Stability Island Contains So Many Chemical Elements with More Neutrons than Protons under Condition that Every Neutron, Free or Bound, Dooms to Decay in about 15 Minutes Due to DS with ENM.
- DS Reveals Nature of the Force that Causes beta Decay. It is Not Weak Force but e.m. Force. Say again, Stable Free Proton Proves Once and for All that d-Q in Free Proton Does Not Decay & W^--Hypothesis $d \to u + W^-$, $W^- \to e^- + \bar{\nu}_e$ is Not Correct. Furthermore, u-Quarks in any Nucleon Never Decay. This Ironic Fact Proves Once and for All that W^+-Hypothesis $u \to d + W^+$, $W^+ \to e^+ + \nu_e$ is Not Correct. It's Correct that W-Particles Do Not Exist, Just like HPs & GBs.

< SA Binding Size and Level Differences >

- Formula of General DBR Tells: IPE Binding Size Inversely Proportional to Mass of Bound Particle. u-Quark's Bohr Radius in Ground State Unknown for Solution of MF-Binding and u-Mass Unknown, Leading to Ambiguous Dynamics Models. We Face Many Different Possible Binding Sizes.
- Roughly, We Have 2 Kinds of Sub-Models:

1. **SQC** (**S**mall **Q**uark **C**ore) Model.
2. **RQC** (Regular Quark Core) Model.

- Small Quark Cores Mean Small Binding Sizes and Small Values of Δr, which Requires Large Values of Δp According to Uncertainty Relation. Large Δp Means Large Momentum p and Large Angle Deviation of Moving Direction. However, It Does Not Necessarily Mean Large ΔE and Large Quantum Level Differences. Uncertainty Relation also Tells that Small Δt Requires Large ΔE, but $\Delta r \sim c\Delta t$ May Not Be Right Quantum Estimation. One May Use $\Delta r \sim c\Delta t$ $\geq c\hbar/2\Delta E \sim \hbar/2mc$ to Conclude: The Heavier the Force Carrier, the Shorter the Force Range. But ΔE in Such Uncertainty Relation Application Carried by **Virtual** Force Carrier. If Carried by a Real Force Carrier such as a HE gamma Photon, then Large ΔE Never Forbids such Real Photon to Travel for Many Light Years before it Hits a Particle. Quantum Level Difference ΔE will Equal Energy Carried by Real Force Carrier or PAP(s) of Real Particles as De-Excitation Occurs to Release Such Real Force Carrier or PAP(s) of Real Particles. Small Radial Excitation ΔE Should Not Necessarily Mean Large Binding Size. We May See this in View of Dynamics, as Described below.
- Dirac Eqn Has Rigorous Solution if PE is Central Time-Independent Inverse PE. In QM, Inverse PE and Square Well PE Yield Similar Result: The Smaller the

Binding Size, the Larger the Level Difference. Once Magnetic & Magnetic-like Forces Enter Stage of Binding Forces that Result May Change Surprisingly.

- An Extremely Serious Question One Can Ask is Why Do MFs Do Work Now That They're Members of Mag-Force Family?

< Reason **MF Does Work** and Changes IPE >

Lorentz Force Formula in CED Indicates that Mag Force Exerted on ***Point-like*** Electric Charge in any Given Mag Field Always Perpendicular to Velocity of the Charge and therefore Does Not Do Work. When 2 Magnetons Such as 2 u-Quarks, 2 DS-Electrons, or u-Quark and d-Electron Interact Each Other via MF, however, They are Not Point-like. Point-like Model is in Severe Conflict with Inherent MM of Magneton. Classical Connection Formula is Deduced under Basic Assumption: Electron's Formed by Many Charged Parts Spinning about Same Axis. CED Proves that $U = -\vec{\mu}\cdot\vec{B}$ is PE of MM $\vec{\mu}$ in Magnetic Induction $\vec{B}$. A Non-Uniform Mag Field $\vec{B}$ May Exert Non-Zero Mag Force on $\vec{\mu}$:

$$\vec{F} = \vec{\nabla}(\vec{\mu}\cdot\vec{B})$$

Of Course, Source MM Produces Non-Uniform Magnetic Field. Thus, MF btw any 2 Magnetons Does Work, Playing Crucial Roles in SA Dynamic World. One May Devise Thought Experiment to Show How Non-Uniform Magnetic Field Can Move Magneton and Do Work. Here We Omit Details of such Discussion.

< **Yukawa** St Force and **Gluon** St Force >

- SQC Requires Extension of Gluon-Force Range and Charge Radius by Virtual Pions btw Internal Double or Single Virtual Gluon Lines. At Least, We Have Seen Reality of MF that Makes Potentials in Rutherford, Mott, and Hofstadter Scatterings Deviate from Coulombian Potentials at Distances Nearby MD. Such Method and Idea are Unable to Get Empirical Sizes of Quark Cores and Hadrons. Established Empirical Upper Bounds of Such Sizes Have No Profound Meaning.
- Putting p^{μ}-Conservation Dirac Delta Factor into Each Vertex of Feynman Diagrams Leads to p^{μ}-Conservation and off-Mass Shell Internal Lines: p^{μ} Transferred in t/s-Channels Spacelike/Timelike. Namely, in p^{μ}-Conservation Events, Transferred 4-Momentum p^{μ} Cannot Be Carried by Individual Virtual Force-Carrier. It's Just Effect of Virtual Force-Carrier Exchange. This View Point of Quantum Field Theory is Supported by Simple Facts that Transferred Energies in So Many Neutral-Current Events Much Less than Rest Energy of Z-Boson. Due to Same Reason, a Transferred Angular Momentum is Dramatically Different from what's Carried by any Individual Virtual Force Carrier. If this is True, Internal Lines of Double Virtual Gluons Not Needed to Couple to Virtual Pion. Besides, Label J=0 for Pions Does Not Prove Their TAM as We Showed in Section "Spin Misconception, TAM Disability, and 4TDSP". Anyway, Yukawa Strong Force of Short Range via Exchanging Virtual Pions of Label J=0 and Strong Force via Exchanging Virtual Heavy Vector Gluons Reconciled.

< **STB** in **SQC** Sub-Model and **RQC** >

- Limiting LV, Proved Overwhelmingly by So Many Facts, Has No Obligation to Produce Non-Limiting LV at SA Scales Ranging from 10^{-13} to $10^{-18}cm$. None of Us Knows at What Distance and Energy Scale Large LV Begins. We are Not Sure How LV Plays a Role in QD. But Two Things are Clear that Importance of MF btw Spinning Electrically Charged Particles for Nuclear or Sub-Nuclear Electron Binding Has Been Ignored for 70 Years Since Neutron was Found to Cause beta Decay Long Before Quark Model was Proposed while Zero LV at Quark Binding Energy Scale and btw $10^{-13}cm$ to $10^{-18}cm$ Cannot Be Accepted as a Scientific Fact Since STD Makes It Impossible for Experimenters to Measure LV at Such Short Distances, or Equivalently, in Corresponding Large γ-Sectors, Even If LV at Such γ-Sector is Indeed Very Large.
- Large LV Might Contribute to QD in SQC Sub-Model Only if Q-Cores Have Small Enough Sizes. STD, Absence of Empirical Curve of LV Factor Function λ at Large γ-Sectors, and Insolvability of SA Quantum Composite Systems Make Accurate Discussion Impossible for Many Basic Issues.
- LV Factor Function λ Determines Potential Energies via Eqn of Motion, Force Splitting, and **ET** (**E**nergy **T**heorem)

(H28a) $$F_{(\mu)} = dp^{\mu} / dt \equiv d(m\lambda\gamma u^{\mu}) / dt$$

(H28b) $$F_{(\mu)} \equiv I_{(\mu)} + S_{(\mu)}, \quad S_{(\mu)} \equiv m\gamma u^{\mu} d\lambda / dt$$

(H28c) $$dE / dt = \boldsymbol{I} \cdot \boldsymbol{u} + m\gamma c^2 d\lambda / dt$$

Very Large $d\lambda / d\gamma$ in **VSS** (**V**elocity **S**hell **S**pace) Nearby $\gamma = \gamma_0 \sim 10^{5\pm\sigma}$ Yields Severe Violation of Momentum Conservation Seen in ***Bizarre Jets*** with Nothing to Balance in Opposite Directions. If No Other Explanation, This Works!
Small Spin-Orbit Coupling at Quasi-Horizontal Bottom of Strong PW Covering MPR of Deeply Bound Quarks, if any, May Yield Spin 3/2 Baryons if Decuplet-Baryon is of $J = 3/2$ in Sense that Non-Radial Excitation with Spin-Flip Takes Place. It also Causes Existence of Many Mesons with Higher J-Labels. Sample LV Model Factor Function (20) Provides with Quasi-Horizontal Bottom of Strong PW in SQC Sub-Model.

Remark: Sample Model (20) was Proposed before Author Recognized MFs and Their Importance. It's Possible that LV Might Play No Important Role in SA Dynamics. We Don't Know whether Large LV Occurs at Bound Quark's GS. In Any Case, Model (20) Shows a Sample of How to Get Limiting LV with Finite N and λ^4-Enhancement of Source 4-Momenta & Energy Density of CHSPs, which is Crucial for Us to Explain ***DM Phenomena***.

- Please Note: Small Non-Radial Excitation Requires Quasi-Horizontal Bottom of Strong PW that Could Yield Small Orbital Excitation in SQC if St MF Does Not Cause Large Orbital Excitation. After Importance of MFs and More and More their Possible Roles Gradually Recognized, We See another Possibility for Small Mass Differences in Hadron Mass-Spectrum: **NCB-States** (see page 74).

- Solutions with Solvable Inverse PEs and Square Well PEs Indicate: the Smaller the Binding Size, the Higher the Energy of Radial Excitation. Mass Spectra of Hadrons Seem, at Least, to Be Empirical Evidences of Small Radial and Orbital Excitations, Ranging from Tens to Hundreds of MeV. Either Quark-Cores Sizes Roughly about 1 Fermi OR Some Forces Ignored before Cause Dramatic Changes in PEs such that New PEs do Lessen Radial Excitations in Small Binding Size Conditions (Smaller or Much Smaller than 1fm). Binding Power and Sticking Ability of MF, Wk MF, and St MF btw Magnetons (Spinning Elementary Particles with Electric, and/or St, and/or Wk Charges) Had Been Ignored for Too Long. NCB-States Open for Future Study.
- Dramatically Different Effects of **Inverse** PE (Coulomb or Coulomb-Type PE) and Deep Square **Well** PE are Seen in Energy Level Formulas: For Inverse PE, Mass Appears in Numerator while it Appears in Denominator for Square Well PE. Dirac Eqn with MFs and LV Space-Force Might Not Be Solvable. But We Still Can See Something:

1. Space Force with Large Jump of LV Factor λ Near $\gamma_0 \sim 10^{5\pm\sigma}$ Produces Well-Type PE Whenever Bound Particle's Speed Becomes Near-c with Large γ-Value Near $\gamma_0 \sim 10^{5\pm\sigma}$ while MF and St MF Not Only Yield Asymmetric PW But Also Widen Width of PW. In Fact, Magnitudes of MFs Proportional to r^{-4} such that Near-c High Speed with Large γ-Value u-Quark Needed for **STB** to Emerge and to Enclose Can Appear at Distance Longer than What We May Expect from Inverse Square Law Force.
2. For RQC, Quarks Cannot Be Bound Deeply. We Have AVI-QNC and NLVUC-QC Scenarios to Explain Ironic FQA Fact as Described before. Besides, Inward Excitation States with Less and Less Binding Sizes Help to Explain FQA Fact. Even if QKoE Not Very Large, below NLVUC as Postulated Upper Bound of All **Elastic Parts** of KkEs, Left Over after PAPP Consumption, if any, and **Absorbed by Unbound Particles to Gain KEs** or **by Bound Particles to Jump to Higher Levels** including Free Regions, Hence u-Quarks are Knocked Out Millions of Times Every Minute in Atmosphere and in Accelerators, DS is Still Able to Explain Ironic FQA Fact in Terms of Absolute Quark-Non-Confinement, Just as Described before. DS Able to Penetrate Dense 'Fog' that Covers Mystery Hidden in Hundreds of Kinds of Middle Stage Particles and Thousands of Decay Modes to Put Spots on and Reveal 4TDSP, So Useful and Powerful Not Only for Explaining Rich and Complicated Collision Events and Decay Modes but Also for Explaining Tough and Ironic FQA Fact.

- From HMC $u\bar{u}$ Ground State Table We See Small Possible Sizes of HMC with True Value Unknown/Undetermined due to Unknown/Undetermined u-Mass.
- BC is 3B-System. BC May Have roughly Same Small Size as that of HMC.
- Small Sizes of HMC and BC Make Them Look like Tiny Particles with Net MMs while DSLs MF-Bound by Effective Magnetic and Magnetic-like Fields Produced by Such Net MMs. Tiny Q-Core with Single Net MM Enable Us to Partially Overcome MMD and Explain Empirical Facts Qualitatively and Half-

Quantitatively. Single Net MM Means Reducing MB-System. Each u-Quark has 3 Kinds of Charge: St, Electric, and Wk Charges. Each Single u-Quark has MM, Wk MM, and Possibly St MM. Usually, Context Determines What Kind of MM and Charge We are Talking about.

< DS-Symmetry, DS-Asymmetry, and Lifetimes >

- DSL MF-Bound by Q-Core May Have Different Spin-Correlation and θ-Zone Location as Quantum State Changes.
- If d-Electron Gets Excited with Spin Vector Flipped Over and Anti-Parallel to Q-Core's Net MM, It Jumps to and Stays in Region 2 Centered at that Net MM, Turning d into s-Quark. Or, d-Electron in Excited State Changes its θ-ZL without Spin-Flipping. Details of $d \to s$ Excitation Not Clear.
- If $\bar{d}$-Positron Gets Excited with Spin Vector Flipped Over and Parallel to Q-Core's Net MM, It Jumps to and Stays in Region 2 Centered at that Net MM, Turning $\bar{d}$ into $\bar{s}$. Or, $\bar{d}$-Positron in Excited State Changes its θ-ZL without Spin-Flipping. Details of $\bar{d} \to \bar{s}$ Excitation Not Clear.
- Neutral K Mesons $d\bar{s}/s\bar{d}$ are <u>***DS-Asymmetric***</u> Regardless which Sample Descriptions of Excitations $d \to s$ and $\bar{d} \to \bar{s}$ are True.
- Charged Pions $u\bar{d}/d\bar{u}$ and Charged K-Mesons $u\bar{s}/s\bar{u}$ DS-Asymmetric, e.g., u-Q in $u\bar{d}$ Contains No any Charged DSL but $\bar{d}$ Contains DS-Positron.
- In Topic < **Slowly and Fast Decaying Mesons** > We've Explained why among Mesons of Label J=0 DS-Symmetric Mesons Live Shorter than DS-Asymmetric Mesons <u>**without Single Exception**</u>, and why all Mesons with Label J=1 Live Extremely Short <u>**without Single Exception**</u>.

<u>**Lifetimes of Neutral Kaons**</u>:

- Asymmetric Distributions and Motions of e^+e^- inside DS-Asymmetric Mesons such as K^0 ($d\bar{s}$) Show: e^- in d is in R1 at Net MM of HMC $u\bar{u}$, e^+ in $\bar{s}$ is in R2 at Said Net MM or is in Different ZL. If e^+ in $\bar{s}$ in R2 at Net MM of $u\bar{u}$, Spin of e^+ Parallel to Net MM of $u\bar{u}$ and Spin of e^- in d while e^+ in $\bar{s}$ Oscillates along Net MM-Line of $u\bar{u}$ with MPR of e^+ away Enough from Center of $u\bar{u}$ so that MF-Attraction btw e^+e^- with Parallel Spins and Anti-Parallel Spin MMs in $\bar{s}$ and d Not Large Enough to Cause e^+e^--Annihilation before e^+ in $\bar{s}$ Steps down to Ground State Leading to De-Excitation of $\bar{s}$. If e^+ in $\bar{s}$ is in Different ZL with Spin and Spin MM Antiparallel to Net MM of $u\bar{u}$, e^- & e^+ Have Parallel MMs so that They Have Much Less Chance to MF-Attract Each Other while Their Electric Attraction Cannot Overcome MF-Binding from u and $\bar{u}$. Then e^+e^- Annihilation Cannot Occur before De-Excitation of $\bar{s}$, while e^- in R1 at Net MM of $u\bar{u}$ Repulses/Attracts $\bar{u}/u$ to Slow down $u\bar{u}$ Annihilations (see page 84), Leading to Lifetimes, such as $5{\cdot}10^{-8}s$ (K_L^0)/$9{\cdot}10^{-11}s$ (K_S^0), much Longer than π^0. Note, e^+ in $\bar{s}$ and in R2 at Net MM of $u\bar{u}$ May Oscillate about

$\bar{u}$ with MPR away from $\bar{u}$. Then e^+ is Mainly in such ZL at u that MF btw e^+ and u Has Much Smaller Magnitude than MF btw e^- and $\bar{u}$.

Faster Decaying of K_S^0 than that of K_L^0 will Be Explained Later.

- Fast e^+e^- Annihilations Occur inside DS-Symmetric, Hence Flavorless Mesons if They Have Quark Contents $d\bar{d}, s\bar{s}, c\bar{c}$, or $b\bar{b}$, Because e^+e^- in Every Such PAP Structure Have Parallel Spins and Anti-Parallel MMs with Their Most Probable Orbits Perpendicular to Spin so that They MF-Attract Each Other Severely. Fast e^+e^- Annihilations in such Mesons Trigger Fast Annihilations of $u\bar{u}$ and Cause Shorter Lifetimes of DS-Symmetric Mesons such as

$$\pi^0, \eta, \eta', \eta_c, \rho^0, \omega, \phi, J/\psi, \Upsilon \quad (\eta_b\text{'s Lifetime Not Measured}).$$

Note: If Neutral Pions Do Not Contain $d\bar{d}$ Mixture Content as OSTP Shows, Fast $u\bar{u}$ Annihilation Inevitable Due to Non-Existence of Resistance from Any Charged DSL.

- In Contrast, Single e^-/e^+ in π^-/π^+ Repels $\bar{u}/u$ while attracts $u/\bar{u}$ to Slow Down Annihilation of $u\bar{u}$, Causing Longer Lifetime of DS-Asymmetric Charged Pions than DS-Symmetric Pion (π^0).
- A **u**-Quark Carries *Strong*, *Electric*, & *Wk* **Charges** while $\bar{u}$ *Opposite* Charges. Neutrinos Carry ***Wk*** Charges **Only** while Other Leptons Electric & Wk Charges. u, e^-, and ν_l Carry **Same Wk Charge** and Their Antiparticles Opposite Wk Charge. **Opposite** Wk Charges **Attract** and **Same** Wk Charges **Repel**, just like Electric Charges.
- In SPS-System $ue^-\bar{\nu}_e$, Wk Repulsive Force btw u & e^- of Same Wk Charge is Dominated by Attractive Coulombian Force and Attractive MF btw Them. They Join Forces to Bind 'Satellite' $\bar{\nu}_e$ which Has Opposite Wk Charge. If Net MM of d-Electron e^- is Much Smaller than Net MM of BC uuu, $\bar{\nu}_e$ Bound Mainly by BC and We Can Expect Many beta-Decay Events with Only Electrons Emitted. If Experimenters Can Find out Percentage of Such Events, It Would Be Indirect Evidence about Net MM of Wd-e in Neutron. It Seems that Experimenters never Discover such Composite Particle that Contains $e^-\bar{\nu}_e$ Only. If They always Fly Out from Neutron Separately, It May Imply that BC Plays Vital Role to Bind $\bar{\nu}_e$. Once Wd-e e^- is Pushed out by Sd-e e^-, It's Unable to Bind $\bar{\nu}_e$ alone whose KE was Just Right for BC & Wd-e e^- together to Bind, Not Right for e^- alone to Bind.
- Color-Particle Index Coupling Factor $(2\delta_{kk'}\delta_{cc'} - 1)kk'/|kk'|$ Put in Strong Force Formula (45) Makes u-Quarks with Same Strong Charge Attracting Each Other and a $u\bar{u}$ Pair with Opposite Strong Charges Also Attracting Each Other.
- K, D, & B-Mesons, both Neutral and Charged Ones, DS-Asymmetric, Decaying Slower than DS-Symmetric (Flavorless) Mesons for Bound d-Electron and/or its Antiparticle ($\bar{d}$ -Positron) Slows Down Annihilation of $u\bar{u}$.
- DS Shows AR (see page 84), such as $e^+ - u/\bar{u}$ Repulsion/Attraction in **<u>Positive</u>**

π, K, D, D_s, B, B_c Mesons and $e^- - \bar{u} / u$ Repulsion/Attraction in Their **Negative** Counterparts, and Non-Existence of e^+e^- Pair inside Such Mesons. Nothing inside Them Could Cut Duration of above AR Short.

< **DS**-Reasons of **Insolvability** of SA Composite Systems >

- One-Body Approximation Good for each Atomic Electron in Atoms with small Z-Numbers where Positive Charges are Concentrated in Small Spaces of Nuclei while Electrons Scattered in Relatively Much Larger Spaces of Atoms. Thereby, Attractive Force Exerted on e^- by Nucleus is Much Larger than Resultant Repulsive Force Exerted on Electron by Other Electrons (see (H21a) on page 63 for detail) so that as an Electron Steps down to Lower or up to Higher Level, Other Electrons' Energy Levels are Not Affected Significantly. This is '**1-Body Excitation**' Approximation in MB Bound Systems where such Approximation's Good. But d-Leptons and Quarks Do Not Form Such Systems. Only Exception is $u\bar{u}$ Bound System without any d-Lepton. Such 2B System is Solvable if its PE in 1BEq is Solvable. Every Its Excited State Has Equivalent State in 1-Body Excitation. Quantum 1B-System in Resultant Force Field of MF-type Force plus Coulomb-type Force Not Solved and May Be Unsolvable. Magnetic Force btw 2 Electric Currents Caused by Fast Orbital Motions of Electric Charges with Near-c Speeds and Magnetic-like Force btw 2 St or Wk Currents Caused by Orbital Motions of St or Wk Charges with Near-c Speeds May Play Important Role in SA World. Such Forces Make Composite Systems Unsolvable.
- Wk Forces Exerted on $\bar{\nu}_e$ by e^- and u-Quark Much Smaller than e.m. Force Exerted on e^- by u-Quark. But Effect it Causes is Not Much Smaller for $\bar{\nu}_e$ Has Non-Zero Mass Much Smaller than that of e^-. **Planet-Satellite Excitation** (**PSE**) is **LE** (**Leptonic Excitation**) in **SA** World.
- Heavy u-Quarks Bound by St Forces (Possibly Including St MF), not Sensitive to LE so that **UE** May Be Independent of LE. It's Not Clear How UE Influences LE in MB-Systems. A Hadron, Roughly, is LE, UE, or Combination of LE and UE State, Called Lepton-u Excitation (**LUE**). Ground States as Zero-Excitation, Included as Special Cases. Any Isolated u-Quark Does Not Exist. Hence UE Always Refers to Excitation of HMC $u\bar{u}$ or BC/AntiBC $uuu / \bar{u}\bar{u}\bar{u}$. For Baryons/Antibaryons, UE's MB-Excitation. With Perturbation of EW Forces from Leptons Neglected, a Meson's UE Has 1BEq. But Its *PE* Might ***Not*** Be ***Solvable*** due to e.m. and/or St MFs btw Spinning Electric/St Charges at Distances Close to **MD**.
- *Necessary* Condition to Better Agree with Empirical ***Deuteron*** Spectrum is to Take Account of MFs, Possibly Including St MF, and Modify Yukawa PE. We Know Nuclear Force No Better than Q-Forces btw Quarks due to Ignorance and Insolvability of MFs and of Nuclei as MB-Systems. Once MF is Considered, Our Basic Understanding of SA World would Surely Be Improved Significantly, Not to Provide Us with Quantitative Solutions but to Reveal Why We Have Such Difficulty or Impossibility to Obtain Quantitative Solutions/Explanations and to Explain Many Experimental Facts Qualitatively/Half-Quantitatively, which are

Difficult or Have No Way to Explain if DS is Not Considered.

- Spin 0 π^0 with $m_{\pi^0} = 134.98$ MeV is 640.51 MeV Lighter than Spin 1 ρ^0 with $m_{\rho^0} = 775.49$ MeV, while $m_\omega - m_\eta = 234.80$ MeV, $m_\phi - m_{\eta'} = 61.785$ MeV, $m_{J/\psi} - m_{\eta_c} = 116.616$ MeV. 'Accidentally' Large Mass Jump from π^0 to ρ^0 is Due to Spin Misconception, TAM Disability, and Ignoring MF-Feature of Excitations. 'Accidentally' Large Mass Jump from π^0 to ρ^0 just Indicates: SM without DS and with Spin Misconception, TAM Disability, and Ignorance of MF-Feature of Excitations Does Not Offer Correct Contents, Spin Labels, and Identifications of Excitations. HMC $u\bar{u}$ Quantum 2B-System in Non-Spherosymmetric MF Fields Not Solved and May Be Unsolvable. DS Tells Crystal Clearly the Reason why We are Unable to Quantitatively Explain Mass Spectrum including Excitations. But as We Pointed Out Before, NCB-States can Explain Small Differences in Mass-Spectrum of Hadrons Qualitatively. May Be, Future Study of NCB-States will Improve Our Understanding of Mass Spectrum in SA World.
- TMET Reveals Dynamic Feature of TME- r Curves with MF-Binding Involved. As Tilting-Angle of OSTP Changes, Degree of MF-Suppression Changes, r_{0TME} and r_c Shift, and Geometric Shape of TME- r Curve Changes. For Each Fixed Tilting-Angle and TME- r Curve, There's a Set of Quantum States. Now We Understand Why So Many New Hadrons Far beyond Original 2 Octets and 1 Decuplet were Discovered soon after Hypothetic Symmetry-Structure Relation was Proposed to Replace Traditional Dynamic Approach plus Structure. It Also Explains why Quantitative Solutions Not Available.

< Miscellany and **Reactions** >

- Non-Existence of $uu/\bar{u}\bar{u}$ 2B-Systems and Existence of $uuu/\bar{u}\bar{u}\bar{u}$ 3B-Systems and of '*Valence*'-Quark Meson Systems Suggest **Magnetism-like** Dynamics: 2 St **Currents** in **Opposite** Directions **Repel** while St Magnetic-like Force is Comparable with Strong Force when Both 2 Interacting St Charges Move at **Near-c** Speeds due to Binding at Small SA Distances, Similar to Magnetic Force that is Comparable with Electric Force in Near-c Speed Cases. This Fact Needs More than Color-Particle Index Coupling Factor to Explain (**Open**). Powerful Magnetic Forces Exerted on Accelerated Charged Particles in All Accelerators Due to Extremely Large Numbers of Slow Electrons inside Wires. But for 2 Charged Particles, Magnetic or Magnetic-like Forces btw Them Comparable with Electric or St Forces btw Them iff Both of Them Move at Near-c Speeds.
- Triplet *uuu* in Each Baryon is Able to Form **Quark Core** Never Affected by Any Decay because **No** Two u-Quarks in Triplet Move in **Opposite** Directions while in CMF **Two** Permanently/Temporarily Bound Particles with Same Mass must Move in **Opposite** Directions as They Revolve about their Center of Mass.
- Without DS, u-Quark in Δ^{++} is Believed to be Able to Make **Eerie Escape** from for New 'Companion' $\bar{d}$ to Form π^+ by Just Emitting Virtual Gluon that Decays into such $\bar{d}$ plus a d . The Latter Occupies Space Escaped u-Quark Left, Fixes uuu-Core Damaged by u's Leaving, and Changes Δ^{++} into Proton.

This Physical Picture without DS is Immediately Disproved by Ironic Fact that No u-Quark in Free Proton Ever Seen to Be Able to Make such **Eerie Escape** for New 'Companion' $\bar{d}$ to Form π^+ by Just Emitting Virtual Gluon that Could Decay into such $\bar{d}$ plus a d while d was Able to Occupy Space Escaped u-Quark Left behind & Fix BC uuu Damaged by u's Hypothetic Leaving under Free-Proton Condition without Being Knocked so that Postulated Process above Could Turn Free Proton into Neutron and Release π^+.

- In Sharp Contrast, Quark Model with DS Offers Natural Explanation for Decay Mode Mentioned above. No u -Quark Able to Escape from Baryon where it was Bound with All-Time Negative ME and Zero-Chance to Escape in any Reaction, Decay, or Event without KkE Received, which is Larger than QKoE. In View of DS, Decay Mode $\Delta^{++} \to p\pi^+$ is Due to Simple DS-Processes below

$$\Delta^{++}(uuu\uparrow) \to uuu + u\bar{u}e^+e^-\nu_e\bar{\nu}_e \to p(uud = uuue^-\bar{\nu}) + \pi^+(u\bar{d} = u\bar{u}e^+\nu_e)$$

1st Step De-Excitation Releasing 3 PAPs. 2nd Step PQTR. So Simple & Natural. Such DS-Reality Picture Fully Consistent with Quark's Bound State.

- As Baryons Switch btw Ground and Excited States (LE or UE), uuu Core-Structure Remains the Same in the Sense that No One Has Ever Been Detached or Replaced. BNC (Baryon Number Conservation) a Book-Keeping Label of Bound Quark-Cores and ERM. PAP Production in ERM Preserves Baryon Number & all Other Quantities of which Each Particle and its Antiparticle Have Opposite Numerical Values. For Example, Every time when a Meson's Found, a $u\bar{u}$ PAP is Produced. Since ERM Makes PAPP and PAPP Conserves Baryon Number, No One in 4TDSP Could Ever Change Baryon Number. Also, Nucleon that Appears among Final State Particles of Every Baryon-Decay Serves as an Empirical **Evidence** of Bound Quark Core in Baryon. Even if Quarks Knocked out Millions of Times Every Minute, DS with 4TDSP Explains FQA Fact and BNC Simultaneously: PAPP and PQTR that Guarantee Instant Stuffing of Every Hole Left behind by Knocked Out u-Quark with One u-Quark Born in PAPP Explain BNC. Instant Capture of One anti-u-Quark by Knocked out u-Quark to Form HMC for a Meson Simultaneously Explains FQA Fact. That anti-u-Quark is Produced in PAPP and its Companion u-Quark Just Fills in Said Hole.
- **Bound Neutrons** in Nuclei **Not Stable** just like Free Neutrons. Repulsion btw Identical d-Leptons & ENM is *Inner Mechanism* inside Neutron, Free or Bound.
- Beta-Decay **Only Channel** for Quasi-Bound d-Leptons to Go under Repulsion in **Free Neutron**. Beta-Decay Pathway **Suppressed** by **QT** in **Elements** Not Above Optimum Line. Larger Z Elements on Stability Island Require N>Z for Stability because Such Larger N is Able to Offset Electric Repulsion btw Protons (Well-Known Explanation). **N>Z** on Stability Island May **Maintain Stability** under Condition that Every Neutron Decays into Proton Releasing $e^-\bar{\nu}_e$ Pair because beta-Decays of Many Neutrons Do Not Occur At Same or Nearly same Time so that No Equal Number of Protons are Needed to Take Released and Relocating d-Leptons.
- In **Beta-plus Decay** and **Electron Capture**, Nuclear *Transmutation* Occurs s.t.

(such that) Mother Nucleus Lost Energy of Amount $E_{mother} - E_{daughter}$. Then Lost Energy Re-Emerges in Form of **PAP**(s). DS Proves Once and for All that **One** e^+e^- **Pair** and **One** $\nu_e\bar{\nu}_e$ **Pair** are Produced for β^+-Decay so that e^+ and ν_e Move away while a u-Quark in Proton Captures e^- and $\bar{\nu}_e$ Turning itself/Proton into d-Quark/Neutron: β^+-Decay is Said to Happen

$$energy + p \to n + e^+ + \nu_e$$

- *Simple DS Forbids Beta-plus Decay if Available Energy is Less than* $2m_ec^2$. It's DS that Tells: 1 Positron and 1 Electron Must Be Produced to Make Reaction to Happen while $\nu_e\bar{\nu}_e$ Pair's Energy May Be Even Less than Experimental Errors to Measure Positron's Energy. A Rigorous Lower Bound Should Be little bit Larger than $2m_ec^2$ but with Difference Smaller than Empirical Errors to Measure Energy of Emitted Positron. Such No Way to Explain Fact Now is Quantitatively and Accurately (within Empirical Errors) Explained in Terms of DS. Once again, DS Demonstrates Extraordinary Power to also Simultaneously Explain why Energy Requirement like this Not Needed at All to e-Capture, as One Can Easily See in the Following:
- In Bound **Electron-Capture**, K-/L-Electron Falls Down to Deep Ground State in e.m. PW, Created by u-Quark to Bind d-Electron. Such Shell-Electron's Stepping down is Part of Nuclear Transmutation. Here Released Energy Takes Form of **One** $\nu_e\bar{\nu}_e$ **Pair without** e^+e^- **Pair** while Three u-Quarks & e^- Join Wk Forces to Capture and Bind $\bar{\nu}_e$ so that u-Quark and Proton Become d-Quark & Neutron respectively, Releasing ν_e: $p + (K/L\text{-}Shell)\, e^- \to n + \nu_e$. It Happens Even if Available Energy Less than $2m_ec^2$ for No e^+e^- Pair is Needed to Nor is to Be Produced to Complete e-Capture Process in View of Quark Model with DS.
- Energy Conservation of Isolated Composite System Means Constant TME (Sum of KEs of Constituent Elementary Particles and PE of Entire MB-System), from Initial to Middle-Stage to Final State, before and after Pair-Production. For $u\bar{u}$ Pair Produced in Pair Production, their *TME* is *Negative* so that Relatively Small Amount of Impact Energy or Energy Released in SA De-Excitation Can Afford to Produce One or Even More $u\bar{u}$ Pairs to Form HMC(s) as Bound System(s), Even If Available Energy Consumed in Pair Production is Much Less than Total Rest Energy of One u and One $\bar{u}$.
- At Least Many Lepton/Anti-Lepton Pairs among Produced PAP-Pairs are Bound Systems. Such Leptons are Born to Be Able to Relocate through QT to Nearby Places with Equal or Close PE Values. As They Relocate in Nearby Nucleons, if any, They Turn Them into Middle-Stage Baryons. As They Relocate in Nearby Produced $u\bar{u}$ Pairs, Various Mesons are Born.
- ERM Seen in Every Decay & High-Energy Collisions. ERM Also Seen in $\nu\bar{\nu}$ Annihilation: e^+e^-, $q\bar{q}$, or Other Type of $\nu\bar{\nu}$ Pair are Found as Products of $\nu\bar{\nu}$ Annihilation though Original $\nu\bar{\nu}$ Particles Only Respond to/Exert Wk Forces.

ERM Applies to Three Types of Processes:

1. De-Excitations of Excited Nuclei and Excited Composite SA Particles.
2. Inelastic Collision with PAPP.
3. PAP-Annihilation.

- **DS** of Quarks/Leptons and 4TDSP (see page 24-28) Have Power to Explain All Reactions. **More Examples** Seen as Follows:
- In Reaction $K^-(\bar{u}s)+p(uud)\to\Omega^-(sss)+K^+(u\bar{s})+K^0(d\bar{s})$, DS-Leptons e^- and $\bar{\nu}_e$ in s-Quark of K^-, Attracted by u-Quark Core in p as Collision Makes Them so Close to One Another, Relocate there (a Place with Lower PE) through QT, Captured by a u-Quark and Turning it into s-Quark & at Same Time u-Quark Core in Proton May Be Pushed to Higher Quantum Level with J=3/2 (see remark below) while $e^-\bar{\nu}_e$ PS-System in p Gets Excited Turning d-Quark into s-Quark so that p Changes to Xi $\Xi^{*0}(uss)$ as Middle-Stage Particle More Spacious than Neutron Ready for its Naked u-Quark to Take $e^-\bar{\nu}_e$ Pair to form Third s-Quark Turning Ξ^{*0} into Ω^-. It Goes Like This: Ξ^{*0} (Lifetime $7.2\cdot10^{-23}$ sec) Makes No Visible Trajectory & Has No Time to Decay since KE Lost in K-p Inelastic Collision Re-Emerges Immediately Right before Ξ^{*0} was Born, Taking Form of Five PAPs: $2e^+e^-2\nu_e\bar{\nu}_e\ u\bar{u}$, Immediately Followed by Leptonic Re-Pairing $e^+e^-\nu_e\bar{\nu}_e\to e^-\bar{\nu}_e e^+\nu_e$ and Capture Process: One $e^-\bar{\nu}_e$ Pair is Captured by Naked u-Quark in Ξ^{*0} and u/Ξ^{*0} Become s/Ω^-. At Same Time, One $e^+\nu_e$ Pair is Captured by $\bar{u}$ in Produced $u\bar{u}$ Pair and $\bar{u}$ Changes to $\bar{s}$ while Another $e^-\bar{\nu}_e$ Pair is Captured by u in Produced $u\bar{u}$ Pair & it Changes to d & K^0 $(d\bar{s})$ is Born. At Same Time, $\bar{u}$ in Naked $u\bar{u}$ Core, Left by K^- after its Original s-Quark Lost its d-Leptons, Captures Another $e^+\nu_e$ Pair/Turns Itself into $\bar{s}$ so that $K^+(u\bar{s})$ is Born. Now Original K^- and Proton Disappear, and Produced Ω^-, K^+, and K^0 Start to Move away Making Tracks of Visible Lengths before They End up with Decays. Ξ^{*0}'s Track Shorter than 22 fm, Invisible and Undetectable.

  Remark: Since Ω^- Only One in Baryon-Decuplet with Lifetime Much Longer than Others, BC in Ω^- May Not Be in Excitation. DS-Explanation of Reaction $K^-(\bar{u}s)+p(uud)\to\Omega^-(sss)+K^+(u\bar{s})+K^0(d\bar{s})$ Truly Independent of whether Ω^- Should Have Label $J=3/2$. For Composite Particles, Spin Misconception and TAM Disability Prove J-Values and TAMs of Constituents Theoretically Unknown and Not Empirically Measurable. **Spin-Labels** and **J-Labels** in SM Tell Qualitatively Excitation Status of Hadrons' Quark-Cores. Quantitatively, They're Values of Neither Spins of Composite SA Particles Nor TAMs of Their Constituents. Heavy Masses and Extremely Short Lives of Hadrons with 3/2 Spin/J-Label in SM Indeed Support Scenario of Non-Radial Excitation of such Hadrons' Cores against St Forces (Possibly Including St MF). Bizarre Long Lifetime of Ω-Baryon is Only Exception. Such Unusual Fact Silently Tells that Dynamics May Not Be Replaced by Symmetry/Group-Representation. Besides,

Extension of Such Idea Led to Super-Symmetry, Predicting so Many Brand New Particles Never Seen Since Beginning of Particle Physics.

- **Kaon**(s) and Leptons Produced in Decay of **D-Meson** Suggest that c-Quark Not Only Contains All Constituent Particles in s-Quark with Higher LE or with 1st or 2nd UE but Also Contains $e^{+}\nu_e$ Pair so that c-Quark & u-Quark Have Same Electric Charge and Spin Label.
- Four-Vane Shaped Attractive Regions R1 & R2 of Source MM's MF Field Give Dynamic Reason Why MF Exerted by Source MM of u-Quark Can Stick Two Oppositely Charged Particles such as $e^{+}e^{-}$ at Same Time Temporarily for Short Period of Time.
- $\Xi_b^- \to \Xi^- + J/\psi$ ($b \to s + c\bar{c}$) Suggests that **b**-Quark is either LE of **s**-Quark, Higher than ***s*** Or UE while J/Psi is Produced Solely in Accordance with ERM.
- Charm and Bottom Related Details are Open for LKoE Issue is Open.
- Lifetimes of Nuclei up to Millions through Billions of Years are **Quantitative** Results of **Unknown** Solutions of MB-Problems with Time-Varying, Non-Central Force Fields Associated with Electric, Wk, & Magnetic/Magnetic-like Forces in Tiny Space Binding without a Rest Source Particle and with Charge Currents Time-Dependent.
- Simple **Extensions** to **Wk** Charge/Magneton/Magnetic-like Force Immediately Show: Wk Magnetic-like Force btw **Wk-Magnetons** (Spinning Wk Charges) is Surely **Attractive** in Some Directions while Repulsive in Some Other Directions No Matter They Have Same or Opposite Wk Charges. Its Magnitude is Much Larger than Wk Force at Every Distance Much Shorter than MD. Thus, Three Wk Charges, Two're Same, Can Stay Together for Very Short Period of Time.
- **Lepton Number Conservation** Book-Keeping Label of DS, ERM and PAPP. PAPP Preserves Lepton Number and Every Such Quantity of which Every Particle and its Antiparticle Have Opposite Numerical Values.

  In Rare Decay Mode $\pi^- \to e^- + \bar{\nu}_e$ (~0.012%), $u\bar{u}$ Pair Quickly Annihilates into Virtual Z^0, Absorbed Immediately by e^- and/or $\bar{\nu}_e$, AND/OR, into Virtual γ Absorbed by e^-, to Gain KEs. This is KC of Annihilation.

<u>**Different Lifetimes**</u> of K_L^0 and K_S^0

- K_L^0 and K_S^0 Have Same Quark Content $d\bar{s}$ Hence Same DS Content but K_S^0 Decays 571 Times Faster than K_L^0. Here is a Sample DS-Explanation:

  Decay Mode $K_S^0 \to \pi^+\pi^-$ Can Be Written in Terms of Quark Model without DS:

$$K_S^0(d\bar{s}) \to \pi^+(u\bar{d}) + \pi^-(d\bar{u})$$

In Terms of DS, It's Written

$$K_S^0(d\bar{s} = ue^-\bar{\nu}_e\bar{u}e^+\nu_e \uparrow) \to \pi^+(u\bar{d} = u\bar{u}e^+\nu_e) + \pi^-(d\bar{u} = ue^-\bar{\nu}_e\bar{u})$$

A Sample DS-Description of Details of This Decay Process is Given as Follows:

Excited DSLs $e^+\nu_e\uparrow$ in $\bar{s}$ Step Down to Ground State, Turning $\bar{s}$ into $\bar{d}$ and K_S^0 into $d\bar{d}$ System and Releasing Energy in Form of a Bound $u\bar{u}$ Pair. $d\bar{d}$ System May Not Exist as a Particle if OSTP of HMC $u\bar{u}$ is Real and Causes Severe MF-Repulsion btw e^- in d and e^+ in $\bar{d}$. But before $d\bar{d}$ Could Be Dismembered, $e^+\nu_e$ or $e^-\bar{\nu}_e$ in $d\bar{d}$ Immediately **QT-Relocate** in Produced Bound $u\bar{u}$, Turning $u\bar{u}$ into $\pi^+(u\bar{d}=u\bar{u}e^+\nu_e)$ or $\pi^-(d\bar{u}=ue^-\bar{\nu}_e\bar{u})$ and $d\bar{d}$ into $\pi^-(d\bar{u}=ue^-\bar{\nu}_e\bar{u})$ or $\pi^+(u\bar{d}=u\bar{u}e^+\nu_e)$. Fast **QT-Relocating** Makes $d\bar{d}$ Middle Stage System Having No Time to Annihilate Nor to Leave Recognizable Track Experimenters Can Measure or See. PAPP of $u\bar{u}$ Bound Pair belongs to Strong Interaction and Goes Fast than e.m. Interaction Process in Different Decay Mode of Neutral K-Meson:

$$K_L^0 \to \pi^+ e^- \bar{\nu}_e \text{ or } K_L^0 \to \pi^- e^+ \nu_e$$

DS-Explanation of above Slower Decay Mode, for Instance, is Written:

$$K_L^0(d\bar{s}=ue^-\bar{\nu}_e\bar{u}e^+\nu_e\uparrow) \to \pi^+(u\bar{d}=u\bar{u}e^+\nu_e)+e^-\bar{\nu}_e$$

Here We Give Sample DS-Description: Excited d-Leptons $e^+\nu_e\uparrow$ in s Return to Ground State, s Changes to d, and K_L^0 Becomes $d\bar{d}$ 2B-System, Releasing Energy in Form of Virtual Force Carriers γ and/or Z-Boson. $e^-\bar{\nu}_e$ in $d\bar{d}$ Absorb Them, Gain KE, and Fly away, Turning $d\bar{d}$ into $\pi^+(u\bar{u}e^+\nu_e)$. Much Slower ($\sim 5.1\cdot 10^{-8}s$) Decaying Due to Much Slower Electro-Weak Interaction.

Note: PDG In Particle Summary Claims a Decay Mode $K_L^0 \to \pi^+\pi^-\pi^0$. π^0 Lives too Short to Show a Recognizable Track. Author Thinks: This Much Slower Mode is Actually Showing Much Slower Non-Strong Interaction Process without π^0 but with Photons and/or Leptons among Final State Particles Assumed by Original Experimenters/Reporters to Be Produced in π^0-Decay. De-Excitation of s-Quark Releases Energies for Production of $u\bar{u}$ Pair and Pushing $\pi^+\pi^-$ to Fly away from Each Other as well as for Production of Photons and/or Some Leptonic PAPs which're Assumed to Come from π^0-Decay by Original Experimenters/Reporters. We Can Expect Larger TKE (Total Kinetic Energy) for $\pi^+\pi^-$ System in CMF and in Decay Mode $K_S^0 \to \pi^+\pi^-$ than TKE for $\pi^+\pi^-$ in CMF and in Decay Mode $K_L^0 \to \pi^+\pi^- +$ Photons and/or Leptonic PAP(s), which are Assumed in PDG's Summary to Come from π^0-Decay. Such Assumption Does Not Match Fact of Slow Decaying now that It Belongs to Fast Strong Interaction Process to Produce π^0, just like Shown in Fast Decay Mode $K_S^0 \to \pi^0\pi^0$. Similarly, PDG Decay Mode $K_L^0 \to 3\pi^0$ May Be Just $K_L^0 \to$ Photons and/or Leptonic PAP(s), which're Assumed by PDG to Come from Decay of Three Neutral Pions. In Such Decay Mode, Fast De-Excitation of s-Quark that Takes Place in Decay

Mode $K_S^0 \to \pi^+\pi^-$ May Halt, Allowing HMC $u\bar{u}$ in K^0 to Annihilate before Excited d-Leptons in s-Quark Could Return to Ground State. As We Said, Anti-d-Leptons in Anti-s-Quark Stay in R2, Allowing d-Electron in d-Quark to Slow down $u\bar{u}$ -Annihilation and Prevent Fast e^+e^- Annihilation and Fast $u\bar{u}$ -Annihilation Taking Place in π^0 -Decay & Other DS-Symmetric Mesons' Decays if They Do Contain PAPs Like $d\bar{d}$ and $s\bar{s}$. Once HMC $u\bar{u}$ in K^0 Annihilate, Neutral Pion Production May or May Not Happen. Due to DS, ERM, & PAPP, K^0 -Decay via $u\bar{u}$ -Annihilation Channel May Produce Same Final State Particles w/o π^0 in Middle. Lifetime of π^0 about $8.4 \cdot 10^{-17} s$, always Leaving Track too Short to Observe. Descriptions of Events w/o π^0 in Many Cases Heavily Depend on Theoretical Assumptions. If DS, ERM, PAPP, and PQTR are Ignored, Is There any Way to Understand Lifetime Difference of K^0 with Same Quark Content? Or, any Explanation of Faster Decay Mode $K_S^0 \to \pi^+\pi^-\pi^0$ with Smaller Branch Ratio than $K_L^0 \to \pi^+\pi^-\pi^0$ of Larger Branch Ratio under Condition that Initial Particles Have Same Quark-Content and Final State Particles Form Two Equal Sets in Two Decay Modes? DS Provides Lots of Brand New Ways to Explain Huge Number of Empirical Events. MB-Systems with Many Kinds of Forces Acting on Each u-Quark and/or Anti-u-Quark and on Each DS-Electron and/or DS-Positron Cause Difficulty to Know Every Detail of SA Events. We All Have Fundamental Difficulty to Figure Out Theoretically and Empirically What Can Happen as a MB-System Bound by More than One Kind of Forces and in Excitation Steps down to Ground State. MB-System Bound by Same Sort of Forces Not Solvable. Details of Quantum MB-System Bound by More than 1 Kind of Forces More Difficult to Figure Out.

- When Energy Released in DS-De-Excitation and/or $u\bar{u}$ -Annihilation in a Meson Decay Large Enough, it May Take Form of **More** PAPs (e.g., $K^\pm \to \pi^\pm\pi^\pm\pi^\mp$).
- In Decays $K^\pm \to \pi^\pm + \pi^0$, Energies Released in Transition of Excited Lepton in $\bar{s}/s$ to Ground State in $\bar{d}/d$ ($\bar{s}/s \to \bar{d}/d$ +Energy) Take Form of **One** $u\bar{u}$ Pair to Form π^0 (or **Two** $u\bar{u}$ Pairs of Lower Kinetic Energies **plus** Some $l\bar{l}$ Pairs to Form π^0 plus $d\bar{d}$ Pair in Mixture of π^0 , if $d\bar{d}$ Can Survive for Long Enough to Justify SM Mixture of π^0). De-Excitation of DSLs in $K^\pm$ Leads to Transition $\bar{s}/s \to \bar{d}/d$ +Energy and Turns $K^\pm$ into $\pi^\pm$. Energy Released in De-Excitation Transferred into Produced π^0 in Decay-Mode $K^\pm \to \pi^\pm + \pi^0$ or $\pi^+\pi^-$ in Decay-Mode $K^\pm \to \pi^\pm\pi^+\pi^-$.
- **Free** ν_e *Rarely* Causes *Inverse* Process of e^- -Capture due to **Repulsion**: u-Quark & d-Electron Bound in Nucleon Carry Same Weak Charge as ν_e and Join Wk Forces to Repel ν_e , Dominating Wk Attractive Force Exerted on ν_e by Bound $\bar{\nu}_e$ and Preventing ν_e from Annihilating Bound $\bar{\nu}_e$ so that ν_e Can **Penetrate Earth** with Little Chance to Interact Even on Their Antiparticles Bound in Nucleons. Such Free ν_e Keeps Distances Larger than Wk MDs Having Almost zero Wk-MF with Those Bound Particles inside Nucleons. No Spin-

Correlation Could Turn Almost Vanishing Wk MFs far beyond Wk MDs to Large Enough Ones so that v_e Could Be Attracted via Wk MF & Move Close Enough to Bound SA $\bar{v}_e$ to Trigger $v_e\bar{v}_e$-Annihilation.

- ***Inverse*** Reactions of e^--Capture May Have Occurred in *Solar*-Neutrino Experiment. Reactions $v_e + {}^{37}Cl \to {}^{37}Ar + e^-$ and $v_e + {}^{71}Ga \to {}^{71}Ge + e^-$ Reveal *Inverse* Process of e^--Capture: $v_e + n \to p + e^-$ (Here e^- is Free while it is Bound in e^--Capture). In Terms of DS, as ***Rare Event***, *e*-Neutrino Hits DSL $\bar{v}_e$ in Neutron and Annihilates it and itself; Annihilation Produces Virtual Photon or Z Absorbed by d-Electron Sending it Free so that Neutron Becomes Proton as It Lost its d-Lepton Pair $e^-\bar{v}_e$. Neutrinos v_e Can Easily Penetrate Earth but They May Be Unable to Penetrate Sun without Observable Loss. Sun and Sun's Core are Much Larger than Earth.

  Note: In View of SM without DS, It's Thought to be an Event of Charged Current: an Intermediate Boson W, Exchanged btw a d-Quark in n & Solar v_e, Turns that v_e into e^- and d-Quark into u-Quark. But Concept of Charged Current with a Charged Intermediate W-Boson Exchanged Could Not Explain Why *d*-Quark in free Neutron Dooms to Decay into u-Quark by Emitting a Charged Intermediate W-Boson while the d-Quark in free Proton Never Found to Emit such Charged Intermediate W-Boson and Decay.

- **Free** $\bar{v}_e$ with Enough Energy May Hit Proton Losing Enough Kinetic Energy and Part of Lost Energy Reappears in Form of an e^+e^- Pair. Then $\bar{v}_e$ & e^- Captured by a u-Quark in Proton Turning it into d-Quark while e^- Moves away so that Proton Captures a Free $\bar{v}_e$ ($p + \bar{v}_e \to n + e^+$) as Relatively Rare Event is Said to Occur. Another Part of Energy Re-Appears in Form of Recoil of Proton.

- **Free Electron** with Enough Energy Hits Proton and Loses KE. Part of It Re-Appears in Form of $v_e\bar{v}_e$ Pair. Then u-Q in Proton Captures $\bar{v}_e$ and e^- Turning itself into d-Q while v_e Moves away: Reaction *free* $e^- + p \to n + v_e$ (Proton Captures a Free Electron) is Said to Occur.

- **Photon** Hits Proton Inelastically, Loses Certain Amount of KE, Gets Absorbed by d-Electron in Proton, Pushing it to Excited State, Turning d-Quark/Proton into *s*-Quark/Σ^+ respectively. While Part of Lost KE Takes Form of 3 PAPs to Re-Emerge: $u\bar{u}e^+e^-v_e\bar{v}_e$. Then PQTR Follows: u Captures $e^-\bar{v}_e$ Changing to d while $\bar{u}$ Captures e^+v_e and Changes to $\bar{s}$ so that $K^0(d\bar{s})$ is Born and So-Called *Photoproduction* is Said to Occur: $\gamma + p \to \Sigma^+ + K^0$. Here e^+v_e Captured by $\bar{u}$ Has Proper Energy to Turn $\bar{u}$ into $\bar{s}$, Not $\bar{d}$. (If Produced Pair e^+v_e Does Not Have Proper Energy to Turn $\bar{u}$ into $\bar{s}$ after Being Captured by $\bar{u}$, then They Simply Turn $\bar{u}$ into $\bar{d}$. Hence, Some Particles Other than K^0 Appear among FSPs.)
  Note: Some Details of MB-PQTR Process are Unknown due to Inherent MMD.

- Free Photon without Interaction Never Make Itself **Disappearing** and Turns

Itself into a **PAP** such as e^+e^-, No Matter How Energetic It is. γ**-Photon** with Energy Larger than $2m_ec^2$ Hit Nucleus, Lost All or Almost All of its Energy, Disappeared or Became a Low-Energy Photon that Eluded Detection. Part of Lost Energy Takes Form of One e^+e^- Pair to Re-Emerge and Photoproduction of e^+e^- PAP is Said to Happen: $\gamma \to e^+e^-$ w/o Elusive Low-Energy Photon.

- More Energetic γ **-Photon** Hit Proton, Lost Nearly All or All of its Energy, Became a Low-Energy Photon that Eluded Detection or Disappeared. Part of Lost Energy Takes Form of One e^+e^- Pair, One $u\bar{u}$ Pair (Definitely Bound with TME Less than $2m_uc^2$), and One $\nu_e\bar{\nu}_e$ Pair to Re-Emerge. Produced $\bar{u}$ in Bound $u\bar{u}$ Pair Captured e^+ & ν_e via QT, Turned Itself into $\bar{d}$ and Produced $u\bar{u}$ and Relocated e^+ & ν_e Became π^+. While a u-Quark in Proton Captured $e^-\bar{\nu}_e$ and Turned Itself into d-Quark and Proton into Neutron. Photoproduction of π^+ is Said to Occur: $\gamma + p \to n + \pi^+$ w/o Elusive Low-Energy Photon.
- Energetic γ **-Photon** Hit Neutron, Lost Nearly All or All of its Energy and Became a Low-Energy Photon that Eluded Detection or Disappeared. Part of Lost Energy Takes Form of One Bound $u\bar{u}$ Pair to Re-Emerge with Quasi-Bound d-Leptons in n Relocating in Produced $u\bar{u}$ Pair so that Neutron Changes to Proton and π^- is Born. Thus, Photoproduction of π^- is Done: $\gamma + n \to p + \pi^-$ w/o Low Energy Photon Eluded.

< Sigma Baryon >

- On Web, Author of "*The Sigma Baryon*" Says: "According to the Particle Data Book, the branching ratio for the decays of the sigma-plus is 51.57% for the $p\pi^0$ pathway and 48.31% for the $n\pi^+$ pathway. This near equivalence is really surprising me–the neutron pathway looks a lot harder." In View of DS We See No Surprise: When Excited d-Electron in s-Quark of Sigma-plus Returns to Ground State Turning s-Quark into d-Quark & Sigma-plus into *Middle-Stage* Particle *proton*, Energy Released in Form of PAPs as long as the Law of Energy Conservation Permits.

In $p\pi^0$ Pathway, $u\bar{u}d\bar{d}$ (2 $u\bar{u}$, $e^+e^-\nu_e\bar{\nu}_e$) are Produced to Form Zero-Pion Mixture, or One $u\bar{u}$ Pair is Produced to Form Zero-Pion without Mixture, while *Middle-Stage* Particle *proton* is Also a FSP.

In $n\pi^+$ Pathway, $u\bar{u}e^+e^-\nu_e\bar{\nu}_e$ PAPs Produced and Re-Paired such that u-Quark in Middle-Stage Particle Proton Captures $e^-\bar{\nu}_e$ Pair Turning Itself/Proton into d-Quark/Neutron. At Same Time Produced $\bar{u}$ Captures an $e^+\nu_e$ Pair Turning itself into $\bar{d}$ Staying with Produced u to Form π^+.

None of Above two Pathways Has a Reason to Dominate the Other. The $n\pi^+$ Pathway is So Natural & Easy in View of Quark Model with DS.

< **CC** (**C**urrent **C**onservation) >

- DS Proves: All Unstable Particles are Composite, not vice versa. All Elementary Particles are Stable. DS Undoes Transition btw Elementary Particles so that Each Elementary Particle Can Only Disappear by Annihilating One of its Antiparticles and Could be Created Only together with One of its Antiparticles. Then, Any 4-Current, if written in Terms of Elementary Particle, is Automatically Conserved in Any Event.
- When One d-Quark in Neutron Changes to u-Quark and $e^{-}\bar{\nu}_e$ are Emitted so that Neutron Becomes Proton, the Current of d Changing to u is Not written in terms of Elementary Particles since d-Quark is no longer an Elementary Particle in View of DS. If d-quark were Elementary just like u-Quark and Transition btw Elementary Particles were Permitted, d-Quark would Have to Emit something to Be Able to Transform itself to u-Quark. But Ironic Fact is: d Quark in Free Proton Never Found to Be Able to Transform itself to u-Quark by just Emitting something. Therefore, Quark Model w/o DS Cannot Co-Exist with W-Particles. Once $e^{-}\bar{\nu}_e$ are Identified as Constituent Particles of Composite d-Quark, 4-Current is Guaranteed to Be Conserved as long as Written in Terms of an Elementary Particle in DS-Model. LV, Massive Force Carriers, and DS Give Coherent Picture of Reality in Physical World. QFT Based on Such Foundation Renormalizable, without IR Catastrophe and with All Quantities Canceled in Renormalization Being Finite.

  More….

Chapter 1 Introduction

Einstein, Dirac, and Feynman [7, 8, and 9] told that the foundation must be changed. Einstein concluded that the field eqs in general relativity are invalid for the very high density of field and of matter due to the well-known singularity problem. His another lifetime conclusion about the age failure seen in cosmology based on general relativity is: "I see no reasonable solution". He had insisted on these conclusions since he wrote them down in a book to explain the meaning of relativity. He never changed them though he did change his attitude towards the probability interpretation in quantum theory. (see III for detail.) The conclusions made by Dirac and Feynman are more explicit. Their far-reaching conclusions can be called **EDF**. Empirical evidences of EDF are called **EDF facts**. Every measurable quantity is finite. This trivial experimental fact disproves the well-known **IR** (infrared) results of massless QED and of other massless gauge theories. It is called **IR fact**.

Renormalization only solves infinity problems partially because not only counter terms are themselves infinities but also, e.g., the *remaining* **IR** infinities counter terms fail to cancel lead to catastrophe. Indeed, $ln(m/K_m)$ remains infinite after IR pole $ln\,\lambda_{min}$ has been canceled [9].

If force carriers were **massless** and infrared cutoffs (IRCs) or regulators were **fictitious**, a process of ***only*** emitting (exchanging) soft force carriers with energies ***greater*** than a ***positive*** value could not exist in this physical world.

Since renormalization is not a complete solution, Feynman concluded:

"*Its solution requires a* **change** *in the* **fundamental laws**."

He told that new laws must provide **convergent factors**. Indeed, at least a massive propagator of photons or any other vector force-carriers automatically give an IR convergent factor to modify massless propagator:

$$1/(q^2+m^2)=(1+m^2/q^2)^{-1}/q^2\ .$$

Iff (if and only if) the force carriers are not massless and IRCs are genuine, all IR infinities become finite and **Yukawa**-type range factors (**YFs**) yield **IR convergent factors** (**IRCFs**) to further reduce the effects of soft force carriers severely.

Zero mass terms in quantum wave eqn for spin 1 force-carriers lead to infinite ranges of the forces and to **inverse potentials** ($\propto 1/r$) without YFs, which are **not square integrable**. The catastrophic IR departure from revolutionary probability interpretation at large distances nearby and after finite ranges of vector forces inevitably leads to IR numerical catastrophe at small enough momenta in quantum theory of massless fields.

As early as in 1980's, T. D. Lee [10] predicted that a **major change** in our basic concepts and theory was more than likely to be undergone. More physicists seem to believe in a radical change. The author does not know the details of what greatest physicists have said about the EDF and about any radical change.

Dirac was very active, repeatedly and explicitly told that the fundamental theory, relativity and quantum theory, must be reformed. Unfortunately, the majority of younger generations have abandoned EDF, '*turning away from the search by their predecessors for a radical solution.*' (Weinberg [3]) Without the "invalidity' statement, relativity is no longer Einstein's but somebody else's. It is a portion of relativity, just like the appeal for reforming relativity and quantum theory and for changing the fundamental laws of physics is a portion of Dirac and Feynman's legacy. Rejecting EDF leads to misinterpretation and distortion of their theories and their creative spirit that will last much longer than the theories they formulated. J.D. Bjorken and S.D. Drell [11][12] indicated the trouble and the doubt caused by divergences in current basic theory.

It is even more serious to face the reality of physical world than the three or more great men in the history of science. The oldest **decisive** experimental evidence of **massive** photons and neutrinos is the **DE fact**: Particles of each sort have different energies.

From Newton to Einstein, kinetic energy has always been a function of mass and velocity. Unchanged fundamental laws claim that each kind of lightlike particles have the **same zero mass** and travel at the **same speed** c. But experiment tells that they have different energies. Einstein relation $E = h\nu$ only circumvented the discrepancy for massless photons. The discrepancy is still there even for photon phenomena.

DE fact proves the **nature** of the **speed** of **light** (**NSL**):

The speed of light is frequency-dependent, frame-variant, and less than c .

Letter c here denotes the finite frame-invariant speed that has been theoretically proved unique. Empirically, the speeds of v-measured photons are very close to c and to each other. Here "v-measured photons" refer to the photons with certain energies and frequencies, whose velocities have been measured experimentally.

Now we prove that the experimental errors in *direct test* of photon speeds are too large to reveal the NSL:

Various experiments since 1930's have provided estimations of a very small photon mass. That is,

$$m_\gamma \sim 10^{-27-z} eV/c^2 = 1.78 \times 10^{-60-z} g$$

for some $z \geq 0$ or $z < 0$. Due to (7), $E = m\not{\gamma}c^2$, with $\not{\gamma} \equiv \gamma$ in relativity and $\not{\gamma} \equiv \lambda\gamma$ in the open theory. Thus, $E_\gamma = \not{\gamma} \times 10^{-27-z} eV$. Suppose $E_{\gamma LB} = 10^{-15} eV$ is the lowest value of the energies of all v-measured photons. Then for all of them, $\not{\gamma} > \not{\gamma}_{LB} = E_{\gamma LB}/m_\gamma c^2 = 10^{12+z}$. Too many choices of λ (including $\lambda \equiv 1$) yield **monotonically increasing** $\not{\gamma} \equiv \lambda\gamma$ so that the speeds of all massive and timelike v-measured photons are **squeezed** in a very **narrow** interval ($u_{\gamma LB}$, c). Here $u_{\gamma LB} = c\sqrt{1-\gamma_{LB}^{-2}}$. Suppose, for example, $\gamma_{LB} = 10^{10+z}$ and $\lambda = 10^2$ at $\gamma = \gamma_{LB}$ so that $\not{\gamma}_{LB} = \lambda(\gamma_{LB})\gamma_{LB}$ holds to indicate that $\gamma_{LB} = 10^{10+z}$ is the lower bound of the γ-values for all v-measured photons. Thus, $u_{\gamma LB} = c\sqrt{1-10^{-20-2z}}$ and all velocities within the interval $(u_{\gamma LB}, c)$ are too close to c and too close to each other to yield positive experimental results of frequency-dependence of the speed of light in vacuum. Clearly,

$$\Delta u_\gamma / c \le (\Delta u_\gamma)_{max} / c = 1 - u_{\gamma LB} / c \approx .5 \cdot 10^{-20-2z}.$$

Experimental error $\Delta c / c \sim 10^{-12}$ for some v-measured photons is too large to reveal the true nature of 'the speed of light'.

The author does not know if military scientists have measured the speed of low-frequency signals sent to and from submarines. The lowest energy of the photons that have been tested for speed may be larger or much larger than $10^{-15} eV$. This would yield a larger lower bound $u_{\gamma LB}$ of the speeds of v-measured photons, which would be closer or much closer to c.

The v-measured photons are **fast-photons**. Even for $\nu \sim 10^{-5} Hz$ photons, $\gamma \ge 10^{7+z}$ and they are fast! It's very clear that slow photons are not detectable even in far future. The only reason is the small value of photon mass. But the DE fact has proved a nonzero mass as well as different speeds of photons once for all. This experimental evidence is completely independent of when one will be able to measure the difference between the speeds of photons with different frequencies.

The motion of virtual soft-photons is quantum and probabilistic. They do not obey classical formula $d = ut$. The range of e.m. (electromagnetic) force is finite but extremely long though real soft-photons are very slow.

Although the finite frame-invariant speed c is unique, the speeds of all v-measured photons are too close to c to yield measurable frame-variance of the speeds of photons. In fact, for any choice of factor functions, the c-postulate yields the same Einstein addition for $f_S(V_{S'S})$ cancels in deducing velocity addition law. Einstein addition $\hat{+}$ says that not only $c \hat{+} v = c$ for all v but also $u_\gamma \hat{+} v \approx u_\gamma$ for all v if u_γ is very close to c. With simple quantitative calculation, one can trivially explain why the frame-variance of the speed of a photon is practically unobservable.

The preferred reference space $\{K\}$ is not such a light-ether that all photons could move at the same isotropic speed relative to it. Even if an observer is at rest in the preferred reference space, the speed of light is still frequency-dependent simply because different frequencies correspond to different energies (see VI) which mean different speeds according to $E = m\lambda\gamma c^2$. Any theory that claims the same zero mass and the same speed for all photons never explains DE fact in any frame. Moreover, the speed of light will be anisotropic even if observed in an isotropic frame K (λ_K is isotropic) as long as the source is moving in K. This is due to Doppler shift. Once the frequency shifts, the energy changes and the speed must change. In fact, $u_\gamma \hat{+} v \ne u_\gamma$ does yield a little frame-difference in photon speed (for u_γ is slightly less than c), which can provide a perfect and rigorous quantitative explanation of Doppler shift via $E = m\lambda\gamma c^2$ to always give the same shift formula as via transformation rule of wave vector (see VI). In any case and in any frame, photons with different frequencies/energies cannot move at the same speed. This is the bottom of line. We treat the formula $E = m\lambda\gamma c^2$ very seriously, no matter what λ would be.

The negative result of Michelson-Morley experiment is just a fact that the velocity-direction dependence of λ at large γ-values in anisotropic earth-systems is always compensated by a change in γ-value for photons traveling in different directions. The change in γ-value causes a change in speed too small to yield positive result in direct testing. The only reason is that the γ-values of the photons in visible light are too large. Indeed, it's easy to prove that

$$0 < u_2 - u_1 < c/\gamma_1^2 \text{ as long as } \gamma_2 > \gamma_1 > 2/\sqrt{3}.$$

Then one can see that $\beta_2 - \beta_1 < \gamma_1^{-2}$ remains too small to measure experimentally for large γ_1 such as $\gamma_1 > 10^{12+z}$ ($\nu > 100Hz$ if $\lambda = 100$ at $\gamma = 10^{12+z}$). For photons in visible lights,

$$\gamma \sim 10^{27+z} >> 10^{12+z}.$$

Thus, a direct test is impossible.

For pulsar and future NASA experiment (U.S. News, Vol. 134, #18, May 26, 2003) that tested or will test frequency-dependence of the speed of light,

$$\Delta t = D/u_1 - D/u_2 = Dc^2(\gamma_1^{-2} - \gamma_2^{-2})/u_1u_2(u_1 + u_2) < \gamma_1^{-2}D/c$$

holds for large γ_1 ($\gamma_1 < \gamma_2$). Even for $D/c = 10^9$ years,

$$\Delta t < \gamma_1^{-2}D/c << 3.15 \times 10^{-12-2z}\, sec$$

is too small to measure because $\gamma_1 >> \gamma_{LB} = 10^{10+z}$ is too large for the photons to be tested in these experiments. (About 2 years later after such predicting, NASA announced the failure to discover a difference in arriving times.)

Experimental values of neutrinos' masses are uncertain. Suppose $m_\nu = .1eV$ for e-neutrinos. Then $\gamma = 10E_\nu/eV$. All e-neutrinos with $E_\nu < 10^{11}eV$ have $\gamma < 10^{12}$ and thereby at least most neutrinos move slower than v-measured photons with $\gamma > \gamma_{LB} = 10^{12+z}$. The γ_5 matrix in interaction term explains parity-violation no matter neutrinos are massless or massive. Small e-neutrino mass indicates the effectiveness of a two component theory. Parity violation with mu and tau neutrinos involved may indicate that the parity violation is indeed due to the γ_5 matrix in the interaction term and is irrelevant to the values of neutrino mass term in Dirac eqn, large, small, or zero.

Relic timelike massive neutrinos were born with high energies. Their speeds were so high that they had not slowed down significantly before embryonic galaxies were formed. Indeed, e.m. interactions could slow down very energetic electrons and protons 'classically' and gradually. But for neutrinos and photons, no physical process could make them slow down gradually though they can be 'stopped' to disappear in quantum absorbing process. Neutrinos rarely interact with other particles. Most relic neutrinos did not slow down after the Big Bang. Massive and timelike relic neutrinos could not lead to a distribution pattern of galaxies different from what astronomers see today.

Hypothetic masslessness of lightlike particles and the choice $\lambda \equiv 1$ with $E = m\gamma c^2$ in relativity squeeze entire velocity spectrum of photons and neutrinos into a singleton $\{c\}$ with $\gamma = \infty$. L' Hôpital limits are correct. But $h\nu = E_\gamma = m_\gamma c^2 \gamma = 0 \times \infty$ has no limiting process. It violates the foundation of all quantitative sciences and cannot serve as the foundation of physics. It's time we were replacing $0 \times \infty$, $0/0$, and infinities canceling infinities with reasonable calculations. It's far more than just a *perfection* issue:

Insisting on tolerating calculations that are not acceptable in view of simple and trivial mathematics inevitably leads to continuously rejecting EDF and tolerating the incorrect foundation and concepts that cause such incorrect calculations and so many experimental disagreements. Then the generations to come and we will have no chance to fix experimental disagreements using changed fundamental laws. Renormalization will shine if infinities are replaced by large finite quantities, no matter how large. The shinning would not be due to the removal of infinities itself but to the new foundation, the shinning renormalization demands. The new foundation that produces such replacement and perfection leads to explanations of so many experimental facts far beyond the perfection issue itself.

The conflict between ***massless*** particles and experimental DE fact is ***fatal*** if the energy formula in classical particle mechanics is **valid** for photons. In VI, we will prove the stunning validity through Doppler phenomenon:

Timelike embryonic photons within vertices of relevant Feynman diagrams were accelerated before emitted as real photons. Using $E = m\lambda\gamma c^2$ and velocity addition law we always get the **same** Doppler formula as from **transformation rule** of **wave vector** regardless whether photons are timelike. (Zero or nonzero mass terms of photons bring about differences in Doppler formula and other fundamental eqs and formulas. But the difference is too small to be practically measurable for most of them.) The match in both massless (with irrational calculations) and massive photon models further proves the wave-particle duality of photons and an emission theory (not Galileo-Newton emission). The match is perfect for massive photon model for all calculations with nonzero photon mass are flawless. This strongly suggests that all **virtual** and **real force carriers** are **massive** and **timelike** (see more evidences later).

The non-timelike separations in 4-dim integrals of Dyson series and spacelike 4-momentum transfers in t-channels will soon be shown fully agreeing with probability interpretation, causality, and the timelikeness of real and virtual force carriers.

IR fact, DE fact, and other EDF facts not only disprove massless particles but also disprove any **0-mass products** such as infinite range of e.m. force, U(1) gauge invariance, its extensions (non-Abelian gauge symmetries), and **gauge byproducts**: gauge-self (GS) couplings, gluon balls, and Higgs particles (HPs).

Experimental records of strong and weak interaction phenomena since 1930s have repeatedly showed nonexistence of gauge byproducts. Among more than hundreds of millions of photos all over the world, no single one is found.

The idea of **massive** vector theory of **Lee-Yang**-type for weak bosons and for photons with non-Abelian and U(1) ***gauge invariance*** **undone** by the mass terms now becomes a *corner stone* for building a simple unified massive vector theory of all non-gravitational forces without gauge invariance and without gauge byproducts.

Charged and neutral currents could be found soon after the quantum theories of weak and of electroweak interactions were proposed because quantization of interactions, exchange of vector force carriers, and the charm quark that indicates the existence of Z_0-mediated weak interactions are revolutionary concepts and are so real. Weak and e.m. forces have the same underlying coupling constant but differ by the masses of their force carriers and by algebraic coupling styles. These ideas hit it and led to the merit of the standard model such as the successful prediction of neutral currents and Cabibbo angles.

Removing **gauge**-couplings (i.e., GS and Higgs couplings), the remaining **algebraic** coupling styles of Yang-Mills, GWS, and Gell-Mann's **non-gauge** coupling terms also lead to success. No any merit of the standard model is attributed to gauge terms. The symbol SU(3)⊗SU(2)⊗U(1) represents algebraic rather than gauge coupling styles. Then it has no conflict with massive Lee-Yang's Lagrangian L_{LY} in a unified massive vector theory that undoes gauge symmetries.

Any **nonzero** mass term of force carriers, no matter how small or how large, not only yields both IRC and IRCF but also is associated with a **finite force-range** in any fixed observation system. That is a **physical reality** not allowing any gauge transformation to change it, regardless whether the Lagrangian can be made gauge invariant and whether the finite range of e.m. force is practically measurable in next 200 or more years. Experimental DE and IR facts prove the reality of a finite range of e.m. force indirectly.

Gell-Mann [13] pointed out that the old track current theoretical development is going on might be wrong. Here is an evidence:

> Gluons were found only in timelike jet-events during 70 years of experiments. If they were ***real*** (not virtual) and ***stable***, some of them could travel longer distances and be found in some of the tens of millions of photos. But none is found to have traveled a longer distance. Thus, either real gluons are not stable or the found ones are virtual gluons. If the found ones were decaying *real* gluons, they had to be **timelike** for the total 4-momentum of the final state particles in each jet was timelike. If the found gluons were **virtual** particles, why have real gluons never been found in the past 70 years now that they are massless or possess a small mass? In all the possible situations, gluons cannot be massless or having a small mass. The **short range** of strong force, according to **revolutionary interpretation** of quantum theory, directly implies **heavy** gluons. For real (not virtual) gluons, the **absence** of **gluon spectrum** in all experiments further proves the heaviness of gluons.

The attempt to explain the light spectra of atoms led to Bohr's great model of atomic structure and revolutionary quantum dynamic theory. The experimental fact of the absence of gluon spectrum at current energy scales is certainly an unusual but precious clue to mysterious structures of hadrons and strong force.

S. Weinberg [14] explained why he and Gross postulated massless gluons: "At first after the discovery of asymptotic freedom it was widely assumed that the gauge bosons in a realistic Yang-Mills theory of strong interactions would have to be quite heavy, to explain why these strongly-interacting bosons had not been discovered long before. Following the precedent of the theory of weak and electromagnetic interactions, it was supposed that the masses of the gauge bosons arose from a spontaneous breakdown of the color SU(3) gauge group, triggered by the vacuum expectation values of scalar fields in a nontrivial representation of this group. But these strongly interacting scalars would contribute positive terms to $\beta(g)$, which could destroy asymptotic freedom. Even worse, in a theory with strongly interacting scalar fields, radiative corrections involving weak interactions would introduce large violations of various symmetries like charge conjugation invariance and flavor conservation, which, as we shall see, would not be violated without the scalars. Then it was suggested to drop the strongly-interacting scalars, and accept the consequence that the gluons, the SU(3) gauge bosons, have zero mass."

In 1981, Weinberg [15] proposed a model without Higgs scalars.

Masses with which non-gravitational force-carriers are naturally endowed do not need hypothetic scalars and make the things Weinberg was worrying about disappearing. In his 1981 model, bound quarks move at the speed of light. This assertion is very close to the reality if quarks are bound within small quark-cores. Small quark cores mean large kinetic energies and large γ values for the confined quarks. For $\gamma \geq 10^2$, the speeds are indeed close to c and to the empirical value of the speed of light.

Gluons found in 3- and 4-jet events should be **virtual** and '**overdrawn**' just like virtual Z-bosons in the events of neutral currents. Then strong coupling constant must be much larger than the one in massless QCD. This yields **LH-model** (very **large** strong coupling constant and very **heavy** gluon mass).

LH-model and possible LV in bound quarks' γ-sector will be seen in II, V, VI and VII. LV may or may not play big roles in SA physics and quark dynamics. But DS (deeper structure) will be seen to be able to explain everything in SA physical world.

For direct experimental tests of photon mass and speed(s) in any future, a negative result will never prove the sameness of the speeds of all photons and an exact zero mass due to the **most fundamental principle** (**MFP**) of physics: *the true* (*exact*) *value of a* ***continuous*** *physical quantity cannot be determined due to inevitable experimental errors*. Counting of **discrete** quantities can be without error if the number is not very large. Two more examples may help to understand the meaning of experimental errors in measuring continuous physical quantities and the MFP: 1. For a free particle, its acceleration and velocity are continuous quantities and experiment with certain error, no matter how small, is unable to prove an exact zero acceleration. Thereby, exact inertial law has not been proved and will not be proved in any future as long as experimental error will always be nonzero in any measurement of continuous quantities. Will an acceleration of $10^{-90} cm / 10^{90} yr^2$ be detected in future by experiment? Nonzero experimental errors in measuring continuous quantities lead to inherent disability of modern physics. Physics will discover and explain more and more phenomena and fine-tune the laws of physics or even make radical and revolutionary changes, but it cannot accomplish everything. The disability is indeed inherent not just for the inertial law. Fortunately, many possible tiny violations of fundamental laws of physics, if any, may not bring about significant effects for the phenomena taking places within laboratories and one may not be bothered by any possible tiny numerical differences. We are only looking for and focusing on the violations that lead to significant effects and are able to explain experimental facts. 2. Parity symmetry claims exact zero-violation that can be disproved experimentally if a nonzero violation larger than experimental errors is confirmed. Counting number difference of the emitted electrons in any two different directions do not have to be accurate. Now that parity symmetry claims exactly zero difference, it has been plainly disproved once for all after a nonzero difference was confirmed. To disprove the symmetry, it is not necessary to count the numbers accurately and to measure the continuous quantity of distribution angles accurately.

To get better estimation or even determine (within experimental errors) the values of the masses of photons and neutrinos and their speeds, one needs future experiments with higher and higher accuracies, though positive results might not be obtained until 24th or much later century. However, to prove that photons and neutrinos are not massless, no any new experiment is needed.

DE fact and IR fact have disproved the masslessness of lightlike particles once for all.

Since 1930's, various experiments have provided estimations:

$$m_\gamma \le 10^{-40\sim-60} g\,.$$

The highest energy of photons found in primary cosmic rays is 20 TeV [1]. Throughout this book, we use these experimental data to write

$$m_\gamma \sim 10^{-27-z} eV/c^2 = 1.78\times10^{-60-z} g$$

$$E_{\gamma,max} \equiv Nm_\gamma c^2 \sim 10^{13+B} eV\,,\quad N = E_{\lambda,max}/m_\gamma c^2 \sim 10^{40+z+B}\,.$$

$$R^f{}_{em} \sim \hbar/m_\gamma c \sim 1.93\times10^{22+z} cm\,.$$

Here, $R^f{}_{em}$ is the **e.m.** force range. The unknown experimental parameters B and z in the exponents are solely due to inaccurate experimental data. Since $E_{\nu,max}$ is unknown for neutrinos, one cannot use $m_\nu c^2 = E_{\nu,max}/N$ to estimate neutrino mass. The most energetic particle of a kind found in primary cosmic rays might differ greatly from the particle of the same kind with $E = E_{max}$ for there is an issue of whether the mechanism to produce such particles with $E = E_{max} = Nmc^2$ exists in the universe.

Measurable physical quantities and phenomena are either sensitive or insensitive to whether N is finite and massless force carriers are actually massive. Some examples are as follows:

1) Atomic physics and numerical success of QED are very insensitive to estimation values of N, as long as N is large enough. A good numerical result in QED was obtained in Bethe's [14] work even with a very small ultraviolet cutoff of order $m_e c^2 \sim 0.5 Mev < 10^{-7-B} \times E_{\gamma,\max}$.

2) IR catastrophe is extremely sensitive to the difference between infinite and finite N. Once N is finite, massless particles cannot exist. The mass $m > 0$ and the YF for massive vector force carriers of each kind remove the IR catastrophe, no matter whether the mass will be practically measurable and what is the value of the mass.

3) Whether the kinetic energies of high-speed neutrinos and photons are given by incorrect operation $E = 0 \times \infty = 0/0$ or by correct operations is extremely sensitive to whether they are massless and whether N is finite. Dark matter problem is extremely sensitive to the difference between infinite and finite N. Claiming infinite N and massless neutrinos and photons led to rejecting them as dark matter and postulating astronomecally large number of stable unknown particles as dark matter. Hundreds of millions of photons of cosmic rays and events in accelerators since 1930's did not show a single one of such unknown stable particles. These overwhelming experimental evidences spot the seriousness of the issue of whether N is finite and whether lightlike particles are massive.

4) Gauge principles are too sensitive to physical reality of nonzero photon mass, finite e.m. force range, IR and DE fact, and the nonexistence of gauge byproducts. In the past 56 years, quantization and renormalization of any massless and massive (with HM) gauge theory of vector fields were done always when arbitrary gauge fields were 'fixed' by certain gauge choices/conditions. Feynman and 't Hooft made this clear long time ago. The non-existence of gauge-independent parameter(s) that would lead to a quantum field theory both quantizable and renormalizable in any inertial frames clearly implies that arbitrary gauge fields are not physical quantities. Namely, one needs '*to abandon part or all of the gauge invariance*' (G. Sterman [4]). He ([4], p. 373) also showed that massive and massless QED satisfy the same Ward identities. To quote, "*So the infrared divergence is indeed associated with the masslessness of the photon.*"

U(1) gauge invariance works in classical e.m. phenomena via massless Maxwell theory not because photon mass is too small and can be neglected, but because classical Maxwell theory does not explain quantum phenomena and a crucial requirement is not under the consideration. That requirement is a must in quantization: All quantum e.m. fields must obey quantum wave eqn for spin 1 particles and must be quantizable in any inertial frame and yield renormalizable theory of quantum fields. All measurable quantities in massless Maxwell theory can be written in terms of $\boldsymbol{E}$ and $\boldsymbol{B}$ and hence arbitrary U(1) gauge fields do not cause contradiction. Since photon mass is so small, the massless theory does not conflict with experiment in classical phenomena at all distances $r << R^f{}_{em} \sim 10^{22+z} cm$. For quantum phenomena, arbitrary gauge fields have no physical meaning, leading to all experimental disagreements with zero mass terms without **HM**

(Higgs mechanism) and with mass terms through HM.

In both massless and massive models, vector fields that obey *both* **classical** eqs of vector fields (massless or massive Maxwell eqs) and **quantum** wave eqs for spin 1 particles (massless or massive) are called **physical vector fields (PVFs)**. Amazingly, in finite-N theory with massive vector force carriers such as massive photons the condition $\partial^{\mu}V_{\mu}=0$ is no longer a gauge condition or a gauge choice but a property of the solutions of massive and generalized massive Maxwell eqs without source or with conserved source currents:

$$\partial^{\nu}\partial^{\mu}V_{\mu\nu}-\mu^{2}\partial^{\nu}V_{\nu}=-\mu^{2}\partial^{\nu}V_{\nu}=-4\pi c^{-1}\partial^{\nu}J_{\nu}=0\,.$$

Then classical massive vector fields automatically obey quantum eqs for massive spin 1 particles. Namely, massive vector fields are automatically 'fixed' and are indeed PVFs. Classical massive vector fields (as the solution of massive Maxwell eqs) are automatically quantizable in any frame at the very beginning without using any gauge choice or generalized gauge condition. This is very easy to see for the vector fields used in perturbation method.

5) Integration constants of the eqs of motion (containing $mu^{\mu}\gamma d\lambda/dt$ term) in finite-N theory for all central and quasi-central forces reveal new potential energies (PEs). The energies $E=m\lambda\gamma c^{2}$ of particles in attracting central force fields do not approach infinities as $r\rightarrow 0$ and the initial kinetic energies of particles in repulsive central force fields do not approach infinities too as the shortest distances r they can reach approach zero. But the new PEs approach (negative and positive) *constants* (**deepest asymptotic freedom**) instead of (negative and positive) *infinities* as $r\rightarrow 0$ (for attractive and repulsive forces respectively). Thus, in view of dynamics, the existence and the nonexistence of spacetime singularities are extremely sensitive to the difference between infinite and finite N.

Finite N is caused by the limiting LV $|\lambda-1|>0$ with $\lambda\gamma\rightarrow N$ as $\gamma\rightarrow\infty$. To accelerate an electron to reach $\gamma=N\sim 10^{40+z+B}$ barely will need an energy-scale of $E=m_{e}c^{2}\lambda\gamma\sim .5\cdot 10^{34+z+B}\lambda$ TeV. This is too large for any possible λ-values between .00001 and 10000 near $\gamma=N$ to be tested experimentally. The experimental way to test zero, small, or large LV directly at that energy scale does not seem to exist even in a far future. Non-limiting LV $\lambda-1=E/mc^{2}\gamma-1$ (zero, small, or large) that takes place at $\gamma<<N$ but with $\gamma\geq$ or $>>10^{5}$ is difficult or almost impossible to check in direct experimental tests for the reason pointed out before (large deviation of the true value of γ from $\gamma=E/mc^{2}$ only causes too little difference in speed in large-γ phenomena). However, severe LV with very large $d\lambda/d\gamma$ within (γ_{0},γ_{20}) with $\gamma_{0}\geq 10^{5}$, called **velocity shell-space (VSS)**, plays a key role to provide a small spacetime bag to assist strong interaction force $\boldsymbol{f}^{st}$ to confine quarks through the ET

$$dE/dt=\boldsymbol{f}^{st}\cdot\boldsymbol{u}+mc^{2}\gamma d\lambda/dt$$

One may read II with model λ (20) for details. **Note**, it was a mistake to think a "bag" of some sort could confine quarks. In cosmic rays, HE particles have energies from few *Mev* to 10^{12} *Gev* (higher or lower). Under such violent knocking, no "bags" can confine quarks. Quark absent fact has been explained in the theory of deeper structure.

'C-particles' or 'c-like' particles, if any, are defined as moving exactly at the frame-invariant speed c. They are called '*lightlike*' in relativity. DE fact not only disproves massless particles but also disproves massive c-particles. To see this, think of this ***subtlety***:

> The particles that are called 'lightlike' in relativity are either **timelike** with their different energies caused by *different* **velocities** OR they are **massive** c-**particles** moving at the same speed c but with their different energies caused by *different* **coefficient masses** proportional to their energies.

Massive c-photons not only lead to the concept of **coefficient masses** that sounds ***eerier*** but also lead to **IR catastrophe** due to their **zero** mass term. In fact, it is $m\lambda$, not m alone enters the **mass term** in every **quantum wave eqn** (see IV). In both infinite- and finite-N theories, $m\lambda$ vanishes for all c-particles (if any): $m=0$ in infinite-N theory while $m \neq 0$ but $\lambda = 0$ in finite-N theory. Indeed, for any continuous function λ, $\lambda = 0$ at $u = c$ is a necessary condition for $N \equiv \lim_{u \to c} \lambda\gamma$ to be finite. Vanishing mass term $m\lambda = 0$ does not lead to YF even if $m > 0$. Mass terms of real photons as well as of virtual photons must be positive in order to yield e.m. YF and IR convergent factors. In IV and V, we will prove:

> Only **timelike** vector force carriers obey **massive** quantum AND classical wave eqs with $m^2\lambda^2 > 0$ (not $m^2\lambda^2 = 0$ or < 0) and therefore have YFs, which reduce soft force carriers' effects severely.

Accordingly, ***lightlike*** particles in this book only refer to massive and timelike photons and neutrinos. A fictitious particle that travels at the frame-invariant speed c now is called a c-like particle or c-particle. As far as **causal character** is concerned, we must replace the term 'lightlike' by either 'c-like' or 'timelike'. For example, 4-vector V_μ or V^μ with $V^\mu V_\mu = 0$ is called lightlike and c-like in infinite- and finite-N theories respectively, and with $V^\mu V_\mu < 0$, timelike in both infinite- and finite-N theories.

There is no kinetic way to reach the speed c due to velocity addition law. There is no dynamic way either, though no infinite amount of energy is needed for a particle to reach the speed c in finite-N theory. In fact, it's easy to prove using finite-N theory that **infinite** amount of **time** is needed for an accelerated timelike particle to reach the speed c as long as the accelerations in all instantaneous frames are *upper bounded*, regardless how large the finite upper bound would be. Classical central interaction force obeys inverse square law (YF equals 1 at all much-shorter than force-range distances). It produces accelerations with no finite upper bounds as $r \to 0$ if LV at short distances is ignored. However, even such central interaction force cannot accelerate a particle to reach the speed c at $r = 0$ in both infinite- and finite-N theory for quantum effect (ground state) at small enough r prevents particle from reaching the state of

$r = 0$ with $u = c$. Now we can conclude:

***All** particles, real or virtual, are **timelike** and **massive**.*

In finite-N theory, the frame-invariant speed c is the lowest unattainable speed or simply the **limiting speed** of photons and all other particles, which can be approached but at which no particle can ever actually travel. It is not '*the speed of light*' anymore. All lights and e.m. waves travel slower than the speed c. A photon with higher energy moves faster than a photon with lower energy as long as the product $\lambda\gamma$ is monotone increasing.

To develop and compare two opposite theories with a postulate preserved and undone respectively and then to see that the said postulate is true or its violation is true, we need a new starting point: a *deeper* and *commoner* assumption shared by the postulate and certain ways to violate it. (Too general discussions are beyond the abilities of individuals.) A deeper and commoner assumption, called ***c-postulate***, claims that *coordinate transformations between inertial frames must leave a finite speed c invariant.* It has been proved that finite frame-invariant speeds are equaled (unique). We assume the frame-invariance of a finite *one-way* speed c since it has been pointed out (W. Rindler [16]) that invariant *two-way* speed would not yield new theoretical results.

Timelike photons and gluons further correspond to violation of the 2nd and the generalized 2nd postulates. However, the c-postulate is obeyed no matter photons, neutrinos, and gluons are massive or massless, and N is finite or infinite.

We now show the consistence of *timelike* virtual force carriers with any *non-timelike* separations $x - y$ and *spacelike* 4-momenta transferred between initial and final states of real leptons and/or quarks, as well as the agreement of *causality* with infinitely many *spacelike* separations in the 4-dim integrals of **Dyson series** and *spacelike* 4-momenta transferred in t-channels.

Virtual vector force carriers' 4-momenta show up in 3-dim expressions of propagators of **massive** vector fields with timelike 4-momenta k^{μ} so that $k^{\mu}k^{\mu} = -m_v^{\,2}\lambda^2c^2 < 0$. The 3-dim expressions of propagators are drawn from the HO (harmonic-oscillator) model of all quantized fields. In 4-dim notation of propagator, however, k_0 is a pure mathematical variable of integration and the 'causal' character of ($\boldsymbol{k}$, ik_0) is irrelevant to the causal character of virtual force carriers. So the '**causal' character** of the 'off-mass-shell' 4-momentum that equals the difference between the 4-momenta of initial and final states, used to label an internal line of Feynman diagram, is *irrelevant* to **causality** for it is not carried by any individual virtual force carrier. This also explains why the energy transferred in a neutral current is much less than heavy Z-boson. The observable effects of exchanging virtual vector force carriers manifest themselves only after the integrations in Dyson series are done. The effects are, for example, certain probabilities to have 4-momenta transferred, which are spacelike in t-channels and timelike in s-channels due to 4-momentum conservation. One can easily check a surprising fact that this discrete feature does not require exact conservation of 4-momentum as long as the violation is small. Furthermore, a

timelike virtual force carrier can be created at x and annihilated at y, *no matter* the **separation,** $x-y$ is **timelike**, **c-like**, or **spacelike**. This is due to the motion of quantum particles, described by probability wave functions: It can appear anywhere next time with corresponding probability and the distance it travels is independent of its velocity. Classical formula *distance=speed×time* does not apply to virtual particles. For virtual force carriers as virtual signals, the only observable effects are those probability amplitudes, where nothing is non-causal though there are infinitely many spacelike separations in the integrals of Dyson series. Two real timelike quantum particles, lepton(s) and(or) quark(s) can exchange a virtual timelike force carrier between any spacetime points x and y. This is because of their quantum motions: Each one can appear at any place at any time to emit or to absorb a virtual force carrier. This never means that they would leave a visible 'spot' at each point where they emit or absorb a virtual force carrier. Every virtual particle is emitted before it is annihilated in any observation system due to the θ–functions originated in Dyson series. It is surprising that the θ–functions are solely caused by elegant iteration method to solve the integration eqs and are not put by hand to meet the requirement of causality. The causality is a direct, quantitative result of the revolutionary concept of quantum motions of real timelike particles and all virtual timelike particles in the revolutionary pictures of quantized interactions.

Massive vector theory **without** gauge invariance/gauge byproducts is **renormalizable** due to the following reasons.

G. Sterman [4] used Lee-Yang's Lagrangian L_{LY} and the definition

$$O_{\alpha\beta}G^{\beta\gamma}(x-y)=-g_{\alpha}{}^{\gamma}\delta^{4}(x-y)\,.$$

He obtained a propagator. The $k^{\beta}k^{\gamma}/m_V{}^2$ term in the numerator vanishes if one chooses such a gauge condition that the chosen parameter in gauge fixing term makes L_{LY} producing quantum eqn for massive spin 1 particles and the fixed gauge field a PVF. In finite-N theory, we will still use the symbol L_{LY} to denote massive Lagrangian of Lee-Yang's type and the corresponding Sterman-type propagator contains no $k^{\beta}k^{\gamma}/m_V{}^2$ term because L_{LY} in finite-N theory is of the form that produces propagator without $k^{\beta}k^{\gamma}/m_V{}^2$ term. Moreover, Weinberg [3] guessed that 4-currents coupled to massive vector force carriers might be conserved ($\partial_{\mu}J^{\mu}=0$). If so, any observable effects caused by $k^{\beta}k^{\gamma}/m_V{}^2$ term would be nullified. The author thinks that this is true not only for the currents photons, gluons, and Z^0 particles are coupled to, but also for the currents $W^{\pm}$ particles are coupled to, since the conservation of lepton and quark numbers requires conserved currents. Electric-charge conservation is sufficient but not necessary for a probability 4-current to conserve.

The classical e.m. coupling constant measured before the last century in a simple laboratory equipped with heavy charged metal spheres could be used to interpret atomic spectra and give numerical success of QED. This is because quantum theory is revolutionary, not phenomenological. HM encouraged many physicists to dedicate themselves to the revolution of generalizing QED for short-range forces at the time when this revolution ran into difficulty. But rejecting EDF means that the HM is not a solution. Experiment has continuously proved the nonexistence of HPs for 40 years. As a price for endowing inherently massive weak force carriers with masses, HM had to endow e, μ, τ leptons with masses and one had to claim that no any structure of leptons will be found in any future. Then too many parameters enter the standard model and it becomes a phenomenological theory. Standard model will shine in further modification without HM and with changed fundamental laws.

The **1st** postulate requires total equivalence among all inertial systems with respect to ***all laws*** of physics, while the **2nd** postulate to ***only*** the propagation speed of light and e.m. waves. It is not consistent to retain the 1st postulate while undo the 2nd one. According to a generalized 2nd postulate, real gluons and real photons move at the speed c and hence are *massless* in view of relativity with infinite N. Because of the DE fact, IR fact, and the experimental facts of strong interactions, the violation of the 2nd postulate and its generalization is obvious.

In view of the more general theory with the c-postulate, which includes both SR (special relativity) and GR as special cases, Minkowski/Einstein spacetimes as 4-dim manifolds with their unique geometric structures provide spacetime frameworks to satisfy the c-postulate globally/locally. Postulates determine the choice of coordinate transformations from the complete atlases in Minkowski and Einstein spacetimes. The chosen transformations shape the covariant laws of physics. Without a change in the transformations, there will be no change in the fundamental laws.

In the complete atlas in Minkowski spacetime, infinitely many linear non-Lorentz transformations preserve uniformity and rectilinearity of the motion of free classical particles, and causal characters of tangent vectors and particles. Denote by S , S' , S'' , K , and K' global **inertial** systems in Minkowski spacetime. To leave c invariant (to obey the c-postulate), linear transformations must satisfy

(1) $$dS'^2 = f_S(V_{S'S})dS^2, \quad \forall S,S'.$$

Here $dS^2 \equiv dx^\mu dx^\mu = d\boldsymbol{x}^2 - c^2dt^2, \forall S$ and $V_{S'S}$ is the velocity of S' relative to (observed in) S. Under the c-postulate, the 1st postulate is true iff $f_S(V_{S'S}) \equiv 1$, $\forall S,S'$. Since (1) holds for arbitrary S and S', we have

$$f_S(V_{S''S})\, dS^2 = dS''^2 = f_{S'}(V_{S''S'})\, dS'^2 = f_{S'}(V_{S''S'})\, f_S(V_{S'S})\, dS^2.$$

The **consistence condition** manifests:

(2) $$f_{S'}(V_{S''S'}) = f_S(V_{S''S}) / f_S(V_{S'S}), \quad \forall S,S',S''.$$

Denote by $\hat{+}$ velocity addition. Then $V_{S''S} = V_{S''S'} \hat{+} V_{S'S}$. With $V = V_{S''S'}$, (2) yields

(3) $$f_{S'}(V) = f_S(V \hat{+} V_{S'S}) / f_S(V_{S'S}).$$

It tells: if $f_S \equiv c_1$ is a constant function for some S, $f_{S'} \equiv 1$ and it is also a constant function for any S'. One can switch S' and S in (3) to prove that $c_1 = 1$.

Consistent **constant** factor functions are all constant 1. Lorentz violation parameters $| f_S - 1 |$ are frame-variant and are **functions** of relative velocity.

For non-constant factor functions, whenever f_S is known for some S, $f_{S'}$ is determined via (3) for any other inertial frame S'.

Factor $\lambda \equiv f_S(\, \boldsymbol{u}\,)^{-1/2}$ enters the expressions for momentum and energy in S ($\boldsymbol{p} = m\lambda\gamma\, \boldsymbol{u},\ E = mc^2\lambda\gamma$) where $\boldsymbol{u}$ is the velocity of a particle/body relative to or observed in S (i.e., $\boldsymbol{u}$ is the velocity of the instantaneous inertial frame $S*$ of a particle/body relative to S). The deviation $|\lambda - 1| > 0$ measures LV. Since the only consistent constant factor function is constant 1, LV $\lambda - 1$ is either identically zero or a non-constant function of velocity. Namely,

λ is either constant 1 or a non-constant function of velocity.

Once the 1st postulate is undone, the complexity of these factor functions hides numerically for experiments with very small errors show that numerical values of f_S for all earth-systems S are simply almost constant 1 for small-γ particles. The interval $10 < \gamma < 10^4$ is within the reach of today's accelerators for the particles other than photons and neutrinos, where possible small LV $|\lambda - 1| > 0$ may already occur. It hides for it is the very difficulty for experiment to measure the high speeds accurately enough to induce very small error in γ-value so that it could be used to prove or to disprove the formula $E = m\lambda\gamma c^2$ with or without LV.

In finite-N theory, photons are massive and timelike. For all the photons with $E_\gamma > 10^{-15} eV$, $\not\gamma \equiv \lambda\gamma > 10^{12+z}$. (For a photon in visible light, one can see that $\not\gamma \equiv \lambda\gamma \sim 10^{27+z}$ is much larger.) They are genuine examples of super-large-γ and super-high-speed phenomena and seem to be able to let experiment to test the factor functions in earth-systems and any LV $|\lambda - 1| > 0$ at super-large γ-values beyond the reach of electrons/protons/ions accelerated in future super-large accelerators built in the coming ten centuries. Unfortunately, the values of $\not\gamma \equiv \lambda\gamma$ for them are too large to tell the values of λ. For example, if $\lambda = 100$ for a photon in red light, the γ-value would be 100 times less than the case of $\lambda = 1$. The change in speed, caused by the difference between $\gamma = 10^{27+z}$ and $\gamma = 10^{25+z}$, is too slight to be practically measurable in a direct way. It makes the region of $\gamma > 10^5$ a 'desert' even in the coming 300 to 30,000 years.

For the high-energy electrons in large accelerators such as 5 MeV~50 GeV electrons, their $\not\gamma \equiv \lambda\gamma$ values are about $10^{1\sim 5}$. We do not know what their λ-values are unless we know the γ-values.

It is difficult to verify the LV at $\not\gamma < 10$ because it is too slight in spite of that the speeds of the electrons and protons with $\not\gamma < 10$ may be measured accurately.

The so-called 'accurate' measurement of speed means small γ-error induced by it. The larger the γ-value, the smaller the error in measuring the high speed required to induce a small γ-error.

The departure of λ from constant 1 at much lower energy scale than $\gamma \geq N$ or at distances much longer than Planck length will be shown to be due to basic experimental facts of strong interaction and dark matter phenomenon (see II, III, V, and VII).

All new eqs and formulas will return to their counterparts in the current basic theory if $f_S(\boldsymbol{V}) \equiv 1$. The open theory with the c-postulate will definitely include the current basic theory as a special case. No exact form of f_S but **model forms** can be guided by experiments due to inevitable experimental errors in all regions and the mysterious 'desert' beyond and within the reach of accelerators. Within the reach, the 'desert' is due to the difficulty to measure the high speeds directly and accurately to induce only very small errors for γ-values and use these accurate γ-values to test Einstein's formula and the LV.

Let K be exactly **isotropic**: $f_K(\boldsymbol{V}) = f_K(V)$. Setting $S = K$ and $S' = K'$ in (3), we see that i) if $V_{K'K} = 0$, then $f_{K'} = f_K$ and $f_{K'}$ is *also isotropic*; ii) Setting $S = K$ and $S' = S$ in (3), we find that $f_S(\boldsymbol{V})$ is **not isotropic** if $f = f_K$ is not constant AND $V_{SK} \neq 0$. The unique set $\{K\}$ of all isotropic systems is called the isotropic or **preferred** reference space. From now on, we always use K to indicate a preferred frame. In case $S' = K$, (3) yields

$$f_K(\boldsymbol{V}) = f_S(\boldsymbol{V} \hat{+} \boldsymbol{V}_{KS}) / f_S(\boldsymbol{V}_{KS}).$$

It can be proved that the solution $\boldsymbol{U}$=$\boldsymbol{V}_{KS}$ that makes $f_S(\boldsymbol{V} \hat{+} \boldsymbol{U})$ independent of the direction of $\boldsymbol{U}$ for any $\boldsymbol{V}$ exists and is unique for any non-constant function f_S as long as f_S is produced by an isotropic function f so that

$$f_S(\boldsymbol{V}) = f(|\boldsymbol{V} \hat{+} \boldsymbol{V}_{SK}|) / f(\boldsymbol{V}_{SK}).$$

It can be proved by (3) with two successive settings $S = K$ and $S' = S$. In general, $\boldsymbol{V}_{KS}$ may not be observable if f_S is too close to constant 1 in a velocity region that covers $\boldsymbol{V}_{KS}$. However, if K is such an inertial system where the **CBR** (cosmic background radiation) is strictly isotropic, then we can find the 'absolute' velocity of the earth by measuring the Doppler shift of the CBR. Earth's motion causes **anisotropy** of the CBR that may be observed from the earth. The technical difficulty is that the CBR is of continuous thermal spectrum with very low temperature and the earth's rotation and revolution speeds are much lower than the moving speeds of distant stars in the expanding universe.

Whenever $V_{S'S}$ =0, translations and spatial rotations are required to form Euclidean group. Therefore,

$$f_S(\boldsymbol{0}) \equiv 1, \forall S.$$

Since $\boldsymbol{0} \hat{+} \boldsymbol{V}_{S'S} \equiv \boldsymbol{V}_{S'S}$, (3) yields $f_{S'}(\boldsymbol{0}) \equiv 1, \forall S'$. Then setting $\boldsymbol{V}$=$\boldsymbol{V}_{SS'}$ in (3) we find that $V_{SS'} \hat{+} V_{S'S} = V_{SS} \equiv 0$ and

(3a) $$f_{S'}(V_{SS'}) = 1 / f_S(V_{S'S}), \forall S, S' .$$

It can also be deduced from (2) by taking $S'' = S$ for $f_S(V_{SS}) = f_S(0) = 1$. Write

(4) $$\Lambda \equiv 1 / f_S(V_{KS}) = f_K(V_{SK}) = f_K(V_{SK}) = f(V_{SK}) .$$

For a fixed K, the frame-invariant

$$dK^2 = f_S(V_{KS}) dS^2 = \Lambda^{-1} dS^2$$

defines **metric**

(5) $$g_{\mu\nu} = \frac{\delta_{(\mu\nu)}}{\Lambda}, g^{\mu\nu} = \Lambda \delta_{(\mu\nu)} .$$

The index quantity $\delta_{(\mu\nu)}$ is the Kronecker *symbol*. Using (5), we find a simple way to lower or raise an index:

$$T_{\mu\nu\ldots} = \Lambda^{-1} T^{\mu}{}_{\nu\ldots}, T^{\mu\nu\ldots} = \Lambda T_{\mu}{}^{\nu\ldots}, \ldots.$$

Since $\delta_{\mu\nu} \neq \Lambda^{-1} \delta^{\mu}{}_{\nu}$ whenever $\Lambda \neq 1$ & $\mu = \nu$, Kronecker symbol $\delta_{(\mu\nu)}$ as an *index quantity* is not a tensor under new transformations with $\Lambda \neq 1$. But, we may write

(5a) $$g_{\mu\nu} = \frac{\delta^{\mu}{}_{\nu}}{\Lambda}, g^{\mu\nu} = \Lambda \delta_{\mu}{}^{\nu} .$$

If we choose another isotropic system K', Λ in (5) will not change for $V_{K'S} = V_{KS}$ due to $V_{K'K} = 0$. Hence, the metric in (5) and (5a) will not change. If J is a *fixed* non-preferred inertial frame and we choose dJ^2 to define metric or distance element, then

$$dJ^2 = f_K(V_{JK}) dK^2 = f(V_{JK}) dK^2 = f(V_{JK}) \Lambda^{-1} dS^2 .$$

The change in scaling occurs if $f(V_{JK})$ is not equal to 1. Nevertheless, one can check that the covariant eqs of laws will remain unchanged. Thus, any theory with the c-postulate is *Weyl-Gauge-invariant*, though the factor functions do not have the physical meaning Weyl once incorrectly assumed.

S^* always denotes *instantaneous rest frame* of a timelike particle. Label

$$\gamma \equiv 1/\sqrt{1 - u^2 / c^2} , \gamma' \equiv 1/\sqrt{1 - u'^2 / c^2} , \gamma_K \equiv 1/\sqrt{1 - u_K{}^2 / c^2} \text{ with}$$

$$\boldsymbol{u} \equiv V_{S^*S},\ \boldsymbol{u}' \equiv V_{S^*S'},\ \boldsymbol{u}_K \equiv V_{S^*K}.$$

Due to (2) and the symbol Λ in (5), we have

(6) $$\lambda \equiv f_S(\boldsymbol{u})^{-1/2} = \sqrt{\Lambda / f_K(V_{S^*K})} \equiv \sqrt{\Lambda}\lambda_K.$$

Then

$$d\tau \equiv \sqrt{-dS^{*2}/c^2} = \sqrt{-\lambda^{-2}dS^2/c^2} = dt/\lambda\gamma$$

is the **proper time interval** of a timelike particle due to (1) and (6). All preferred frames in $\{K\}$ provide the same frame-invariant $dK^2 = f_S(V_{KS})dS^2$ determining metric (8) while instantaneous rest frames S^* of timelike particles provide frame-invariants dS^{*2} that determine frame-invariant proper time intervals $d\tau$ crucial to the formulas of momentum and energy and to the dynamics.

Since $U^\mu \equiv dx^\mu / d\tau$ and $p^\mu \equiv mU^\mu \equiv (\boldsymbol{p}, iE/c)$ we have

(7) $$\boldsymbol{p} = m\lambda\gamma\,\boldsymbol{u}, \qquad E = m\lambda\gamma c^2,$$

$$c^2\boldsymbol{p}^2 - E^2 = -m^2\lambda^2c^4, \qquad p^\mu p_\mu = -m^2\lambda_K{}^2c^2.$$

Here $p_\mu = p^\mu / \Lambda$. Infinitely many forms of λ yield finite limits of $\lambda\gamma$ as $\gamma \to \infty$. This is the origin of ultraviolet cut-offs, the non-existence of massless particles and of any zero-mass products.

The identity $\beta^2 \equiv 1 - \gamma^{-2}$ yields $\Delta\beta \approx \Delta\gamma / \gamma^3$ for all high-speed particles with large γ and $\beta \approx 1$. Any departure of λ from 1 causes departure of γ-value from what is determined by Einstein's formula $E = m\gamma c^2$. Even if $\lambda(\gamma) = 100$ at the γ-values of *visible photons* with

$$E_\gamma \sim 1eV,\ m_\gamma = 10^{-27-z}eV/c^2,\ \gamma = E/\lambda m_\gamma c^2 \sim 10^{25+z},$$

a super large γ-departure $\Delta\gamma = E/m_\gamma c^2 - E/\lambda m_\gamma c^2 \sim 99\gamma \sim .99\cdot 10^{27+z}$ only causes a super tiny speed-departure. In fact, $\Delta\beta \approx \Delta\gamma/\gamma^3 \sim 10^{-48-2z}$ would be too small to measure experimentally. The experimental errors in testing the speed of light have been $\Delta\beta/\beta \sim 10^{-5\sim-12}$ since 1940's. The errors are too large to reveal the said severe departure $\lambda - 1 = 99$. Then that severe departure is completely hidden behind the well-tested photon phenomena. This clearly proves that large-γ region beyond today's accelerators is indeed a 'desert' though so many important properties of photons and e.m. waves have been well tested since 1860's and intensively since 1930's. It is very difficult or even impossible for experiment to tell directly (with necessary accuracies much higher than 10^{-12}) about the values of the high speeds, γ, and the corresponding λ of high-speed particles with

$\gamma >> 10^5$. However, the departures within (γ_0,∞) with $10^5 \le \gamma_0 << N$ will be shown to be very helpful to explain DM (dark matter).

Atomic clocks are very accurate. But they moved too slow to verify any deviation from Einstein time expansion. We will not mention any low speed phenomena and corresponding accurate experimental facts.

General linear transformations obeying the c-postulate obey (1) with consistence conditions (2) and (3) and yield metric (5). The velocity addition law for *parallel transformations* remains

(8) $$\boldsymbol{u} = \boldsymbol{u}' \hat{+} \boldsymbol{V} = [\frac{1}{\Gamma}\boldsymbol{u}' + \boldsymbol{V} + \frac{\Gamma-1}{\Gamma V^2}(\boldsymbol{V}\cdot\boldsymbol{u}')\boldsymbol{V}](1+\boldsymbol{V}\cdot\boldsymbol{u}'/c^2)^{-1}$$

with $\Gamma \equiv 1/\sqrt{1-V^2/c^2}$, $\boldsymbol{V} = \boldsymbol{V}_{S'S}$. Write $f_S(\boldsymbol{V}_{S'S})^{-1/2} = \lambda(\boldsymbol{V}_{S'S}) = \lambda(\boldsymbol{V})$. For a moving rod and clock, we have

(9) $$\Delta x = \lambda(\boldsymbol{V})\Delta x'/\gamma, \quad \Delta y = \lambda(\boldsymbol{V})\Delta y', \quad \Delta z = \lambda(\boldsymbol{V})\Delta z'$$

$$\Delta t = \lambda(\boldsymbol{V})\gamma\Delta t'.$$

(9) and (3a) imply that an observer at rest in S can see a **transverse *contraction*** of a moving rod at rest in S', iff an observer at rest in S' can see a **transverse *extension*** of a moving rod at rest in S. The net effect, caused by $\lambda(\boldsymbol{V})/\gamma$ or $\lambda(\boldsymbol{V})\gamma$, is still relative for the relative effect of the Lorentz factor γ dominates.

The new transformations have matrix representation

$$\{ f_S(\boldsymbol{V}_{S'S})^{1/2}\alpha(\boldsymbol{V}_{S'S}) \mid \forall S,S' \}$$

where $\alpha(\boldsymbol{V}_{S'S})$ is the Lorentz transformation matrix whose velocity parameter is the vector value of $\boldsymbol{V}_{S'S}$. We have suppressed the parameters for translations and spatial rotations in $\alpha(\boldsymbol{V}_{S'S})$. Any *non-equivalence* among inertial frames does not allow removing **frame indices**/*subscripts* since f_S are not the same function for all S. Denote

$$f_S(\boldsymbol{V}_{S'S})^{1/2}\alpha(\boldsymbol{V}_{S'S}) = T(SS')$$

where $\boldsymbol{V}_{S'S}$ now is treated as an *indexed quantity*. There is a unique 3-vector $\boldsymbol{V}$ (velocity) corresponding to each indexed quantity $\boldsymbol{V}_{S'S}$. Now $T(S'S)$ are treated as induced indexed quantities corresponding to matrices

$$T = f_S(\boldsymbol{V})^{1/2}\alpha(\boldsymbol{V})$$

with $\boldsymbol{V}$ being the velocity parameter. It can be proved that all $T(S'S)$ form a **groupoid** [17] for each given factor function $f = f_K$ that generates a corresponding consistent set $\{f_S\}$ for all S. It can be called the f **-groupoid**.

When f is identically 1, the **1-groupoid** can be parameterized with frame indices removed. Then it becomes the Lorentz group:

$$\{T(S'S)\}=\{\alpha(V_{S'S})\}\Rightarrow\{\alpha(V)\}$$

A groupoid possesses a common algebraic structure of the transformations between nonequivalent as well as equivalent objects. The writer prefers to use term 'physical group' for the product of two non-successive transformations, say the one from the sun to the earth and the one from the Mars to the moon makes no sense. (This physical fact is invisible in abstract groups since the transformations between physical objects have been parameterized due to the equivalence among transformation objects.) In all numerical calculations, $\boldsymbol{V}_{S'S}$ is regarded as a 3-vector velocity.

The identities $\gamma\equiv 1/\sqrt{1-v^2/c^2}$ and $\beta\equiv v/c$ allow free switches between speed-variables v, β, and γ. For any exact isotropic function $f(v)$ one can always write

$$f(v)=1+h(\gamma)+\beta^2\gamma^2/N^2=1-N^{-2}+h(\gamma)+\gamma^2/N^2,$$

as long as $h(\gamma)=f(v)-1-\beta^2\gamma^2/N^2$. One can directly work on selecting $h(\gamma)$ to shape a working model of factor function $f(v)$ that agrees with experimental facts. Set $f_K(v)=f(v)$. Since $f(0)=1$, $h(1)=0$. In finite-N theory, all $h(\gamma)$ with $\lim_{\gamma\to\infty}[h(\gamma)/\gamma^2]=0$ yield **finite** limits:

(10) $$N\equiv\lim_{v\to c}[f(v)^{-1/2}\gamma]=\lim_{v\to c}\frac{\gamma}{\sqrt{f(v)}}<\infty\ .$$

Using (6) and (8) with $f_K(v)\equiv f(v)$, we find direction dependent limit in terms of non-isotropic frames

(11) $$N_S(\boldsymbol{n})\equiv lim\,\lambda\gamma=\sqrt{\Lambda}N\Gamma^{-1}(1+\boldsymbol{A}_S\cdot\boldsymbol{n})^{-1}\ \text{as}\ \boldsymbol{u}\to c\,\boldsymbol{n}\ \ (|\boldsymbol{n}|=1)\,.$$

Where $\boldsymbol{A}_S\equiv\boldsymbol{V}_{SK}/c$, $\Gamma\equiv 1/\sqrt{1-A_S^2}$. The direction dependence of the limit in terms of non-isotropic frames is caused by and only by the limiting direction dependence of λ in terms of non-isotropic frames. Obviously,

$$N_K(\boldsymbol{n})=N$$

is exactly isotropic. Since (8) is obtained via parallel transformations,

$$\boldsymbol{V}_{SK}=-\boldsymbol{V}_{KS}\,.$$

Evidently, N_S is <u>finite</u> for any S iff N_S is <u>finite</u> for some S. The *contravariant* **ultraviolet cut-offs** for *timelike* particles' 4-momenta are

(11a) $$\Lambda^{\mu}(\boldsymbol{n}) \equiv \lim p^{\mu} = mcN_S(\boldsymbol{n})(\boldsymbol{n},\, i) \quad \text{as } \boldsymbol{u} \to c\,\boldsymbol{n} \ \ (|\boldsymbol{n}| = 1).$$

Any particle of mass m with $E \approx N_S(\boldsymbol{n})mc^2$, if ever exists in the universe, is called **most energetic** particle of its kind. From (11) we get the largest difference caused by non-isotropy of any fixed non-preferred frame S

(11b) $$N_S(\boldsymbol{n})_{max} - N_S(\boldsymbol{n})_{min} = 2\sqrt{\Lambda\Gamma}NA_S.$$

Here the maximum and the minimum are understood as due to the difference in direction. It is very large but not practically observable in direct test for two reasons. 1) The numerical results of quantum theory of fields are not sensitive to the values of the ultraviolet cut-offs as long as they are large enough in all directions. 2) We do not have most energetic particles to verify the maximum anisotropy (11b) in a non-isotropic frame S. Even if 20 TeV photons found in primary cosmic rays are the most energetic photons, the sources are far and unknown. Then the anisotropy is no way to check directly. Besides, it can be proved that the S-depending cut-off in (11) just leads to frame-variant energy for any same c-photon (if any) observed in S and S' respectively. This will, after a 4-wave vector is defined in (59), leads to a frequency-shift formula for any same fictitious c-photon observed in S and S' respectively. We'll use transformation rule for the 4-wave-vector to deduce the shift formula for timelike photons ($\beta_\gamma < 1$) and get (59c). The two formulas become the same as one sets $\beta_\gamma = 1$ in (59c) for c-photons. It becomes the Doppler formula for c-photons in current theory if one sets $f_S(\, V_{S'S}\,) \equiv 1$. This match proves that even the severe anisotropy in the phenomena involving most energetic photons shown in (11) and (11b) agrees with experimental facts. For timelike particles, the match will be perfect without using the eerier concept of coefficient masses for c-photons.

There is another perfect agreement with experiment indicated before: The anisotropy of λ in anisotropic earth-systems plus direction dependent speed of the photons emitted from a source at rest and observed in anisotropic earth-systems can yield the observed isotropic atomic spectrum. Energy differences between quantum levels determine the energies and frequencies of emitted photons. They are independent of moving direction of emitted photons. Namely, the spectra of E_γ, ν, and $\lambda\gamma = E_\gamma / m_\gamma c^2$ are isotropic for any source at rest and observed in any anisotropic frame. But both λ and γ for the emitted photons are not isotropic in earth-systems. The direction-dependence of the speed of these photons is not practically measurable because of their large γ-values. The reason is simple and trivial:

The direction-dependence of the γ-values of these emitted photons is significant so that the differences in γ-value could compensate the significant anisotropy of λ in anisotropic earth-systems at large γ-values. The large γ-values, however, make the corresponding differences in speed too small to measure directly. This argument rigorously explains the **empirical isotropy** under the condition that λ in terms of earth-systems is significantly anisotropic at large γ-values. It is too obvious and any detailed quantitative calculation is trivial and redundant.

As $u \to c$, finite limit $\lambda\gamma \to N_S$ requires $\lambda \to 0$. This produces an infinite limit $f_s(\boldsymbol{u}) = 1/\lambda^2 \to \infty$ as $u \to c$. However, for any S, S', we have $V_{S'S} < c$ and $f_s(\boldsymbol{V}_{S'S}) < \infty$. Limiting frame S' with $V_{S'S} = c$ for any realistic frame S does not exist in the universe. In (7), (9), and other expressions of physical quantities, $f_s(\boldsymbol{V}_{S'S})$ appears with negative exponent. No infinity appears in any expression of physical quantity. In (9), the limiting Lorentz contraction is still zero just as in special relativity. Einstein time expansion approaches infinity as $\gamma \to \infty$. In finite-N theory, however, time expansion is always finite even as $\gamma \to \infty$.

The infinite limit ($f_S(\boldsymbol{u}) = 1/\lambda^2 \to \infty$ as $u \to c$) does not cause any trouble while the infinity $\lambda\gamma \to \infty$ as $u \to c$, such as caused by $\lambda \equiv 1$ in relativity, leads to a series of experimental disagreements and irrational calculations.

In terms of an isotropic frame K,

$$\lambda_K \equiv f_K(u_K)^{-1/2} = f(u_K)^{-1/2} = \lambda_K(\gamma_K),\ \gamma_K \equiv 1/\sqrt{1-u_K^{\ 2}/c^2}\ .$$

We now *suppress subscript* K for simplicity of notation and write

$$\lambda \equiv f(u)^{-1/2} = 1/\sqrt{1-N^{-2}+h(\gamma)+N^{-2}\gamma^2}\,. \qquad (12)$$

Due to velocity addition,

$$\gamma_K = \Gamma(1+\boldsymbol{A}_S\cdot\boldsymbol{\beta})\gamma\ .$$

Since (12) is just λ_K with subscript K suppressed, (6) and (12) lead to the general form of λ in terms of any non-preferred system S:

$$\lambda = \sqrt{\Lambda}/\sqrt{1-N^{-2}+h\varpi\gamma+N^{-2}\varpi^2\gamma^2}\ ,\quad \varpi \equiv \Gamma(1+\boldsymbol{A}_S\cdot\boldsymbol{\beta})\,. \qquad (13)$$

As the earth moves, $A_S \neq 0$ all the times for any earth-system S. At most there will be only one instant time when $A_S = 0$. Therefore, for earth-systems, λ in (13) is indeed **anisotropic**. Infinitely many choices of $h(\gamma)$ and λ induced by it via (13) make LV too slight for $\gamma < 10$. Then the anisotropy of λ in terms of earth systems with $A_S \neq 0$ is severely suppressed for $\gamma < 10$.

For $10^{1\sim2} < \gamma < 10^4$, LV $|\lambda - 1| > 0$ might be significantly nonzero and the anisotropy shown in (13) might be appreciable. The difficulty to confirm such LV and anisotropy experimentally is due to the difficulty to measure the speeds higher than .99*c* and the γ-values larger than 10 with high accuracy.

In V, one will be able to check that the curvature of the trajectory of a high-speed charged particle in a uniform magnetic field is unable to tell the speed accurately. In fact, $R = mc\lambda\gamma u/(qB\sin\theta)$. The accurate measurement of R will not tell accurately about the speed and γ since λ appears in the formula.

For earth-systems, $A_S << 1$ is supposed to be a good approximation so that $\varpi \equiv \Gamma(1+\mathbf{A}_S\cdot\boldsymbol{\beta}) \approx 1$ in (13) for earth-systems. In this book, we will use (12) as a good approximation for the general form of a model λ in terms of earth-systems. A very special setting $h(\gamma)\equiv 0$ in (12) yields a simple model λ and a total differential for $\gamma d\lambda/dt$

(14) $$\lambda = 1/\sqrt{1-N^{-2}+N^{-2}\gamma^2} = N/\sqrt{N^2-1+\gamma^2},$$

$$\gamma d\lambda/dt = d[\lambda\gamma - Nth^{-1}(\lambda\gamma/N)]/dt.$$

A total differential of $mc^2\gamma d\lambda/dt$ term in the ET (19) is needed to write down the integration of the eqn of motion as the conserved TE and then the PE can be written explicitly. In this book TE may mean TME or total energy including rest energy. Context determines what exactly TE means. TME always means total mechanical energy without rest energy.

In general, a function $f(x)$ defined on $(\sqrt{|a|},\infty)$ ($a<0$) or (0, ∞) ($a\geq 0$) is called **quasi-inverse** if it can be written as

$$f(x) = b/\sqrt{a+x^2},$$

where a and $b>0$ are constants. For $a\neq 0$, it becomes very close to $f(x)=b/x$ for and only for all $x >> \sqrt{|a|}$. We call b the **roof** of the quasi-inverse function. If $a=0$, the function is strictly proportional to $1/x$. The model λ (14) is *quasi-inverse* and is so far the simplest non-constant factor function found to yield an explicit expression of conserved TE in finite-N theory (see II). It is useful for establishing the concept of λ-dependence of PE. We say that (14) is a **limiting-departure** model without measurable non-limiting departure $|\lambda-1|>0$ for $\gamma << N$. In general, for any *quasi-inverse* piece λ_i within an interval (γ_{i1},γ_{i2}), the term $\gamma d\lambda_i/dt$ is a total differential:

(14a) $$\lambda_i = \lambda_0(a+\gamma^2)^{-1/2}, \gamma\in(\gamma_{i1},\gamma_{i2}),$$

$$\gamma d\lambda_i/dt = d[\lambda_i\gamma - \lambda_0 th^{-1}(\lambda_i\gamma/\lambda_0)]/dt.$$

Here $\lambda_0>0$ and $a\neq 0$ are constants. It is remarkable that the total differential in (14a) does not contain a explicitly. But one can check later that a must be positive in order to obey the ET (19). Numerically, λ_i in (14a) is close to $\lambda_i = \lambda_0/\gamma$ iff $|a|/\gamma^2 << 1$. Now that the expression of the total differential in

(14a) is independent of the value of a, the expression of the integration of the ET (19) and the PE will be the same no matter a is very large or very small. But the $\gamma - r$ relation depends on a so that the same expression of PE does not mean the same thing for different values of a. We will see that the value of the roof λ_0 is crucial to determine the **characteristic distances (CDs)** where the PE starts to turn to be quasi-constant to yield asymptotic freedom. Clearly, (14) is a special case of (14a) with $\lambda_0 = N$ and $a = N^2 - 1$.

In terms of all earth-systems, $\Lambda = 1$ is too accurate for many choices of $h(\gamma)$ in (12). However, we will keep Λ to maintain strict covariance.

Chapter 2 Particle Mechanics

Label $F_{(\mu)} \equiv F^\mu / \lambda\gamma$. The **eqn of motion** $F^\mu = dp^\mu / d\tau$ leads to its usual form

$$(15) \qquad F_{(\mu)} = dp^\mu / dt \equiv d(m\lambda\gamma u^\mu)/dt.$$

From (15) we get $F_{(\mu)}u^\mu = -mc^2\gamma^{-1}d\lambda/dt$. Here $u^\mu \equiv U^\mu / \lambda\gamma = (\, \boldsymbol{u},\, ic)$. For any contravariant 4-vector V^μ other than p^μ and U^μ, we will use notation $V_{(\mu)} \equiv V^\mu / \lambda\gamma \equiv (\, \boldsymbol{V},\, V_{(4)})$ throughout this book. Then one can check that

$$(16) \qquad dE/dt = \boldsymbol{F} \cdot \boldsymbol{u} + mc^2\gamma^{-1}d\lambda/dt.$$

Lorentz force formula mismatches (16) whenever λ is not constant. To obey (16), we use a **force splitting**

$$(17) \qquad F_{(\mu)} = I_{(\mu)} + S_{(\mu)};\; S_{(\mu)} \equiv m\gamma u^\mu d\lambda/dt.$$

Here $I_{(\mu)}$ is called the **interaction force** and $S_{(\mu)}$ **space force**. In general, $I_{(\mu)}$ can be any kind of interaction forces or their resultant, while $S_{(\mu)}$ is universal.

In Minkowski spacetime, non-constant λ causes departure from Minkowski metric. This brings about non-vanishing space force and causes dynamic effect just like the departures from Minkowski metric in Einstein spacetime cause dynamic effects in both general relativity and in new gravity theory with finite N. We will show that geometrized gravity cannot be materialized and quantized and gravitons do not exist (see III). Similarly, the space force in (17) causes dynamic effect but it cannot be materialized and quantized to give a quantum field of new force carriers. Space force carriers do not exist. Essentially, the term $m\gamma u^\mu d\lambda/dt$ is called space force for it joins forces with interaction forces to produce dynamic effects in the sense that it changes the old ways in which interaction forces cause dynamic effects.

An eqn of motion **equivalent** to (15) can be drawn from (15) and (17):

$$(17a) \qquad I_{(\mu)} = m\lambda d(\gamma u^\mu)/dt.$$

Now the ***inertial law*** reads $\boldsymbol{u}$ is a constant 3-vector iff the interaction force $\boldsymbol{I}$=**0**. The e.m. (interaction) force is

$$(18) \qquad I^{em}{}_{(\mu)} = \Lambda q F_{\mu\nu} u^\nu / c,\; F_{\mu\nu} \equiv \partial_\mu A_\nu - \partial_\nu A_\mu.$$

It becomes the Lorentz force formula (LFF) if $\Lambda \equiv 1$. We still call (18) **Lorentz**

force formula in new physics in honor of great H. A. Lorentz. The force splitting (17) leads to this minimum change in the e.m. force formula. It is almost the LFF for $\Lambda = 1$ is too accurate in earth-systems. However, the Lorentz force now does not produce the same dynamic effect as in Lorentz invariant CED (classical electrodynamics) since (17a) contains factor λ. In general, since

$$1/\gamma + \gamma u^2/c^2 \equiv \gamma ,$$

(17) and (16) prove the **energy theorem (ET)**:

(19) $$dE/dt = \boldsymbol{I} \cdot \boldsymbol{u} + mc^2 \gamma d\lambda / dt \, .$$

The inertial law shown in (17a) tells that any nonzero term $mc^2\gamma d\lambda/dt$ in (19) causes dynamic effect only if an interaction force $\boldsymbol{I}$ is present $(\boldsymbol{I} \neq \boldsymbol{0})$.

Experiments may have not verified whether $mc^2\gamma d\lambda/dt$ term is significantly nonzero within $10 < \gamma < 10^4$. It's quite sure that experiments have not verified whether $mc^2\gamma d\lambda/dt$ term is significantly nonzero as $\gamma > 10^4$. Particles (not neutrinos and photons) in accelerators move with $\gamma < 10^6$.

QM cannot explain high-energy phenomena where particle-transitions occur. However, for stationary bound states, QM and PEs play fundamental roles. To explain Lamb shift, QED must use **QM** with **classical force fields** to match the **large portion** of an empirical energy value, leaving only a tiny portion for radiative correction to explain. Even radiative correction itself must use **classical external field** to modify the propagator so that the 2nd order approximation could be treated as a 1st order approximation.

Feedback from quantum experimental facts and basic features of quantum field theory is called **Q-feedback**. CED can be regarded as a classical theory of e.m. force **without** Q-feedback from radiative corrections. An idea classical theory **with** Q-feedback would make a feature of a new e.m. PE as the solution of *inverse Dirac problem* (**IDP**) where the spectrum of energy levels and corresponding probability wave functions are known quantities determined either by experimental values or by successful numerical results of QED. At the present, we do not know if IDP is solvable mathematically to tell the **exact PE** or its expansion to any desired order. Practically, the experimental facts and QED numerical results of energy levels can be very accurate for lowest energy levels but not for wave functions. Thus, we have to consider feedback from radiative corrections. Although IDP is not practically solvable to get an idea PE, we may postulate that an idea PE for each bound state exists and is able to explain relevant spectrum via QM. The effect of all radiative corrections is contained in the idea PE. This is called the **PE-Postulate**. If it is true or approximately true, we are allowed to shape the graph of an idea or even just a working PE directly from experimental facts.

In finite-N theory, the concept of electrical charges is generalized while classical vector fields of all non-gravitational forces obey massive Maxwell eqs (see (31), (46), (47)) and quantum wave eqn for massive spin 1 particles simultaneously. Lorentz condition for each massive vector field is a property drawn from the solutions of massive Maxwell eqs. The scalar potentials of point-like rest charges are all of **Yukawa**-type:

$$\phi = \Lambda q' e^{-\mu r} / r \, .$$

For the Yukawa-type potentials of **strong charges**, $q' = k_{eff} e_s$ is the effective strong charge of the source that includes a three-body factor particularly for bound quarks within baryons and a Q-feedback factor due to heavy gluon mass and the strong force mechanism in LH-model. Numerically, they are **inverse potentials** (a potential is not a potential energy) at all the distances much shorter than the corresponding finite ranges of all non-gravitational forces, though they remove IR catastrophes at the distances near and longer than the finite force-ranges.

However, there is **NO** PE that is **entirely inverse** at shorter than force-range distances in finite-N theory since the eqn of motion contains $mc^2\gamma d\lambda / dt$ term that changes the ways how inverse potentials cause dynamic effects at short distances.

Any PE needed to match empirical facts is produced by and corresponds to certain form of λ through the ET (19) (see the PE-Theorem below).

The e.m. force formula (18) is generalized to **strong force formula** (45) with the Q-feedback. In general, an anti-quark can also be called a quark. However, we must distinguish the difference between a quark and its anti-quark and only particular ones are called quarks in order to understand **strong coupling style** in (46), which yields repulsive force only between two identical quarks or anti-quarks. It can be modified so that the strong force is always attractive under the condition that all quarks have strong charge e_s while all anti-quarks $-e_s$. Then we have conservation of strong charges and conserved 4-currents coupled to gluons that do not carry strong charges. The said conserved 4-currents yield renormalizable non-gauge theory for massive quantum vector gluon-fields without HM.

LH-model indicates that the observed strong force is caused by virtual pions between double-gluon lines and/or by quark loops. Therefore, the observed strong force so far can be called **pion-force**. Gluons are very heavy and **gluon-force** caused by single virtual gluon exchange without anything in the middle is of very short range beyond the reach of today's accelerators. The effective coupling constant η_s^2 in terms of pion-propagator factor $1/(m'_\pi + k^2)$ then is determined by a power of integrals with massive gluon propagator

$$e_s^2 / (M^2 + k^2)$$

that is not small but less than 1 to validate perturbation method. Here e_s is the strong charge carried by quarks. The author does not know if the perturbation method is allowed to be invalid within certain energy region. We may keep this open. (**Note**, this issue is resolved via the topics < NLVUC > in the Highlights).

In terms of massless propagator factor $1/k^2$, the effective coupling constant η_s^2 is running slightly at $m'^2_\pi << k^2 << M^2$ for

$$e_s^2/(M^2+k^2)=\eta_s^2/k^2,$$

$$\eta_s^2=e_s^2k^2/(k^2+M^2)=e_s^2/(1+M^2k^{-2})=e_s^2[1-1/(1+M^{-2}k^2)]=e_s^2M^{-2}/(k^{-2}+M^{-2})$$

is running **slightly** at $k^2 << M^2$. Moreover, when rewrite pion and/or loop quark propagators in LH-model in terms of massless gluon propagators in massless QCD,

$$1/(m'_\pi+k^2)=(1+m'^2_\pi/k^2)^{-1}/k^2,$$

the factor $(1+m'^2_\pi/k^2)^{-1}$ is also running but **slightly** at $k^2 >> m'^2_\pi$ and it increases from 0 to ½ and then to the maximum value 1 as k increases. This alone is not the empirical running. Experiment shows decreasing running as k increases. The empirical running is the overall consequence of multiple reasons. However,

$$(1+m'^2_\pi/k^2)^{-1}\approx 0 \text{ at } k^2 << m'^2_\pi.$$

At the distances $r>10^{-13}cm$, this together with the Yukawa factor suppresses all the powerful strong forces. This has been proved by all the experimental facts of nuclear force between nuclei and strong force between nucleons since 1930's and intensively since 1960's. The very large strong interactions at small momenta or equivalently, at distances longer than $10^{-13}cm$ have never been observed in a single event of nucleon-nucleon and nucleus-nucleus interaction phenomena. At distances longer than hadrons' sizes, short range strong force fades automatically. The automatic fading is fully agreeing with the overwhelming evidences.

Remark: FQA fact has been explained via theory of DS.

In view of LH-model, the Feynman diagrams of exchanging virtual pions between virtual double-gluon lines (with opposite spins) and/or virtual u/d quarks in loops determine dynamics and effects via Feynman rules. This is called **LH mechanism** with **double-gluon diagrams** and/or **loop diagrams**. An obvious effect of LH mechanism is that the effective coupling in terms of fictitious massless gluon propagator is running smoothly. No any divergent pole in the entire energy region. The running is small between the energy scalars m'_π and M. For loop diagrams, $m'_\pi\approx m_\pi$. For double-gluon diagrams, $m'_\pi=m_\pi$. At sufficiently high-energy scales, η_s^2 becomes almost constant. The gluon mass M is very heavy but its value is unknown.

On the contrary, if we write massive photon propagator in terms of fictitious massless propagator, there is no running for $(1+m_\gamma^2/k^2)^{-1}=1$ is too accurate at all the energy scales experimenters have been dealing with. Possible virtual $e^{\pm}$ pairs with opposite spins between virtual double-photons lines with opposite spins and/or $e^{\pm}$ loops have never been treated as massless photon-propagators in massless QED. Loops have always been calculated in radiative corrections in massless QED. In view of LH-model, however, the virtual $q\bar{q}$ pairs with opposite spins between virtual double-gluon lines with opposite spins and/or $q\bar{q}$ loops have always been treated as massless gluon-propagators in massless QCD.

Smooth LV somewhere at $m'^{\,2}_{\pi} << k^2 << M^2$ is possibly significant. It also contributes to smooth running according to the PE theorem that will be stated and proved soon.

For a (e.m., weak, or strong) charge q in a central static fields of a rest similar charge in K, $\boldsymbol{I} \cdot \boldsymbol{u} = -d(k_{eff} q\phi)/dt$ where $k_{eff} = 1$ for **e.m.** force.

In terms of any preferred frame, λ is exactly isotropic. We may write $\lambda = \lambda(u) = \lambda(\gamma)$. The relation between PEs and λ is shown in the following theorem.

PE−Theorem: As long as $\int \gamma d\lambda / d\gamma d\gamma$ is **integrable** and

$$\boldsymbol{I} \cdot \boldsymbol{u} = -d(k_{eff} q\phi)dt$$

is at least approximately true, we have an **integration** of the *eqn of motion* and a PE $\underline{U}(r)$.

Proof: If $\boldsymbol{I} \cdot \boldsymbol{u} = -d(k_{eff} q\phi)dt$ is at least approximately true, the integration of (19) is

(19a) $$E + k_{eff} q\phi - mc^2 \int \gamma d\lambda / d\gamma d\gamma = c_0.$$

The explicit expression of the corresponding PE $\underline{U}(r)$ can be obtained if $\int \gamma d\lambda / d\gamma d\gamma$ is integrable and (19a) is **solvable** for

$$E \equiv mc^2 \lambda\gamma = E(k_{eff} q\varphi, c_0).$$

It yields

$$\underline{U}(r) = -E(k_{eff} q\phi, c_0) + \underline{C}.$$

Weaker conditions may lead to nonexistence of mathematical expression of PE $\underline{U}(r)$ though it exists and $\underline{U}(r) + E = constant$. In every possible case, (19a) determines the γ - r relation for each set of initial conditions. ■

The simplest example is $\lambda \equiv 1$ in current basic theory that produces

$$\underline{U}(r) = -k_{eff} q\phi + constant.$$

Examples of PEs produced by **non-constant** factor functions λ will be given soon.

For massive vector fields of any non-gravitational force carriers with rest sources, used to interpret atomic spectrum, Lamb shift, and quark dynamics, ϕ is a Yukawa-type scalar potential that is an inverse potential at all the distances much shorter than the range of force. The e.m. PE with $k_{eff}=1$ is also an inverse PE except for the distances near and beyond the force range or shorter than e.m. **CDs** (**characteristic distances**). The strong PE is a quasi-inverse PE roughly between the ranges of strong force and small quark cores, where the Yukawa-type scalar potential is almost proportional to r inversely and the LV is quasi-zero. With this feature, DS with MFs has explained the masses of charged and neutral pions. Please see the Highlights and the Appendix.

One can see **three** reasons to cause effective couplings **running**:

1. The **mechanism** of strong force in **LH-model** that modifies massless propagator with large strong coupling constant and heavy gluon mass.
2. **LV** $|\lambda-1|>0$ that produces departures from inverse PEs for all forces.
3. **Yukawa**-type range factors with significant effect only near and beyond the finite ranges of non-gravitational forces.

Model λ (20) is an example for severe LV near and after the characteristic γ_0 where $10^5 \le \gamma_0 << N$. LV is said to be **severe** if the deviation $|\lambda-1|>10$. In VIII we will see that DM needs severe LV to explain.

A simple example of non-constant λ that produces a PE through the PE-theorem is the limiting departure model λ (14). It is quasi-inverse and is numerically almost proportional to γ inversely for all $\gamma >> N$. This yields almost a constant energy $E = mc^2\lambda\gamma \sim Nmc^2$ for all $\gamma >> N$. Then an almost constant PE $\underline{U}(r)$ is expected at short enough distances, say $r << R^c$. It is called the **deepest asymptotic freedom**. In fact, (14) and (19a) yield an **integration** of (19):

(19b) $$Nmc^2 th^{-1}(\lambda\gamma / N) + k_{eff} q\phi = c_0.$$

Its 1st order approximation is

$$mc^2\lambda\gamma + k_{eff} q\phi = E + k_{eff} q\phi = c_0$$

with extremely high precision at all accessible energy scales with $\lambda\gamma << N$. Eqn (19b) is indeed **solvable** for $\lambda\gamma / N$:

$$\lambda\gamma / N = [\Omega(r)-1]/[\Omega(r)+1], \quad \Omega(r) \equiv \exp[2(c_0 - k_{eff} q\phi)/Nmc^2].$$

Since $E = Nmc^2 \cdot \lambda\gamma / N$ and $E + \underline{U}(r) = \underline{C}$ ($\underline{C}$ is a constant determined by choice), we get a **PE**

$$\underline{U}(r) = -Nmc^2[\Omega(r)-1]/[\Omega(r)+1] + \underline{C}.$$

With massive scalar potential $\phi = q' e^{-\mu r} / r$ and the symbol $Y \equiv e^{-\mu r}$ one can rewrite

$$\text{(19c)} \qquad \underline{U}(r) = 2Nmc^2 / [1 + \Omega(r)] + \underline{C} - Nmc^2 = Nmc^2 \frac{1-\Omega(r)}{1+\Omega(r)} + \underline{C} \,,$$

$$\Omega(r) \equiv a_0 \exp(-\underline{R}Y / r) \,, \quad \underline{R} \equiv 2k_{eff} qq' / Nmc^2 \,,$$

$$a_0 \equiv \exp(2c_0 / Nmc^2) = k_0 \exp(\underline{R}Y_{i0} / r_{i0}) \,,$$

$$k_0 \equiv (1 + \lambda_{i0}\gamma_{i0} / N)/(1 - \lambda_{i0}\gamma_{i0} / N) \,.$$

Initial conditions r_{i0}, γ_{i0}, and $\lambda_{i0} \equiv \lambda(\gamma_{i0})$ influence the shape of the graph of the PE. (On the contrary, for a PE such as the Coulomb PE in current theory with $\lambda \equiv 1$, its graph cannot change in shape though it moves parallelly as initial conditions change.)

For *attractive* force, $\underline{R} < 0$. At $r << R^c \equiv |\underline{R}| \equiv 2k_{eff} |qq'| / Nmc^2$, where R^c is called the **characteristic distance (CD)**, the PE becomes almost **constant**. This CD and the corresponding deepest asymptotic freedom is caused by the limiting LV at $\gamma >> N$ shown in (14), and is proportional to coupling strength $k_{eff} |qq'|$. But it is too short to yield observable effects for non-gravitational forces. We will see that model λ (20) is of non-limiting LV and the 1st asymptotic freedom takes place at $R^c{}_1 >> R^c$.

To get the gravitational PE, we use (19b, c) with replacements:

$$k_{eff} q\phi \to GMm / r \quad \text{and} \quad k_{eff} qq' \to GMm \,.$$

It does tell that particles within black holes, if any, are asymptotically free and will not be pulled into any spacetime singularities. This deepest asymptotic freedom holds no matter one applies (14) or (20). Singularity theorems become invalid once the strong force between nucleons and quarks are no longer negligible such as in the physical processes taking place in a neutron star. This is because any theory of geometrized gravity such as general relativity and the new gravity theory with local LV explicitly tell that the trajectories of classical particles severely depart geodesics (as the solutions of the field eqs) for very high density of field and of matter since the strong interactions cannot be neglected in the phenomena. This justifies Einstein's conclusion [7] about the invalidity of his field eqs for those phenomena. The above gravitational PE simply indicates that the limiting LV and the corresponding factor functions cause dynamic effects to prohibit gravity force from pulling particles inside black holes, if any, into spacetime singularities. It is fully consistent with the minimum spacetime sizes of the particles in the sense of uncertainty coexisting with the genuine ultraviolet cutoffs in finite-N theory.

Each $\lambda(\gamma)$ that leads to the existence of a PE $\underline{U}(r)$ via (19a) is called a **PE-**

producer. The constant sum $\underline{U}(r)+E$ is called the TE. Here E includes rest energy and TE>TME. In general, if $\int \gamma d\lambda / d\gamma d\gamma$ is approximately *piecewise* integrable then we may have approximately *piecewise* expression of the PE for an interaction. For instance, any *quasi-inverse* **piece** shown in (14a) together with

$$\boldsymbol{I}\cdot\boldsymbol{u} = -d(k_{eff}q\phi)dt$$

yields an integration of (19):

(19d) $$\lambda_0 mc^2 th^{-1}(\lambda\gamma/\lambda_0)+k_{eff}q\phi = cons\tan t\,,\ \gamma\in(\gamma_{i1},\gamma_{i2}).$$

It can be obtained from (19b) by replacing N with λ_0. Also, it is solvable for $\lambda\gamma/\lambda_0$ and one can get the corresponding **piece** of the PE from (19c) by replacing N with λ_0:

(19e) $$\underline{U}(r)=\lambda_0 mc^2\frac{1-\Omega}{1+\Omega}+\underline{C}\,,\qquad \Omega\equiv a_0\exp(-\underline{R}Y/r)$$

$$\underline{R}\equiv 2k_{eff}qq'/\lambda_0 mc^2\,,\qquad a_0\equiv\frac{1+\lambda_{i0}\gamma_{i0}/\lambda_0}{1-\lambda_{i0}\gamma_{i0}/\lambda_0}\exp(\underline{R}Y_{i0}/r_{i0})\,.$$

This piece holds within (r_{h1} , $r_{h2}\approx R^{c}{}_{1}$). Denote

$$R^{c}{}_{1}\equiv|\underline{R}|\equiv 2k_{eff}\,|qq'|/\lambda_0 mc^2$$

which are such four characteristic distances in the four kinds of forces, called the **1st CDs**, where the PE in each sort of interactions turns to be *quasi-constant* within four intervals ($r_{h1},r_{h2}\approx R^{c}{}_{1}$), where r_{h1} corresponds to γ_{i2} in (19d) and r_{h2} to γ_{i1}. The values of $r_{h1,2}$ vary with the strength of interaction force. Within the interval $(\gamma_{i1},\gamma_{i2})$ where $\lambda(\gamma)$ is quasi-inverse of the form of (14a), the γ - r relation implied in (19d) becomes explicit once we solve

$$E+\underline{U}(r)=\underline{C}\ \text{ for }\ \gamma\in(\gamma_{i1},\gamma_{i2})\,.$$

Indeed, from (19d), (7), and (14a) we get

(19f) $$\gamma=\frac{\sqrt{a}(\Omega-1)}{2\sqrt{\Omega}}\,,\quad \Omega\equiv a_0\exp(-\underline{R}Y/r)\,.$$

Here a_0 and $\underline{R}$ are given in (19e). It requires a positive a. Connections with neighboring pieces are not considered and (19f) is invalid nearby and at end-points of the interval $(\gamma_{i1},\gamma_{i2})$.

The deepest asymptotic freedom takes place at

$$r < R^c \equiv k_{eff} \mid qq' \mid / Nmc^2$$

while the 1st asymptotical freedom occurs at the distances shorter than the 1st CDs

$$r < R^c{}_1 \equiv k_{eff} \mid qq' \mid / \lambda_0 mc^2 .$$

Since $\lambda_0 << N$, $R^c{}_1 >> R^c$. Appropriately chosen λ_0 will greatly enlarge the strong CD so that the quarks could be bound in small cores with sizes somewhere between 10^{-15} and 10^{-20} cm. According to model λ (20),

$$\lambda_0 \sim \text{Ħ}\gamma_1 / \sqrt{2} .$$

Those parameters in (20) can be estimated and are adjustable.

A best strong PE that best matches the reality is unknown because of the 'desert' without direct experimental data. Fortunately, in many cases, quantum energy levels are **not sensitive** to the differences among the working potential models. Let's see the following examples.

A proton is not point-like. The amount of deviation from Coulomb PE at short enough distances is so big, like filling up infinitely deep PW with infinitely high potential peak so that the potential of a non-point-like proton does not approach $-\infty$ as $r \to 0$. We know that such severe deviation does not affect the numerical results of atomic physics and QED since the probability for a bound electron to 'visit' such short distances is too small, and the bound electron is not sensitive to such large amount of deviation taking place there. Another example is quark model with potential models of strong force. Large numbers of papers were published using various potential models that may differ from one another at long and short distances by huge differences. But the spectra of excited states are close to each other as long as those potentials have small differences in the main regions of probability wave functions.

Atomic spectra are not sensitive to a deviation from Coulomb PE at distances shorter than 10^{-13} cm. High-energy e.m. interaction phenomena may not be sensitive to a deviation from Coulomb PE at distances shorter than $10^{-16\sim-18}$ cm. Then a best PE-producer λ is also unknown even through intensively tested e.m. interaction phenomena. The major obstacle is the difficulty to measure the speeds accurately for high-energy electrons/protons with large γ-values. For detectable photons, accurate measurement of the speeds with extremely small errors that would induce small errors in γ-values of the photons (in visible lights through low-frequency e.m. waves) are and will be impossible in 200 years. For any high-speed particle of large γ-value, a very tiny difference in speed would cause big difference in γ-value. We may know the energy $E = m\lambda\gamma c^2$ and even the product $\lambda\gamma$ accurately, such as for gamma photons in Mossbauer effect. Please note, to know $\lambda\gamma$ accurately one must have known the mass accurately. But in any case, we may not know λ and γ separately simply because the only way to know the large γ-value is to know the speed accurately, not through a well-measured energy or $\lambda\gamma$. Before we discuss model (20), we state a theorem that reveals certain relations between the graphs of $\lambda(\gamma)$ and $\underline{U}(r)$.

Graph-Theorem: Suppose a PE $\underline{U}(r)$ exists with

$$E+\underline{U}(r)=constant\ .$$

For each interval (r_{i1}, r_{i2}), if any, where the graph of $\underline{U}(r)\approx -g^2/r$ is *quasi*-**inverse**, there is an interval (γ_{h1}, γ_{h2}) where the graph of $\lambda(\gamma)\approx\lambda_1$ is *quasi*-**horizontal**, and vice versa. For each interval (r_{h1}, r_{h2}), if any, where the graph of $\underline{U}(r)\approx \underline{C}-E_1=constant$ is quasi-horizontal, there is an interval (γ_{i1}, γ_{i2}) where the graph of $\lambda(\gamma)$ goes down quasi-inversely and vice versa. Moreover, if the graph of $\lambda(\gamma)$ goes **up** *quasi*-**vertically** within a narrow interval ($\gamma_0,\gamma_0+\varepsilon$), if any, the graph of $\underline{U}(r)$ drops **down** *quasi*-**vertically** within a narrow interval ($r_0-\delta$, r_0), and vice versa.

Proof: The only reason for the above relations is (19) and a constant sum of $E=m\lambda\gamma c^2$ and $\underline{U}(r)$. ■

Various potentials have been used in phenomenological potential-models for quark-dynamics, such as linear plus inverse plus constant potentials, harmonic-oscillation plus inverse potentials, square well potentials, and so on. The PE theorem tells that for every PE there is a PE producer λ corresponding to it. The absence of gluon spectra and the existence of hundreds of middle stage unstable particles shown by PDG (particle date group) on their website make it impossible to get one that corresponds to the reality because many body systems and MF-binding systems are not solvable. Any one such as (20) will be just a model or an approximation. Now that quantum levels are not sensitive to many differences among working PEs, any efforts to confirm an exact form of λ and the PEs it produces is not only impossible but also not needed to explain empirical facts.

New development in subatomic and particle physics reveals the stunning effects of magneton forces. LV is assumed to be negligible for bound quarks and for the DS leptons bound or quasi-bound in SA composite particles. In order for DM to be explained, extremely small or very small rest masses of CHSPs must receive large $\lambda^5\gamma$-enhancement. The formula $T^{\mu\nu}=\rho^*U^{\mu}U^{\nu}=\lambda^5\rho\gamma u^{\mu}u^{\nu}$ can be seen in VIII. It's equivalent to large λ^4-enhancement of their energy densities. A sample of such λ will be given in (20) with adjustable parameters.

Label

$$I_{(a,b)}\equiv I(a,b,\gamma)\equiv 2^{-(a/\gamma)^n-(\gamma/b)^n}\quad (a<b)$$

Evidently, $I(a,b,\gamma)\approx 0$ for $\gamma<a$ with $(a/\gamma)^n>>1$ as well as for $\gamma>b$ with $(\gamma/b)^n>>1$. The cut is sharp if n is very large. Within (a , b), $I(a,b,\gamma)\approx 1$. At $\gamma=a$ and $\gamma=b$, $I(a,b,\gamma)=1/2$ with very small error. Write

(20) $$\lambda = 1/\sqrt{1-e^{-(\gamma_0/\gamma)^n} + 2\mathcal{H}^{-2}I(\gamma_0,\infty) + N^{-2}\gamma^2 - \varepsilon}\,,$$

$$I(\gamma_0,\infty) \equiv I_{(\gamma_0,\gamma_{20})} + I_{(\gamma_{20},\gamma_1)} + \gamma_1^{-2}\gamma^2 I_{(\gamma_1,\gamma_2)} + I_{(\gamma_2,\infty)}\,.$$

Here, $\gamma_0 < \gamma_{20} < \gamma_1 < \gamma_2 << N$, $\varepsilon \approx 0$. (The exact value of the constant ε is solely determined by the condition $\lambda(\gamma) = 1$ at $\gamma = 1$. Assume γ_1 are close to γ_{20}. With appropriately chosen $h(\gamma)$, (12) becomes (20). Model λ (20) and $d\lambda / d\gamma$ is smooth. The anti-derivatives $\int \gamma d\lambda / d\gamma d\gamma$ exist. At least the integrand is approximately piecewise integrable. The isotropic (20) is for an isotropic frame K but with the frame-subscript K omitted. To get the factor function λ for anisotropic earth-systems, one can simply replace γ in (20) by $\Gamma(1+ \boldsymbol{A}_S\cdot\boldsymbol{\beta})\gamma$.

Let's describe the values and functional forms of smooth λ in (20) piecewise. $\lambda(\gamma) = 1$ at $\gamma = 1$ and $\lambda(\gamma) = 1$ with high precision for $1 < \gamma < 0.99\gamma_0$. $\lambda(\gamma_0) \sim \sqrt{2}$. There is the 1st (large) quasi-**vertical** jump from $\lambda(\gamma_0) \sim \sqrt{2}$ to $\lambda(\gamma_{20}) = \lambda_{max} \sim \mathcal{H} / \sqrt{2} \sim 10^{2\sim4}$. $\lambda(\gamma) \sim \mathcal{H} / \sqrt{2}$ for $\gamma_{20} \le \gamma \le \gamma_1$. ($\gamma_{20}$ and γ_1 can be very close.) There is the 1st quasi-**inverse** piece of $\lambda(\gamma)$:

$$\lambda(\gamma) \sim \frac{\mathcal{H}\gamma_1 / \sqrt{2}}{\sqrt{\varepsilon(\gamma) + \gamma^2}} \quad \text{for } \gamma_1 < \gamma < \gamma_2\text{, where}$$

$$\varepsilon(\gamma) \equiv [1 - e^{-(\gamma_0/\gamma)^n} - \varepsilon]\mathcal{H}^2\gamma_1^2 / 2$$

is very small for all γ with $\gamma^n >> \gamma_0^n$ so that $\lambda(\gamma) \sim \gamma_1\mathcal{H} / \sqrt{2}\gamma$ is quite accurate for $\gamma_1 < \gamma < \gamma_2$. There is the 2nd (small) quasi-vertical jump from $\lambda(\gamma_2) \sim \gamma_1\mathcal{H} / \gamma_2$ to $\lambda(\gamma_2 + \Delta\gamma) \sim \mathcal{H} / \sqrt{2}$. (Please note, $2^{-(1+\Delta\gamma/\gamma_2)^n} \approx 0$ even for very small $\Delta\gamma$ as long as n is very large). The 2nd quasi-inverse piece is

$$\lambda(\gamma) \sim \frac{\mathcal{H}}{\sqrt{2}} / \sqrt{1 + N^{-2}\mathcal{H}^2\gamma^2 / 2} = N / \sqrt{2N^2\mathcal{H}^{-2} + \gamma^2}$$

for $\gamma_2 < \gamma < \infty$. Then, $\lambda(N) \approx 1$ and $\lambda(\gamma) \sim N / \gamma$ numerically for $N\mathcal{H}^{-1} << \gamma < \infty$.

Let $\lambda_{min} = \lambda(\gamma_{10})$ be the local minimum within $1 < \gamma < \gamma_0$. With $n > 10^3$ and $10^5 \le \gamma_0 << N$, one can check that $\gamma_{10} \approx \gamma_0 \approx \gamma_{20}$ though $\gamma_{10} < \gamma_0 < \gamma_{20}$. For $1 \le \gamma < \gamma_{10}$, the graph is almost **horizontal** with $\lambda_{min} = \lambda(\gamma_{10}) \approx 1$. Possible measurable LV for $10 < \gamma < \gamma_0$ is not considered in model (20). Within the narrow interval $\gamma_0 < \gamma < \gamma_{20}$, the graph is rather **vertical** like a 'cliff' of height $\lambda_{max} \sim 10^{2\sim4}$. The larger the n, the steeper the 'cliff'.

The 1st large LV occurs near γ_0, and we call it the **characteristic** γ_0. By assuming $\lambda_{max} \approx H/\sqrt{2} \sim 10^{2\sim4}$ we see $\lambda \sim 10^{1\sim4}$ for $\gamma_{20} < \gamma < N/10 \sim 10^{39+z+B}$. That is, most v-measured photons with such γ-values have large valued λ and receive very large extra λ^5-enhancement in addition to γ-enhancement of their masses. Although λ in (20) is not monotone, $\lambda\gamma$ is still monotone. Large collection of possible forms of λ yield graphs similar to (20).

The spherical shell $\gamma_0 < \gamma < \gamma_{20}$ in velocity space is called the **velocity shell-space (VSS)**. The shell is very thin for any $n >> 10^3$. The smaller the n, the thicker the VSS. Positive derivative $d\lambda/d\gamma$ in the narrow interval (γ_{10}, γ_{20}) makes the graph of $\lambda(\gamma)$ climbing up from $\lambda(\gamma_{10}) \approx 1$ to $\lambda(\gamma_{20}) \sim 10^{2\sim4}$. The larger the derivative, the steeper it goes up. Correspondently, the graph of the PE for any interaction force in a narrow interval (r_{10}, r_{20}) drops down from an inverse PE (as r becomes smaller and larger for attractive and repulsive interaction force respectively). The steeper the graph of $\lambda(\gamma)$ goes up, the steeper the graph of the PE drops down from an inverse PE. The function $\gamma d\lambda/d\gamma$ with λ given by (20) is piecewise integrable if one uses approximate expressions piecewise. Then each piece of integration

$$E_i + k_{eff} q\phi - mc^2 \int \gamma d\lambda_i / d\gamma d\gamma = c_i$$

is found to be solvable for the product $\lambda_i\gamma$ and for $E_i = mc^2\lambda_i\gamma$ with an explicit solution

$$E_i = mc^2\lambda_i\gamma = f_i(k_{eff} q\phi, c_i).$$

The PE then can be written piecewise:

$$\underline{U}_i(r) = -f_i(k_{eff} q\phi, c_i) + \underline{C}_i.$$

There are four major pieces in model (20) and alike: horizontal piece within $1 \le \gamma < \gamma_0$, quasi-vertical piece at γ_0, the 1st quasi-inverse piece within $\gamma_1 < \gamma < \gamma_2$, and the 2nd quasi-inverse piece within $\gamma_2 < \gamma < \infty$. Writing the two quasi-inverse pieces in the form of (14a), we get $\lambda_0 \sim H\gamma_1/\sqrt{2}$ and $\lambda_0 \sim N$ for them respectively and the corresponding two CDs:

$$R^c{}_1 \equiv 2\sqrt{2} k_{eff} \,|qq'|/H\gamma_1 mc^2, \quad R^c \equiv 2k_{eff}\,|qq'|/Nmc^2.$$

They produce two pieces of the PE that can be written from (19e) by setting $\lambda_0 \sim H\gamma_1/\sqrt{2}$ and $\lambda_0 \sim N$ respectively. We omit the details here.

Complicated strong interaction phenomena tell that strong PE is complicated and may not be solvable. The formula $E_n = \hbar^2\pi^2 n^2/8ma^2$ for infinitely deep square PW is very useful in atomic/nuclear physics. It led to $E_2 - E_1 = 28.2eV$

for atomic binding energy with $a = 10^{-8} cm$ and $m = .511 MeV$. For nuclear binding energy, $a = 2\times10^{-13} cm$ and $m = 938.3 MeV$. Then the formula leads to $E_2 - E_1 = 38.4 MeV$. These results are in order-of-magnitude agreement with experiment though the atomic Coulomb PE differs from square PW greatly. Indeed, to the order of magnitude, the solution of quantum wave eqn is not sensitive to the functional form of the PE and the geometric shape of its graph as long as the PEs in various models are numerically close to each other in the main probability region of the wave function of the constituent particle.

A small mass term for e-neutrinos with energy less than $10 MeV$ would require no λ-enhancement and hence $\gamma_0 > 10^7$ if their mass is $1 eV$. If the mass term $\mu \equiv mc\lambda_K / \hbar$ in the new Dirac eqn (27) can be of large values for high-energy neutrinos in weak decays while the parity violation can be interpreted by γ_5 matrix in H_I , the parity violation does not impose any restriction on the value of γ_0 . If $\gamma_0 \sim 10^5$, then from model (20) one can see that all neutrinos with energy between $m_\nu c^2 \cdot \lambda_{\max} 10^5$ and $N m_\nu c^2 / 10$ will receive very large λ^5-enhancement of their effective gravity mass.

Chapter 3 Gravitation

The curved spacetime is still assumed a connected time-oriented 4-dim **LM (Lorentz manifold)** whose complete atlas contains Lorentz transformations as well as non-Lorentz transformations. Its geometric structure distinguishes itself from other 4-dim manifolds and makes it able to obey c-postulate locally. Some mathematicians, such as B. O'Neill [18], are specialists in the theory of manifolds and its applications to relativity theories. One may read their books to figure out why they call it LM. Empirical recording of events is to tell where and when they did and will happen so a theory of the laws of physics can be disproved or proved within experimental errors. The definition of 4-dim manifold built an outstanding bridge between empirical data and theoretical/mathematical structures of physics. Lorentz coordinate systems in Minkowski spacetime are *isometries* and the new coordinate systems connected by new transformations obeying (1) can be called **f-homotheties** (see [18] for definition of **homothety**), which are conformal maps with *constant* conformal factors. '*Constant*' means that they're independent of spacetime coordinates x^{μ} though they do depend on relative velocities btw coordinate systems. Since dS^2 is no longer an invariant, we must use local instantaneous rest frame of a timelike particle to define proper time interval and the 4-momenta of particles in curved spacetime. If *local* inertial observation systems have **local f-homotheties** (local homothety coordinate systems selected from the complete atlas by factor function f) as their mathematical & theoretical models, so does the local instantaneous rest frame of a timelike particle in curved spacetime that is still Einstein spacetime. The models for factor function λ in Minkowski spacetime are guided by experiment, which can be used to determine the choice of local inertial frames in curved Einstein spacetime.

The same hypothesis of the **frame-invariance** of ***mass*** and ***charges*** (FIMC) since Newton and Einstein ($dm^* = dm$, $dq^* = dq$) now gives **proper densities** of mass and charge from (9). For mass density,

$$\rho^* \equiv dm^* / dV^* = dm/(\lambda^{-3}\gamma dV) = \lambda^3\gamma^{-1}\rho\,.$$

The hypothesis means and guarantees the frame-invariance of mass/charge conservations: If the violation of conservation is zero, it is zero in all frames, and if the violation is nonzero, it is nonzero in all frames. Conservation or non-conservation itself is another postulate, independent of the FIMC. **Mass/charge conservation (MCC)** postulate claims that for a continuous flow of particles of a kind (stable, of course), $(\rho u^{\nu})_{;\nu} = 0$ in all local frames in Einstein spacetime and $\partial_{\mu}(\rho u^{\mu}) = 0$ in all global inertial frames in Minkowski spacetime. Whether or not the conservation holds exactly, one can always construct two kinds of 4-vectors of mass/charge currents under the FIMC:

$$j^{\mu} = \rho^* U^{\mu} = \lambda^4 \rho u^{\mu} \text{ and } \underline{j}^{\mu} = \lambda_K^{-4} j^{\mu} = \Lambda^2 \rho u^{\mu}\,.$$

(They are the same in current basic theory with $\lambda_K \equiv 1$.) Due to (6), λ_K is frame-invariant. Hence, both j^μ and $\underline{j}^\mu$ are 4-vectors. The latter is the simplest 4-vector that manifests the frame-invariance of the MCC postulate: $\partial_\mu \underline{j}^\mu = \Lambda^2 \partial_\mu(\rho u^\mu)$ and $\underline{j}^\mu;_\mu = \Lambda^2(\rho u^\mu);_\mu$. Although $\partial_\mu(\rho^* U^\mu)$ and $(\rho^* U^\nu);_\nu$ are frame-invariant in Minkowski and Einstein spacetime respectively, their values (zero or nonzero) are independent of whether conservation postulate is true. For example,

$$\partial_\mu(\rho^* U^\mu) = \partial_\mu(\lambda^4 \rho u^\mu) \neq \lambda^4 \partial_\mu(\rho u^\mu)$$

if λ is not constant and its argument, the velocity field ***u***=***u***(***x***, *t*) associated with the continuous media of moving particles, depends on x^μ. Thereby, λ_K^{-4} is the simplest factor to yield a frame-invariant eqn of mass/charge conservation to express (the frame-invariance of) the MCC postulate explicitly. We call λ_K^{-4} **conservation factor**. It is hidden in current basic theory for it appears there as a factor identically equal to constant 1.

We just have shown that the 4-vector $\underline{j}^\mu$ itself is a conserved 4-current iff the MCC postulate does match the reality. Under the MCC postulate, we still can construct both two types of mass/charge 4-currents in new physics with Lorentz violation, such as j^μ that does not conserve and $\underline{j}^\mu$ that does.

Conserved probability 4-currents of the elementary particles coupled to any massive non-gravitational force carriers are needed to insure renormalizability of massive vector fields without HM. For e.m., weak, and strong 4-currents (see V), we will use λ_K^{-4} and write $\underline{j}^\mu = J^\mu \equiv \lambda_K^{-4} \rho^* U^\mu$. However, we'll soon see that geometrized gravity cannot be materialized and quantized to make any quantum theory of gravitational fields renormalizable. Hypothetic existence of gravitons contradicts the principle of equivalence. Now that quantization/renormalization do not apply to geometrized gravity, a conserved 4-current of mass is not needed and we now write the stress-energy-momentum tensor for perfect fluid as

(21) $$T^{\mu\nu} = c^{-2} P^* g^{\mu\nu} + c^{-2} \rho^* U^* U^\mu U^\nu + j^\mu U^\nu, j^\mu \equiv \rho^* U^\mu .$$

Here $U^\mu \equiv dx^\mu / d\tau = \lambda\gamma dx^\mu / dt \equiv \lambda\gamma u^\mu$. New gravitational field eqn is

(22) $$G^{\mu\nu} = 8\pi G T^{\mu\nu} / c^4 .$$

Here $G^{\mu\nu} = R^{\mu\nu} - R g^{\mu\nu} / 2$ remains Einstein gravitational tensor, where $R^{\mu\nu}$ is the *Ricci curvature* tensor and R is the *scalar curvature*. The **vacuum eqn** $G^{\mu\nu} = 0$ is the same as in GR (general relativity). The simplest vacuum solution remains Minkowski spacetime, which is both flat and Ricci flat. Here we leave it open for future study whether a cosmological term should be included.

Remark: (22) without **CT** (cosmological term) is one of the historical records of mistakes and struggles seen in the 1st edition. Corrections and advancement are shown in the highlights and in chapter 8, including the CT-Theorem. In general, if one finds something in chapter 1 through chapter 7 that contradicts DS and later development in gravitation and cosmology seen in the Highlights, chapter 8, and the Appendix, that would be a mistake the author made before 2007. It's not removed for the author believes that the young generations need to see not only merits and advancement of science but also the mistakes and struggles made by individuals and by scientific community.

In Einstein spacetime, the requirement $T^{\mu\nu};_\nu = 0$ had shaped $G^{\mu\nu}$. (In 1937, 21 years later after general relativity was published, Einstein together with Infeld and Hoffman, guided by the requirement $T^{\mu\nu};_\nu = 0$, deduced the eqn of motion from field eqn itself. In the deduction, mass conservation postulate had played a key role to shape the eqn of motion). For a medium composed of dust particles,

$$T^{\mu\nu} = j^\mu U^\nu = \rho^* U^\mu U^\nu .$$

Here $j^\mu = \rho^* U^\mu = \lambda^4 \rho u^\mu$ is not conserved 4-current under mass conservation postulate. Then the requirement $T^{\mu\nu};_\nu = 0$ leads to the eqn of motion

$$(23) \qquad \rho^*(U^\nu \partial_\nu U^\mu + U^\tau U^\nu \Gamma^\mu{}_{\tau\nu}) = -U^\mu (\rho^* U^\nu);_\nu$$

For stars, planets, and light beams passing by the sun, the RHS is almost zero since

$$(\rho^* U^\nu);_\nu = (\lambda^4 \rho u^\nu);_\nu \approx \lambda^4 (\rho u^\nu);_\nu = 0 .$$

If we set $T^{\mu\nu} = \underline{j}^\mu U^\nu = \lambda_K^{-4} \rho^* U^\mu U^\nu$ using replacements $j^\mu \to \underline{j}^\mu$ & $\rho^* \to \lambda_K^{-4} \rho^*$, then we will have an eqn of motion with vanishing RHS but its LHS is the same as in (23). Indeed

$$(\lambda_K^{-4} \rho^* U^\nu);_\nu = (\lambda_K^{-4} \lambda^4 \rho u^\nu);_\nu = (\Lambda^2 \rho u^\nu);_\nu = \Lambda^2 (\rho u^\nu);_\nu = 0 .$$

Here $(\rho u^\nu);_\nu = 0$ due to MCC postulate in Einstein's spacetime. Then there will be no extra λ^5-enhancement of the masses of CHSPs and DM remains a myth.

The new field eqn contains λ & Λ being almost constant 1 up to extremely high speeds. Then Einstein field eqn, where λ and Λ become identically 1, remains extremely accurate up to extreme high-speed regions of heavy bodies, and its three experimental proofs (bending of light rays, advance of the perihelion for Mercury, gravitational red shift) are valid with high precision in view of the new gravitational theory. However, we will see very large difference regarding the gravitation contributions of astronomically large number of CHSPs (neutrinos and photons) that pervade the gigantic spaces within and wrapping clusters and super clusters.

To interpret dark matter as mainly high-speed neutrinos and photons, the setting $T^{\mu\nu} = j^\mu U^\nu = \rho^* U^\mu U^\nu$ is crucial since $\rho^* U^\mu U^\nu = \lambda^5 \rho \gamma u^\mu u^\nu$ contains λ^5, a large-valued factor for fast-photons/fast-neutrinos with $\gamma_{20} \le \gamma < 0.1N$, according to model λ (20) and a huge collection of similar model factor functions. This roughly covers photon-frequencies $10^2 Hz < \nu < 10^{27} Hz$ and neutrino-energies between $m_\nu c^2 \cdot \lambda_{\max} 10^5$ and $N m_\nu c^2 / 10$. In fact, (20) gives $\lambda_{\max} = \lambda(\gamma_{20}) \approx H / \sqrt{2}$, estimated $10^{2\sim 4}$ and $10 < \lambda < \lambda_{\max}$ for those photons and neutrinos that receive large λ^5-enhancement in addition to the γ-enhancement. It can be adjusted for a better explanation of dark matter. This enhancement of effective gravity masses by large values of λ^5 is called the λ^5 **-enhancement** of fast moving masses. Energetic photons and neutrinos with both γ- and λ^5-enhancement very large are effectively the 'super-heavy' particles postulated in some models to explain dark matter phenomena. However, 'super-heavy' particle models with unchanged fundamental laws exclude the contribution of neutrinos and photons and hence require astronomically large number of such super-heavy stable particles, which is disproved by experimental records cumulated in 70 years: no single one has been found in all the photos (hundreds of millions to billions).

The λ-enhancement of photon mass terms yields frequency-dependent e.m. YF. However, exponent of λ there is 1 and the e.m. YF values remain almost constant 1 up to very long distances even for the largest λ-enhancement of photon mass terms.

Neutrinos and photons in universe are 'dust particles' with $P^* = 0$ and $T^{\mu\nu} = \rho^* U^\mu U^\nu = \lambda^5 \rho \gamma u^\mu u^\nu$. Most galaxies are not spherically symmetric. Space distributions of neutrinos and photons are more uniform than other matter (see VIII). Besides, they have irregular moving directions. We may estimate

$$T^{44} = -\lambda^5 \rho \gamma c^2 .$$

For low speed particles as well as for all the stars and planets in the universe, $\lambda^5 \gamma \approx 1$ and $T^{44} = -\rho c^2$. This is the origin of a popular concept saying that for any particle or body, 'the mass determines the gravitational field it produces'. Model (20) and many possible similar models make a wide velocity space $(\gamma_0, N/100)$ within which the high-speed photons and neutrinos receive large λ^5-enhancement in addition to the large γ-enhancement of their masses.

Photons are visible iff they are of the frequencies of visible lights and reach our eyes or apparatus. They come directly from the original sources or from non-original sources after being reflected. Detectable photons with frequencies outside visible lights can be detected directly in the labs iff they reach apparatus on earth. Most photons (much more than 99.999999%) in the universe never reached and will never reach our earth. To us, they are obviously dark, invisible.

The vertical cut in a Kurie plot determines $E_{\min} = m_\nu c^2 (\lambda\gamma)_{\min}$ (the energy of the **softest/slowest timelike neutrinos** of a kind with m_ν being their rest mass). It has nothing to do with the TE of neutrinos. Neutrinos and photons in a space that wraps a cluster are significant elements of dark matter if their TE is large enough. Astronomers observe most powerful explosions, such as γ-explosions, in the universe almost every day. One such explosion may release a TE larger than the TE emitted by the sun in 10 billion years. All such explosions might have produced astronomically large number of super energetic photons and neutrinos. Large number of less and much less violent explosions within the centers of galaxies may not be visible for the background is too 'bright' or 'noisy'. As long as the energy radiation constants per unit mass of the centers of galaxies are much larger than solar constant per unit mass, all the neutrinos and photons emitted and cumulated in hundreds of thousands of years have large TE. The accumulation agrees with the 2nd law of thermodynamics. The gravitational fields produced by these invisible particles are 'detectable' only collectively via their astronomically large number with the γ- and λ^5-enhancement.

An approximately uniform distribution of matter in the shape of a sphere, ellipsoid, or disk produces gravity force approximately proportional to the cube or square of the distance from the center. This causes the observed fast orbiting of the stars in the *peripheries* of galaxies about the centers of galaxies, fast orbiting of galaxies about the centers of clusters, and fast orbiting of clusters about the centers of super-clusters.

Accurate quantitative calculation is impossible. We do not know an accurate value of the TE and the space and energy/frequency distribution pattern of neutrinos and photons in a space that contains and wraps a galaxy, cluster, or super cluster. The absence of gluon spectrum and unsolvable SA MF-binding systems and MB systems that make any attempt to find out theoretical relation btw empirical masses of SA composite particles and λ-depending PEs doom to fail also bring uncertainty to details of the functional form of an idea factor function λ. Thereby, it is very difficult to know the value of H in (20) and exact form of λ in general even if the orbiting motions of the stars in a galaxy can be measured as accurately as the motions of the planets in our solar system. Since $\lambda_{max} \sim H/\sqrt{2} \sim 10^{2\sim4}$ in (20) is adjustable, the difficulty to explain dark matter phenomenon is gone in principle.

Extra **λ^5-enhancement** of **effective gravity masses** for most fast-photons and energetic neutrinos is needed because ***relic*** neutrinos/photons have uniform distribution within galaxies/clusters/super-clusters. They cannot produce central forces to cause fast orbiting about any center. Most non-relic neutrinos and photons emitted since the formation of embryonic galaxies have fled into the vast spaces between galaxies, clusters, and super clusters. Only those cumulated in t years for a galaxy of t light-year diameter may produce central force to cause fast orbiting motion of the stars in the periphery of that galaxy, and so on.

The λ^5-enhancement of effective gravity masses for most fast-photons and energetic neutrinos is another consequence of such λ as (20) and large number of similar models. The range of e.m. force depends on photon mass only and is irrelevant to the λ^5-enhancements.

After the revolution of QM and the numerical success of QED, Einstein's approach to unify gravitational and non-gravitational forces was abandoned by most physicists. They have gone off in another direction to try to make the dream come true: to quantize gravity. A wave eqn was deduced and seemed to tell the existence of spin-2 massless particles, called '***gravitons***'. However, some small terms of the weak field were thrown away in the deduction. They were not infinitesimals. The existence and the non-existence of gravitons are of radical, not just numerical difference. Furthermore, if one puts an electric charge, a film and a meter, used to record radiation, into an Einstein Elevator, the same reasoning Einstein used in his celebrated thought experiment to predict bending of light rays tells that an electric charge does not radiate when it is freely falling and it does when at rest in an external gravitational field. (We omit the detail of the reasoning in this thought experiment leaving as an exercise for readers.) Coordinate transformations are unable to change the reading of a meter nor the chemicals of any film. Herrera at Brookhaven National Laboratory published an article in Phys. Rev. D in 1980's, where a Lorentz invariant formula was obtained by generalizing the Larmor formula and then a generally covariant form was written through Weinberg Rule. It drew the same conclusion. Putting gravity quanta into Feynman diagrams cannot explain this: Why exchanging virtual photons may cause emitting real photons while exchanging virtual gravitons cannot cause emission of real photons. Moreover, the principle of equivalence is unable to make existence of gravitons and nonexistence of them 'equivalent'.

Glashow once gave a very nice explanation of why renormalization is needed in QED. He referred to the existence of clouds of virtual photons. His argument would be perfect if the quantities canceled by counter terms in renormalization are actually finite, no matter how large. The genuine physical picture he described implies that a quantum theory of fields is acceptable only if it is renormalizable. More than likely, the gravity that has been **geometrized** cannot be quantized, because such quantized theory is not only ***unrenormalizable*** but also leading to ***contradiction*** at the very beginning, before any sophisticated mathematical formulation was done. Therefore, the departure from Minkowski metric does not correspond to quantum field of force-carriers though it causes dynamic effects in both Minkowski & Einstein spacetime. Since the gravitational acceleration on earth is too small, one needs extremely fast charges to verify experimentally whether a freely falling charge radiates. The technical difficulty would be how to avoid the interference from other charged particles. To put huge amount of electrical charges on a metal sphere of huge size would not help to check whether charges at rest in gravitational fields radiate since it is too difficult to cool a huge metal sphere to almost $0^{0}K$ and to get rid of the interference from the thermal motions of the charged particles within the sphere. Usually, no real experiment would disprove a conclusion of a thought experiment. Extremely violent explosions in the universe have been observed, but no gravitational waves and gravitons have ever been found. This is the experimental evidence of the nonexistence of gravity quanta and any radiation caused by exchanging such quanta. The **crucial unlikeness** of geometrized gravity force and materializable, quantizable, and renormalizable non-gravitational forces may imply far more things than we see right now. See more proofs of the nonexistence the so-called gravitational waves (pages 160,161). Weak force doesn't obey equivalence principle. It's not geometrized but materializable/quantizable/renormalizable. Electrons accelerated by weak force do radiate! Experiment plainly proves radiation of photons in β decay.

Remark: In the Highlights we see that DS undoes W-particles and beta-decay is mainly due to ENM with e.m. interaction forces btw two identical d-electrons. As historical record of mistakes, the author does not remove the above mistake.

Charged particles falling into a neutron star or a black hole (if any) are of course expected to radiate before they absorbed by the neutron star or disappear beyond event horizon. The radiation is not caused by large acceleration of gravity but by furious e.m. interactions between high-speed electric charges.

Acceleration of an electric charge must cause radiation only in the absence of gravity. The true reason of real-photon emission is exchanging non-gravitational force-carrier, no matter there is gravity to enlarge or reduce (even cancel) the acceleration. In view of Maxwell theory, acceleration must cause radiation and an electric charge without acceleration must not radiate for Maxwell eqs are valid only in the absence of gravity. All detected radiations such as emitted from antennas are caused by non-gravitational interactions, without a single exception.

As GS coupling terms brought in **nonlinearity** similar to what is seen in geometrized gravity, they brought in '**unrenormalizable**' **infinities** including ***collinear divergences*** and the gauge theory was in severe conflict with trivial experimental facts at the very beginning. HM retains gauge invariance that does not exist as a physical principle according to DE fact, IR fact, and a mass term in L_{LY}. It allowed gauge transformations to change the reality of unchangeable finite force-range by changing mass term in any fixed inertial frame, and brought as many Higgs coupling parameters as leptons and quarks that were endowed with masses and postulated as structureless within and beyond any energy scales. The nonexistence of HPs and gluon balls features extra experimental evidences to support non-gauge massive vector theory with LV.

If geometrized gravity cannot be quantized, it does not mean there are no discrete energy levels of quantum particles in a gravitational field. We do not have a super heavy particle to bind an electron via gravity force to verify this experimentally. But from quantum mechanics, one can see that the reality of discrete quantum levels of energy does not depends on whether the potential fields can be materialized and quantized.

We do not need to know what kind of singularity theorem might be deduced from (22) with (21). Both general relativity and the new gravity theory tell that geodesics can be used to describe the motion of classical particles iff non-gravitational forces are negligible. They are not negligible between nucleons within the range of strong force and between quarks after nucleons are smashed in further gravitational collapse, if any. Then any theory of geometrized gravity explicitly indicates severe deviation of the motions of classical particles from geodesics and therefore *singularity theorems* **do not apply** to the **physical word** whenever the existence of strong force is no longer negligible. Moreover, in view of quantum theory, geodesic-description of the motion of particles is valid only if the effects of quantum motion are negligible. Quantum eqs in curved spacetime with a well-established potential model based on a radical theory of strong force and on a radical form of λ should be considered to investigate gravitational collapse and black holes. Singularity theorems are very beautiful and provide elegant description of physical process of gravitational collapse until strong interactions make trajectories departing from geodesics greatly.

When all nuclear energies are exhausted, plasma and the core of a star cool down, and annihilation (not $e^{\pm}$ annihilation) of electrons into the core becomes irreversible, gravitational collapse occurs and a neutron star is born. During this physical process, nucleons stop following geodesics as they are close to one another simply because of the existence of strong force. For the limited applicability of geodesics, the existence of ground states in quantum theory is a reason that is even more radical.

The **ground states** (the lowest among discrete energy levels) and the corresponding **minimum average distances** between interacting particles in bound systems are the basic feature of quantum phenomena, which are the necessary consequence of square-integrable wave functions with probability interpretation. Usually, the larger the coupling constant, the shorter the minimum average distance. Electrons can be knocked out from atoms and ions or captured by nucleons by chance and destroyed (annihilated), but there is no way to force them stay closer to the nucleons in stable states. When atoms and ions are smashed as bound electrons are annihilated in gravitational collapse, Bohr-type radii of atoms and ions are crossed over. Then the grand states and the minimum average distances (much shorter than Bohr radius) between nucleons as well as between quarks prevent further collapse. At least, further gravitational collapse of particles that only follow geodesics could never happen in the universe after neutron stars collapsed and black holes were born. Schwarzschild radius is proportional to mass M of a star but the radius of a neutron star is proportional to $M^{1/3}$. This is the true reason a black hole may be born if the mass M of a star just before collapsing into a neutron star exceeds certain amount.

From viewpoint of energy conservations (as the integrations of the eqs of motion), the TME as the sum of kinetic and potential energies of the particles falling into a neutron star and then into a black hole are conserved. The furious radiation of particles such as observed around a supernova takes away large amount of kinetic energy so that the particles falling into a black hole would have a negative TME and evaporation of black hole would not occur. (The evaporation of a black hole occurs iff a larger black hole exists nearby.) Quantum tunnel is a term used to interpret the probability motion of a quantum particle. For example, a quantum particle with negative TME and bound in a PW with finite depth can appear (with certain probability) outside the PW while a bound classical particle within the PW can never appear outside without having received enough energy. However, a quantum tunnel for a quantum particle with negative TME and bound within a PW to escape from the PW and become a free particle with non-negative TME does not exist in nature. To escape and become free, enough amount of energy must be received by the bound particle. Quark evaporation has never been observed in tens of millions of photos. Correct understanding of quantum tunnel is fully agreeing with overwhelming experimental evidences. Cold electron emission and α-decay are two typical examples of quantum tunnels. In the former, a strong electric field must be directed towards the surface of the metal so that at a short distance from the surface the PE is less than the energy of the electron inside the metal. In the latter, only those α particles that have positive energy can escape from the nucleus (Quantum Mechanics, I. M. Rae, 1986). Quantum tunnels obey energy conservation. Note: Here ‘energy’ should be ‘ME’.

It is sufficient for further collapse to occur if quarks can be annihilated to produce unknown particles with new type of **stronger** force so that the minimum average distance between newborn particles can be smaller to lead to further collapse. High-energy head-on collisions of protons never produced such unknown particles and stronger force. Therefore, we are not sure whether there may be further collapse.

The behaviors of geodesics and singular or nonsingular solutions of field eqs do not prove nor disprove the existence of spacetime singularities (SS) in nature. The existence of SS can be excluded in a theory iff there are minimum spacetime sizes of quantum particles in the sense of uncertainty, which can only be obtained in a theory without ultraviolet divergences and without massless particles.

The minimum spacetime sizes of particles in the sense of uncertainty will be deduced in IV. In V, we will see a dynamic reason for the impossibility of collapsing into spacetime singularity: The universal space force

$$S_{(\mu)} \equiv m\gamma u^{\mu} d\lambda / dt$$

balances any interaction force $I_{(\mu)}$ at distances $r < R^{c} \equiv 2g^{2} / Nmc^{2}$. This leads to deepest asymptotic freedom and prevents gravity force from pulling all particles into any spacetime singularity.

Einstein [7] left us with his lifetime conclusions: "*singularities must be excluded* " and "*One may not therefore assume the validity of the eqs for very high density of field and of matter*". We cannot blame Einstein for not telling us what theory might be valid. Everybody has limited time and energy. He did too many things for us.

'***God does not play dice***' may not be a lifetime conclusion of Einstein. According to a story by a Russian physicist/historian who wrote a book about the relativity principles in classical and quantum physics, Einstein said, one year before he passed away, that he thought he might look like an ostrich burying his head in the sand of relativity not to see the disgust quanta. What is the sand of relativity? Einstein did not see it. His spacetime theory was so good for $\gamma < 10^{1\sim2}$ phenomena that even numerical success of massless QED was obtained. No single eqn in quantum theory could be deduced if quantum theory did not marry any spacetime theory (Newton's or Einstein's) and an *E-**p*** relation or energy formula in classical particle mechanics was not used. Obviously, spacetime theory and classical theory of heavy bodied are basic portions of quantum theory. Separation is impossible. No quantum theory can be 'pure' for it must return to and recover the classical theory for heavy particles and bodies, as quantum particles are replaced by heavy particles and bodies in the real world. Quantum theory does not fail in classical limit. The requirement of a single theory also requires that classical theory be a special case of quantum theory. Lab systems and recording devices are heavy. In view of quantum theory, we must treat them as classical and regard transformations and classical theory as a part of quantum theory. In this sense, relativity principles are not the sand that hurts the probability interpretation of the foundation of quantum theory. The only sand visible at Einstein's time should be the determinacy assumption of motion: Motion should be described in terms of trajectories, not probability wave functions. With this sand, it was thought that hidden parameter(s) should put the foundation of quantum theory on a position of temporary form of the true mathematical principle of natural philosophy. In this book, we have shown that to explain the experimental facts the current basic theory has difficulty to explain we need to change fundamental laws. In this sense, relativity does contain sand (besides the determinacy assumption of motion) that hurts. With changed sets of elementary/composite particles and fundamental laws of physics, one will be able to see whether there is a single experimental fact so far quantum or classical theory based on changed fundamental laws and DS could not explain. Please read the Highlights, Chapter 8, and the Appendix to see more details.

No single step can be made without mistake in any attempt to make a major change without being lightened by all the works done by physicists in the past century, regardless of their validity or invalidity, accuracy or inaccuracy, correctness or incorrectness. To probe the reasons of the observed phenomena has always been a breath-taking activity of mankind, right or wrong.

< Non-Uniform Structures and Cosmological Principle >

From hadrons to galaxies, we see that everything has ***structure*** and every observed uniform distribution of matter is 'local': water in the sea, air in the atmosphere, stars in a uniform portion of an arm of galaxy.... The observed part of the universe and beyond cannot be exceptions. This is author's opinion in the 1st edition of this book published online in 2005. Many thought that the largest size of structure could be about 1.2 billion *ly*. Then cosmological principle could still be true. On 8/6/2015, "Epochtimes" reports: an astronomer just discovered a group of gamma-stars forming a circle with size much larger than 1.2 billion *ly*.

Stars and space btw them contain charged particles radiating photons such as radio waves, microwaves, infrared photons, visible lights, ultraviolet photons, ... Moreover, planets, celestial bodies, dusts, and molecules are '***converters***'. They also radiate microwave photons (see pages 229, 230). Planck institution made a nice photo about 2 years ago, showing the brightest area of CBR matching the same distribution pattern of the stars, dusts, and hydrogen gases in our galaxy. It proves once for all an earlier conclusion the author made more than 4 years ago that many CBR photons are not relic, produced by '***convertors***' in star era (see pages 229, 230).

Extra evidences of structure at largest scales can be seen in the ***CBR holes*** discovered some years ago. We now show that *CBR holes are not holes lacking CBR photons but holes lacking such CBR photons that are moving towards our earth due to lacking of baryonic 'convertors' radiating CBR photons*:

In the universe, relic and non-relic CBR photons move in all directions, just like other CHSPs. Only extremely small portion of CBR photons could and will reach the earth/be detected and called 'CBR' seen on our earth. CBR holes are, 100% sure, not the holes lacking CBR photons. They just contain little of stars and 'convertors' as CBR source matter radiating CBR photons. CBR holes, 100% sure, contain many CBR photons moving in all the directions not towards our earth. In each CBR hole, the lack of the CBR photons moving towards us is due to the lack of stars and 'convertors' as the CBR source matter. In any corner of the universe, as long as it contained ***enough*** matter that radiated CBR photons, ***enough*** CBR photons could move towards us and many observers on earth could detect them coming from that corner. CBR holes are the extra evidences of nonuniform distribution of stars, galaxies, and other baryonic matter such as dusts and molecules at largest scales.

The EF-cosmology and the cosmology in new physics with LV are based on the same assumption, called 'cosmological principle': The major source matter of GF at large scales in the universe is homogeneously distributed. CHSPs are the major source matter and DM at large scales in new physics with LV. (See chapter 8 for details.) CBR photons are part of CHSPs. CHSPs, including both relic/non-relic CBR photons, move faster than the expanding metric space, support the cosmological principle for their fastest speeds and motions in all directions make them enter and fill every hole in the Universe during a very long period of time since the beginning of star era while stars, galaxies, dusts, and nearly all baryonic DM, comoving with the expanding metric space, are unable to enter and fill all the holes. ***Holes grow with the expansion*** of metric space. ***Initial condition*** determined the ***earliest nonuniform distribution*** of matter and the ***tiniest holes*** at the non-singular beginning. ***Only CHSPs***, moving faster than the expansion of metric space, all galaxies, and all other non-CHSP matter, ***could catch/overtake receding galaxies and holes btw them and distribute uniformly*** in the Universe.

Now we see that approximately uniform distribution of galaxies and CBR source matter without structure wouldn't be a reality while cosmological principle holds due to CHSPs and cannot be violated by nonuniform distribution of galaxies nor by the newly discovered reality of the existence of '***CBR holes***'.

If geometrized gravity cannot be materialized and quantized, its range may not be finite and then the gravity can provide dynamics to maintain non-uniform structures in spaces of **any large sizes**. The law of gravity together with ***initial*** and ***boundary*** conditions determine structures in small spaces such as our solar system as well as in spaces of **any large sizes**.

< Gravitational Wave and Graviton >

Two well-known theoretical features of field eqs are:

1. Field eqs of ***strong*** GF depart ***wave eqs*** severely.

Field eqs in GR, EF-cosmology, and EFD-type gravitation and cosmology in new physics with LV are NOT wave eqs. No one could ever show that strong GF could satisfy wave eqs as approximations so that ***strong*** gravitational waves could be produced in some places in the universe and we could detect them on the earth.

2. Under and ONLY under ***weak*** GF condition, field eqs lead to ***wave eqs*** as approximation after some quantities of weak GF are thrown away. The condition is that GF must be so weak that the departure from Minkowski metric is very small: $g_{\mu\nu} = g^{(0)}{}_{\mu\nu} + h_{\mu\nu}$, where $g^{(0)}{}_{\mu\nu} = (1,1,1,-1)$ ($g^{(0)}{}_{\mu\nu} = 0$ if $\mu \neq \nu$) and $h_{\mu\nu}$ are all small quantities.

The above '*weak GF condition*' leading to wave eqs is not weak at all in real life situation. The gravity forces on our earth can be so ***strong*** that we all have experiences to feel/see their consequences without doing any lab researches. But GF on earth is too ***weak*** in the sense of tiny departure from Minkowski metric. In fact, the solution of field eqs for GF on earth is the *Schwarzschild* metric:

$$dS^2 = (1-2GM/c^2r)^{-1}dr^2 + r^2(d\theta^2 + \sin^2\theta d\varphi^2) - (1-2GM/c^2r)c^2dt^2$$

We can calculate the departure from Minkowski metric and see that $|h_{\mu\nu}|<10^{-8}$. So GF on earth is supposed to obey wave eqs approximately and everything on earth should approximately vibrate all the times. Is this a joke? What's wrong?

A legendary deduction from field eqs to wave eqs under weak GF condition is shown on page 326 in the literature [20]. The starting point's seen on page 298 in [20]: An approximate formula for Christoffel symbol (11-59) replaces the rigorous formula (11-23) on page 288 in [20], with $g^{(0)\beta\mu}$ replacing $g^{\beta\mu}$ (= $g^{(0)\beta\mu}$

$+h^{\beta\mu}$). The reason is interpreted as "***$h_{\mu\nu}$ are small quantities whose products are negligible***". All the quantities of the form $h\partial h$ are treated as higher order small quantities and are removed/abandoned. Suppose this is OK, we go along with it. But then the Riemann-Christoffel tensor (11-60) is simplified to an approximate formula (11-61) as one "***neglecting nonlinear terms in the*** $h_{\mu\nu}$". The removed terms are of the form $\partial h\partial h$. The author just (8/18/15) found an example showing how a nonlinear term can make the genuine solution departing wave solution:

Consider a 1-dim weak vibration $x=\varepsilon sin\omega t$ ($0<\varepsilon<<1$) which is a special solution of the 1-dim oscillation/vibration eqn $d^2x/dt^2 + \omega^2 x = 0$. Here $|x|$ can be as small as one wants. Once a nonlinear term $(dx/dt)^2$ enters, we have

$$d^2x/dt^2 + (dx/dt)^2 + \omega^2 x = 0$$

Its solution is

$$t = \pm \int e^x[\omega^2 e^{2x}(0.5-x) + c_1]^{-1/2}\, dx + c_2$$

The weak vibrations all disappear completely without trace. This simple example shows that a nonlinear term $(dx/dt)^2$ causes radical departure.

Small quantities x & ε are algebraic quantities. Higher order small quantities x^n ($n\geq 2$) are algebraic quantities too. But $(dx/dt)^2$ is not a regular algebraic term. Let's see this example: $|x|=\varepsilon|sin\omega t|$ can be as small as one wants at all times, but $|dx/dt|=\varepsilon\omega|cos\omega t|$ can be as large as one wants at many or most times as long as ω is large enough to make $\varepsilon\omega$ as large as one wants....

For a 1-dim non-wave eqn $\partial^2\varphi/\partial x^2 + (\partial\varphi/\partial x)^2 - c^{-2}\partial^2\varphi/\partial t^2 = 0$ that differs from the well-known 1-dim wave eqn in a nonlinear term $(\partial\varphi/\partial x)^2$, its numerical solutions might further reveal how such nonlinear term can make the genuine solution radically departing the wave solution of the 1-dim wave eqn for $\varphi(x, t)$, that is $\partial^2\varphi/\partial x^2 - c^{-2}\partial^2\varphi/\partial t^2 = 0$.

The author does not know if gravitational wave, originally, was Einstein's idea. If it was, he never proved using field eqs that when GF gets stronger and stronger the gravitational wave would get stronger and stronger too. He might have missed the crucial difference btw e.m. field and GF. When written in terms of 4-potentials, classical massless/massive Maxwell eqs without source lead to exactly the same kind of eqs as quantum wave eqs for massless/massive free spin 1 particles, no matter the e.m. fields are weak, strong, or super strong. Because of such stunning feature, quantization of e.m. fields and existence of radio waves, microwaves, and light waves become so natural and reasonable.

It's a baseless/unreasonable imagination to think that super-strong GF at and after the time of the super explosion of the Big Bang had produced gravitational waves so strong that their relic more than ten billion years later could be detected by some physicists on earth after they all found no any gravitational waves in 70 years of intensive searches while many big explosions in the universe have been detected. For example, γ-explosions are detected by astronomers nearly every day. One such explosion may release a TE larger than the TE emitted by the sun in 10 billion years. No gravitational waves nor gravitons have been found by anyone.

The author would be very happy to point out that a thought experiment has revealed the contradiction btw the equivalence principle and the existence of graviton. It is left as an exercise for readers.

Chapter 4 Quantum Eqs for Free Particles

Feedback from quantum phenomena and theory (**Q-feedback**) will be useful to establish **classical theory** of interactions. The thought experiment J.D. Jackson [19] used to show why the **wave-phases** should be **frame-invariant** is perfect for both e.m. and probability waves. The new de Broglie wave of a free particle is

$$\psi = | p > \equiv A\, exp(i\, \hbar^{-1} \Lambda^{-1} p^{\mu} x^{\mu}).$$

With $\underline{\hbar} \equiv \Lambda \hbar$ the new **uncertainty relations** are

(24) $$| \Delta x || \Delta p | \geq \underline{\hbar} / 2, | \Delta t || \Delta E | \geq \underline{\hbar} / 2 \ .$$

Here, $\underline{\hbar} \equiv \Lambda \hbar$ and $\hbar$ are practically the same in all S up to very high absolute speed V_{SK}. For such systems, $N_S = N$ is very accurate. The **minimum length** and the **minimum time interval** of a timelike real particle in the sense of uncertainty are

$$l_{min} \approx \hbar / 2Nmc \text{ and } t_{min} \approx \hbar / 2Nmc \ .$$

Since, $N \sim 10^{40+z+B}$, for *electrons* we get

$$l_{min} \approx 1.93 \times 10^{-51-z-B} cm \text{ and } t_{min} \approx 6.44 \times 10^{-62-z-B} sec \ .$$

Without knowing the exact values of N and of those minimum spacetime sizes, finite-N theory has finally excluded the existence of spacetime singularities in nature. Now Einstein can rest.

The new uncertainty relations in (24) are deduced from the same kind of rigorous deduction as in regular quantum theory with zero LV and with Einstein and de Broglie relations, where $\Lambda \equiv 1$. In view of new physics with LV, the regular uncertainty relations are just special case of (24). Now we deduce new quantum wave eqs containing Λ and λ. Label

$$\lozenge \equiv \partial^{\mu} \partial_{\mu} - \mu^2 \text{ and } \mu^2 \equiv m^2 c^2 \lambda_K{}^2 / \hbar^2 ,$$

where $\lambda_K{}^2 = \Lambda^{-1} \lambda^2$ due to (6). Obviously, μ^2 is frame-invariant. The new **Klein-Gordon eqn** for spin 0 free particles is

(25) $$\lozenge \phi = 0 \ .$$

Here ϕ is a scalar. The basic operators for free particles are

$$p^{\mu} \rightarrow -i\, \hbar \partial^{\mu} ,\ \hat{\boldsymbol{u}}^{n} | p > = c^{2n} E^{-N} \boldsymbol{p}^{n} | p >, n = 1, 2, 3 \ldots .$$

Define operator $\hat{\lambda} \equiv f_S(\,\hat{\boldsymbol{u}}\,)^{-1/2}$. We see that $\hat{\lambda}\psi = \lambda\psi$ holds for free particles. The new quantum **Proca eqn** for free spin 1 particles is

(26) $$\Diamond V_\tau = 0 \,.$$

Inhomogeneous quantum Proca eqn is

(26a) $$\Diamond V_\tau = -\frac{4\pi}{c} J_\tau \,.$$

The new Dirac eqn for spin ½ free particles is

(27) $$(\gamma^\mu \partial_\mu + \mu)\psi = 0 \,, \quad (\gamma^\mu = \sqrt{\Lambda}\gamma_\mu) \,.$$

Lorentz invariant Dirac eqn (LIDE) and (27) lead to the same **continuity eqn**. One can check that (27) is covariant under the new transformations if ψ obeys the same transformation rule as in LIDE. In fact, the coordinate transformation factor f_S or $f_{S'}$ cancels when we transform the eqn between S and S'. This implies independence of the **spinor space** from spacetime **coordinate space**. In general,

$$\mu^2 \equiv m^2 c^2 \hat{\lambda}_K{}^2 / \hbar^2 \equiv m^2 c^2 \Lambda^{-1} \hat{\lambda}^2 / \hbar^2$$

in (25) and (26) and μ in (27) are **operators** so a plane wave expansion also obeys (25), (26), or (27). All real and virtual particles are '**on-shell**'. It means on positive mass shell with positive μ because $m^2\lambda^2 > 0$ and that their 4-momenta are timelike with

$$p_\mu p^\mu = -m^2 c^2 \Lambda^{-1} \lambda^2 < 0 \,.$$

As we pointed out in I, the 3-dim expressions of propagators drawn from the HO (harmonic-oscillator) model of all quantized fields show that every virtual timelike particle is on positive mass-shell. The 4-momentum used to label an internal line of Feynman diagram is indeed 'off'-mass shell but it is not carried by any individual virtual force carrier. The e-ν pair produced in a weak decay has a TE much less than the rest energy of a W boson. Overwhelming experimental facts of weak decays indicate that the 4-momenta transferred are the effects of exchanging virtual force carriers, not the ones carried by those force carriers.

Photons are almost on 0-mass shell for $m_\gamma \lambda_\gamma$ is too small. The sun neutrinos are produced inside the sun and the conditions there are not clear. The author does not know if the temperature in sun's core is lower than expected. The neutrino production rate is sensitive to temperature, not to its mass.

With $V_{\mu\nu} \equiv \partial_\mu V_\nu - \partial_\nu V_\mu$, the Lagrangians of (26) and (26a) are

(28)
$$L_{LY0} = -V_{\mu\nu}V^{\mu\nu}/4 - \frac{1}{2}c^2 m_v^{\ 2}\lambda_K^{\ 2}\hbar^{-2}\ V^\mu V_\mu - \frac{1}{2}(\partial_\mu V^\mu)^2,$$

$$L_{LY0} = -V_{\mu\nu}V^{\mu\nu}/4 - \frac{1}{2}c^2 m_v^{\ 2}\lambda_K$$

$$L_{LY} = L_{LY0} + \frac{4\pi}{c}J^\mu V_\mu .$$

The symbols L_{LY0} and L_{LY} stress their similarity with the original Lee-Yang's massive Lagrangian. In (28), L_{LY0} returns to the original Lee-Yang's Lagrangian if one sets $\lambda_K \equiv 1$ and multiplies $(\partial_\mu V^\mu)^2/2$ by an arbitrary parameter ξ to form what is called *gauge fixing* term. The particular setting $\xi = 1$ yields (28) that produces (26), (26a), and Sterman's propagator without $k^\beta k^\gamma / m_v^{\ 2}$ term in its numerator. For massive photons we write

(28a)
$$L_{LYem0} = -F_{\mu\nu}F^{\mu\nu}/4 - \frac{1}{2}c^2 m_\gamma^{\ 2}\lambda_K^{\ 2}\hbar^{-2}\ A^\mu A_\mu - \frac{1}{2}(\partial_\mu A^\mu)^2,$$

$$L_{LYem} = L_{LYem0} + \frac{4\pi}{c}J_{em}^{\ \ \mu}A_\mu .$$

Chapter 5 Classical Interaction Theory with Q-Feedback

For classical e.m. fields, define $A_\mu \equiv (\boldsymbol{A}, i\phi^{em})$ so that

$$\boldsymbol{E} \equiv -\nabla\ \phi^{em} - \partial_t \boldsymbol{A}/c\ , \boldsymbol{B} \equiv \nabla \times \boldsymbol{A}\ .$$

Write $F_{\mu\nu} \equiv \partial_\mu A_\nu - \partial_\nu A_\mu$. Since photons are massive and timelike and spin 1 **massive** timelike particles obey massive quantum eqn (26), classical massive e.m. fields must obey **massive Maxwell eqs** (tensor form):

(29) $$\partial^\mu F_{\mu\nu} - \mu_\gamma^{\ 2} A_\nu = -4\pi c^{-1} J_{em\nu}, J_{em\mu} \equiv \lambda_K^{\ -4} \rho * U_\mu = \Lambda \rho u^\mu\ .$$

where $\mu_\gamma^{\ 2} \equiv m_\gamma^{\ 2} c^2 \lambda_K^{\ 2} / \hbar^{\ 2} \equiv m_\gamma^{\ 2} c^2 \Lambda^{-1} \lambda^2 / \hbar^{\ 2}$. For a ***rest*** point-like electric charge q at the origin of S, (29) gives

$$A_\mu = (\boldsymbol{A}, i\phi^{em}) \text{ with } \phi^{em} = q\exp(-\mu_\gamma r)/r\ .$$

Coulomb potential $\phi^{em} = \phi_C \equiv q/r$ is extremely accurate at

$$r << R^f{}_{em} \sim \hbar / m_\gamma c \sim 1.93 \times 10^{22+z} cm\ .$$

$\lambda_K^{\ -4}$ is the simplest factor to yield a frame-invariant eqn of charge conservation and to insure conserved e.m. source 4-current:

(29a) $$\partial^\mu J_{em\mu} = \partial^\mu (\Lambda \rho u_\mu) = \Lambda \partial^\mu (\rho u^\mu) = 0$$

The **massive** e.m. **Lagrangian density** with **given source** $J_{em}^{\ \mu}$ is

(30) $$L_{em} = L_{LYem} + \frac{1}{2}(\partial_\mu A^\mu)^2$$

Here massive L_{LYem} is given by (28a). The Lagrangian eqn

$$\partial_\mu (\partial L / \partial \partial_\mu A_\nu) - \partial L / \partial A_\nu = 0$$

yields massive Maxwell eqn (29) if $L = L_{em}$, and it yields inhomogeneous quantum Proca eqn (26a) for massive spin 1 photons if $L = L_{LYem}$. It has been pointed out in section I that the solutions of massive Maxwell eqs for all classical massive non-gravitational vector fields automatically satisfy the generalized Lorentz condition and the quantum wave eqn for massive spin 1 particles. But not vice versa. However, the well-known setting $V_{\mu 0} p^\mu = 0$ used in perturbation approach for vector fields guarantees that the solutions of quantum Proca eqs for spin 1 particles or quantum vector fields satisfy the general Lorentz condition $\partial_\mu V^\mu = 0$ and Maxwell eqs for classical vector fields. It is true for both massless and massive cases. This is the very feature of PVFs. The setting

$$a_\mu p^\mu = A_{\mu 0} p^\mu = 0$$

and the mass term without HM make L_{em} and L_{LYem} equivalent. This equivalence gives an explanation of field strength (crucial in classical theory and phenomena) as well as the scalar potential of a rest charge with revolutionary probability interpretation of pure quantum wave functions of spin-1 force carriers. The classical theory of vector fields such as Maxwell theory indeed has solid foundation and is supported by probability interpretation, which is a corner stone of Lorentz invariant as well as the modified quantum theory with LV.

Classical and quantum Proca eqs in current basic theory have nonzero mass terms too. But they are Lorentz invariant. Massive Maxwell eqs (31) and quantum wave eqn (26) and (26a) for massive spin 1 particles are not Lorentz invariant. Lorentz invariant Proca eqs recorded the 1st attempt to realize the revolutionary idea of massive photons. The mass term in *Lorentz invariant* classical and quantum Proca eqs is $\mu^2 \equiv m^2c^2 / \hbar^2$, while it is $\mu^2 \equiv m^2c^2\Lambda^{-1}\lambda^2 / \hbar^2$ in new physics with LV. For all earth systems $\Lambda = 1$ is too accurate. However, as the speeds of timelike photons approach c, there are no upper bounds for photons' energies and frequencies in any Lorentz invariant theory while there are in finite-N theory with (29).

Using Lorentz invariant Proca eqs to support massive photons and the full meaning of massive photons and to reject LV, we will be unable to get ultraviolet cutoffs and the minimum spacetime sizes. The EDF does not take Lorentz invariant Proca eqs as a starting point to get radical solutions.

Since massless particles do not exist due to the DE fact and IR fact, gauge invariance is not a principle of physics. We should fully accept the far-reaching meaning of a mass term in L_{LY}. Vector fields of all vector force carriers including photons obey the **same type** of **differential eqs**

(31) $$\partial^\mu V_{\mu\nu} - \mu^2 V_\nu = -4\pi c^{-1} J_\nu, J_\mu \equiv \lambda_K^{-4} \rho * U_\mu = \Lambda \rho u^\mu,$$

where $V_{\mu\nu} \equiv \partial_\mu V_\nu - \partial_\nu V_\mu$ with $V_\mu \equiv (\boldsymbol{V}, i\phi) = (\boldsymbol{A}, i\phi^{em})$, $(\boldsymbol{W}, i\phi^{wk})$, and $(\boldsymbol{G}, i\phi^{st})$, $\mu^2 = \mu_\gamma^2$, μ_{wz}^2, and μ_{gl}^2, while $\rho \equiv dq/dV$, dq_w/dV, and dq_s/dV for e.m., weak, and strong forces respectively. We call q, q_w, and q_s electric, weak, and strong (interaction) charge respectively. In view of quantum theory of fields, charges directly relate to coupling constants. For leptons and quarks, $q = \zeta e$, $\zeta = 0, \pm 1, \pm 1/3, \pm 2/3$, depending on the settings of various quark models, $q_w \equiv e$, $q_s = 0$ (for leptons), $q_s = e_s$ (for quarks), $q_s = -e_s$ (for anti-quarks). Here $q_w \equiv e$ means weak force is always repulsive with coupling style containing

Weinberg angle and weak coupling is unified with e.m. coupling on the fundamental level. It does not mean a neutral particle without electrical charge like a neutrino will have e.m. interactions. The strong coupling style (45) with flavor and color indices will show attractive strong force between every pair of non-identical quarks/anti-quarks. For photons and gluons, $q = q_w = q_s = 0$. For weak force carriers, $q = 0, \pm e$ and $q_w = q_s = 0$.

Note: **DS** proves that W-particles are fictitious and beta decay is caused by e.m. interaction (ENM), not by repulsive weak force. It is wrong to think that weak force is always repulsive. Such wrong idea is not removed as a historical record to show the struggles and mistakes of physicists including the author.

We call (31) **universal massive Maxwell eqn** or simply **massive Maxwell eqn** in honor of great J.C. Maxwell. The universal Lagrangian density for dynamic systems with massive vector fields of non-gravitational forces and with given source J^μ is written

(32) $$L = L_{LY} + \frac{1}{2}(\partial_\mu V^\mu)^2$$

where L_{LY} is given in (28). Again, L and L_{LY} are equivalent due to the mass term and the setting $V_{\mu 0} p^\mu = 0$ in perturbation approach.

The following conclusions can be made from (17a).

I1) For any timelike particle, $\boldsymbol{u}$ is constant vector iff $\boldsymbol{I=0}$ (**inertial law**).

I2) u is a constant iff $\boldsymbol{I} \cdot \boldsymbol{u} = 0$.

I3) In case $\boldsymbol{I} = -h(r)\boldsymbol{r}$ is any exact central interaction force in a non-preferred (non-isotropic) frame S ($V_{SK} \neq 0$), an exact circle cannot be the exact trajectory of a classical particle in the central force field as long as λ is not a constant. In any slow motion around a center, λ is too close to constant 1 to make the deviation observable.

I4) $\boldsymbol{r} \times \boldsymbol{u}\gamma$ is a constant 3-vector iff $\boldsymbol{r} \times \boldsymbol{I} = \boldsymbol{0}$.

The **theorem of angular momentum** is written from (15) and (17)

(33) $$d(\boldsymbol{r} \times \boldsymbol{p})/dt = \boldsymbol{r} \times (\boldsymbol{I} + m\gamma\, \boldsymbol{u}\, d\lambda/dt)\ .$$

For all the observed periodic motions about centers (stars, planets, electrons in atoms), $\gamma < 2$ with $d\lambda/dt$ almost vanishing. Then the observed conservations of angular momenta in central interaction force fields with $\boldsymbol{r} \times \boldsymbol{I} = \boldsymbol{0}$ are true with extremely high precisions.

For a **two body problem** in K, the opposite Coulomb-type interaction forces obeying inverse-square law (YFs are almost 1 at distances much shorter than force-ranges) are $\boldsymbol{I}_1 = -k_{eff} q_1 q_2 (x_2 - x_1)^{-2}\, \boldsymbol{i}$, and $\boldsymbol{I}_2 = -\boldsymbol{I}_1$ (Newton's third law applies only for interaction force $\boldsymbol{I}$ in the force-splitting $\boldsymbol{F} = \boldsymbol{I} + \boldsymbol{S}$). We find

(34) $$\boldsymbol{I}_1 \cdot \boldsymbol{u}_1 + \boldsymbol{I}_2 \cdot \boldsymbol{u}_2 = -dU_{12}/dt, U_{12} \equiv k_{eff} q_1 q_2 / |x_2 - x_1| .$$

With (19) and (14), one can check that

(35) $$U_{12} + Nm_1c^2 th^{-1}\xi_1 + Nm_2c^2 th^{-1}\xi_2 = constant ,$$

where $\xi_i \equiv N^{-1}\lambda_i\gamma_i$. If q_2 is <u>at rest</u> at the origin before the collision, $q_1 = q$, and $m_1 << m_2$, then $U_{12} = k_{eff} q q_2 / r = k_{eff} q \phi$, $Nm_2c^2 th^{-1}\xi_2 \approx 0$, and (35) returns to (19b).

The Coulomb-type interaction forces $\boldsymbol{I}_i$ (not $\boldsymbol{F}_i$) obey Newton's third law: $\boldsymbol{I}_2 = -\boldsymbol{I}_1$. Consequently, the total momentum does not conserve exactly. Let's prove that the violation is negligible up to extremely high-energy region and up to extremely short distance, while we do have another **conserved quantity of motion** that departs total momentum in extreme high-speed region. Now we apply (17a) to the two charged particles colliding along x-axis, using β, u, a, and p to indicate the corresponding **signed** quantities. With (14) and $\xi \equiv N^{-1}\lambda\gamma$, we find that

$$\gamma^2 = (N^2-1)\xi^2/(1-\xi^2), \ \beta = \pm\sqrt{1-1/N^2\xi^2}/\sqrt{1-1/N^2} ,$$

(36) $$p = m\lambda\gamma u \equiv Nmc\,\xi\beta = \pm Nmc\sqrt{N^2\xi^2-1}/\sqrt{N^2-1} .$$

From (14) one can check that

$$a = du/dt = \pm(d\xi/dt)cN^{-2}\xi^{-2}/\sqrt{(1-N^{-2})(\xi^2-N^{-2})} .$$

Since $\boldsymbol{I}_2 = -\boldsymbol{I}_1$, (17a) yields $m_1\lambda_1\gamma_1^{\,3}a_1 + m_2\lambda_2\gamma_2^{\,3}a_2 = 0$. Hence, the ***conserved*** total **quantity of motion** is

(37) $$m_1 th^{-1}\delta_1 + m_2 th^{-1}\delta_2 = constant \quad (\delta_i \equiv p_i / Nm_ic, i = 1,2)$$

As $r \to 0$ in head-on collisions of two attracting particles, $th^{-1}\delta_1$ and $th^{-1}\delta_2$ approach $\pm\infty$ respectively, but the sum in (37) remains constant all the times and at all distances. The *1st order approximation* of $th^{-1}\delta$ with extremely high precision gives the conservation of total momentum:

$$m_1\delta_1 + m_2\delta_2 = constant \text{, i.e., } p_1 + p_2 = constant .$$

If the collision is head-on and is *symmetric* when observed in a preferred system K, i.e., $m_1 = m_2$, $u_1 \equiv -u_2$, then the conservation of total momentum is exact even in finite-N theory. For earth systems, head-on e-e and p-p collisions in large accelerators are almost symmetric for the absolute velocity of the Earth is not high. The numerical difference between δ and $th^{-1}\delta$ is almost zero for all non-

most energetic particles with $\xi \equiv N^{-1}\lambda\gamma << 1$ and the larger the limit N , the higher the speed needed for the violation of the conservation of total momentum to be practically measurable. For all high-energy particles in accelerators and at least most in cosmic rays, $\xi \equiv N^{-1}\lambda\gamma << 1$. We do not know whether there are some most energetic particles ($E \approx Nmc^2$) other than most energetic photons ($E_\gamma \approx Nm_\gamma c^2$) ever exist in the universe. In terms of any earth-system, the eqs of conservation laws are numerically almost the same as the deduced conservation laws in a preferred system.

If λ is given by (20), it has four major pieces. The 1st piece is almost constant within $[1,\ \gamma_0)$. Then the 4-momentum conservation is numerically as usual as in current basic theory. The 2nd piece is quasi-vertical within a narrow interval (γ_0, γ_{20}), called VSS.

In rare events of collisions whenever one particle's velocity is within the thin VSS and the other's is not, the total 3-momentum of two colliding particles will not conserve. This might be the reason of the rare events at CERN (bizarre behavior of p^+p^- collisions) where one jet shoots off on one side, with nothing visible balancing its flight in the opposite direction. The energies, velocities, and γ-values of p^+ p^- particles in collisions can be well controlled in the experiment. But those events are caused by interactions between quarks, not between an entire proton and an entire antiproton. The motion of bound quarks is very probabilistic. It is rare for a bound quark having the γ-value within the thin VSS at the instant time of collision.

The 3rd and the 4th pieces are quasi-inverse within (γ_1, γ_2) and (γ_2, ∞) respectively. The minor quasi-vertical jump at γ_2 is not treated as a major piece for the purpose to interpret quark confinement. The minor jump helps to retain large λ-values for photons and neutrinos with $\gamma_2 < \gamma < N\hbar^{-1}$. Within (γ_1, γ_2), the expression of conserved quantity of motion is different from (37) that is deduced from model λ (14). It is left for future discussion.

Due to the very difficulty of accurate speed-measurement in large-γ motions, any direct accurate measurement of γ-value is very difficult or almost impossible for high-speed particles with $\gamma > 10^{2\sim5}$. Then the theoretical calculation of both energy and momentum using a γ-value lost its empirical foundation even if the model λ used in the calculation indeed matches the reality. The only way to measure the high-energy and large-momentum of a large-γ particle accurately is to make a process with all initial and final state particles moving slowly except for the one under the investigation. Namely, all other initial particles and all final state particles are slow moving or at rest so that their speeds can be measured accurately. This means that the errors for induced γ-values are extremely small and their energies and momenta can be calculated accurately due to their well-known and well-measured masses and a reliable fact that λ is too close to constant 1 at low speeds. A simple example is the absorption of a photon by an atom. However, once there is any violation of conservation, some details in the conclusion drawn from the same empirical data would be different. However, if the violation of momentum-conservation manifests itself in the form of unbalanced shoot, the said quantum phenomenon becomes a decisive evidence of the violation of momentum-conservation.

In summary, the conserved **quantity** of **motion** is

$$\underline{p} \equiv Nmcth^{-1}(p/Nmc) \text{ and } \underline{E} \equiv Nmc^2th^{-1}(E/Nmc^2)$$

with model (14) so that

$$\underline{p}_1 + \underline{p}_2 = constant \text{ and } \underline{E}_1 + \underline{E}_2 + U_{12} = constant$$

for an isolated two body system in head-on collision. An **equivalent** form of the latter is

$$E_1 + E_2 + \underline{U}_{12} = constant .$$

Since the differences $\underline{p} - p$ and $\underline{E} - E$ are too small up to extreme high-speed regions, the conservation delta factor at each vertex of Feynman diagrams produced numerical success of massless and massive QED.

With model (20) and within the interval (γ_0, γ_2), the above result should be modified by the replacement $N \to H\gamma_1/\sqrt{2}$. Although the two roofs have large difference ($H\gamma_1/\sqrt{2} << N$), $H\gamma_1/\sqrt{2} \sim 10^{7\sim12}$ is still very large. Then no violation of the conservation of total 3-momentum in collisions would be observed as regular events in today's large accelerators.

For any <u>non-constant</u> function λ, $\boldsymbol{F}$ in (37) always contains a component $\boldsymbol{S} \equiv m\gamma\, \boldsymbol{u}\, d\lambda/dt$ that is always in the same direction with either $\boldsymbol{u}$ or $-\boldsymbol{u}$. This means that for a charged particle moving in a given e.m. field, the dynamic system is **not** exactly a Lagrangian system whenever λ is not constant.

A classical dynamic system with exactly zero damping does not exist on earth. The reason Lagrangian systems work in both classical and quantum theories is not because there is no damping in nature and only Lagrangian systems match the reality but because those terms that make real physical dynamic systems non-Lagrangian are too small to consider.

Exact Lagrangians in the current basic theory, however, can be obtained by neglecting some quantities that are extremely small up to extreme high-speed regions.

Denote gravity force by $\boldsymbol{I}_g$. Set $\boldsymbol{I} = \boldsymbol{I}^{em} + \boldsymbol{I}_g$. From (37) we get

(38) $$d\,\boldsymbol{p}/dt - \Lambda q(\,\boldsymbol{E} + \boldsymbol{\beta} \times \boldsymbol{B}\,) = \boldsymbol{S} + \boldsymbol{I}_g .$$

It is equivalent to a **quasi-Lagrangian eqn**

(39) $$d(\partial L/\partial\, \boldsymbol{u}\,)/dt - \partial L/\partial\, \boldsymbol{r} = \boldsymbol{S} + \boldsymbol{I}_g,$$

$$L = mc^2 \int \varsigma\lambda(c\varsigma)(1-\varsigma^2)^{-1/2} d\varsigma + \Lambda q(\,\boldsymbol{\beta} \cdot \boldsymbol{A} - \phi^{em}\,),$$

which is **integrable** if $\lambda \equiv 1$ (current theory) OR λ is given by (14). It is piecewise integrable if one uses model λ (20). Using (14) for $S=K$ with $\Lambda=1$ and omitting the frame subscript K for all frame-variant quantities we get

(40a) $$L=\frac{mc^2}{1-N^{-2}}(1-\frac{1}{\lambda\gamma})+q(\boldsymbol{\beta}\cdot\boldsymbol{A}-\phi^{em}).$$

(40b) $$L=-\frac{mc^2}{1-N^{-2}}\frac{1}{\lambda\gamma}+q(\boldsymbol{\beta}\cdot\boldsymbol{A}-\phi^{em}).$$

They are equivalent. If we neglect both gravity force $\boldsymbol{I}_g$ and space force $\boldsymbol{S}$, the quasi-Lagrangian eqn (39) turns to be a Lagrangian eqn and the quasi-Lagrangian (40a) and (40b) become Lagrangians.

The numerical success of QED is obtained by neglecting gravity. It is difficult to complete a theoretical treatment that would include gravity in Feynman diagrams and Feynman rules. Numerically, there is no difficulty at all. One can simply neglect gravity force in a quantum theory of non-gravitational fields on earth. Similarly, we may neglect space force for the purpose to apply Lagrangian method. In current basic theory, the ratio of the gravity force to the Coulomb force between the same two charged particles is *very small* **at any distance**, though *both two kinds of forces approach infinities* as $r\to 0$. However, from (14) and (17) we can prove that

(41) $$\boldsymbol{S}=-\xi^2\,\boldsymbol{I}\cdot\boldsymbol{\beta\beta}\quad(\xi\equiv N^{-1}\lambda\gamma).$$

The magnitudes of both interaction force and space force approach infinity as $r\to 0$ but the *ratio is not small* **within the CDs**. Then the space force is able to balance the interaction force at the distances shorter than CDs to provide with deepest asymptotic freedom shown explicitly in the new generalized PE obtained from (19c) with $R^c{}_{em}\equiv|\underline{R}^{em}|\equiv 2|\zeta\zeta'|e^2/Nmc^2$ replaced by

$$\underline{R}^c\equiv 2\underline{g}^2/Nmc^2.$$

From (20) and (17) we can get the expressions of space force $\boldsymbol{S}$ piecewise. We omit the details in this book. Can we omit space force for the purpose to get a Lagrangian system and a Lagrangian that immediately recovers all the numerical success of QED? The answer is yes:

The numerical success of QED is obtained by stopping calculating at short enough distances. Beyond those short distances, the e.m. force that approaches infinity (as $r\to 0$) gives no numerical contribution to the 1st nine or more after decimal digits of observable quantities, neither the space force, included or neglected.

Due to **frame invariance of an action**, $L'dt'=Ldt,\forall S,S'$. Since

$$d[\partial(\boldsymbol{A}_S\cdot\boldsymbol{\beta})/\partial\,\boldsymbol{u}]/dt=c^{-1}d\,\boldsymbol{A}_S/dt=\boldsymbol{0}$$

we can omit the term $\boldsymbol{A}_S \cdot \boldsymbol{\beta}$. Put back subscript K in (40b) to get L_K. Then, in a **non-preferred** system S

(42) $$L = -\Lambda \frac{mc^2}{1-N^{-2}} \frac{1}{\lambda\gamma} + \Lambda q(\, \boldsymbol{\beta} \cdot \boldsymbol{A} - \phi^{em}\,)\,,$$

(42a) $$d(\partial L/\partial \boldsymbol{u}\,)/dt - \partial L/\partial \boldsymbol{r} = d\,\boldsymbol{p}/dt - \Lambda q(\,\boldsymbol{E} + \boldsymbol{\beta}\times\boldsymbol{B}\,) - d\,\boldsymbol{Q}/dt\,.$$

(42b) $$\boldsymbol{Q} \equiv \frac{mc}{N^2-1}\Gamma^2\lambda\gamma(1+\boldsymbol{A}_S\cdot\boldsymbol{\beta})\,\boldsymbol{A}_S\,.$$

The Lagrangian eqn with the Lagrangian (42) gives

(43) $$d\,\boldsymbol{p}/dt = \Lambda q(\,\boldsymbol{E} + \boldsymbol{\beta}\times\boldsymbol{B}\,) + d\,\boldsymbol{Q}/dt\,.$$

If $S = K$, then $\Lambda = 1$, $A_S = A_K = 0$, and therefore $Q = 0$, while $\boldsymbol{u} = \boldsymbol{u}_K$, $\boldsymbol{p} = \boldsymbol{p}_K = m\lambda_K\gamma_K\,\boldsymbol{u}_K$, $\boldsymbol{E} = \boldsymbol{E}_K$, $\boldsymbol{B} = \boldsymbol{B}_K$. So in terms of a preferred frame K, (43) and (42a) becomes

$$d\,\boldsymbol{p}_K/dt_K = q(\,\boldsymbol{E}_K + \boldsymbol{\beta}_K\times\boldsymbol{B}_K\,)$$

which is the eqn of motion in K with space force omitted. For all earth-systems, $\boldsymbol{Q}$ and $d\,\boldsymbol{Q}/dt$ is negligible. In fact, for all S with $A_S < 1-10^{-30}$, one can see that

$$N^{-2}\Gamma^2 < .5\times 10^{-50-2z-2B}\,.$$

Moreover, from (43) and (42b) one can see that the departure

$$d\,\boldsymbol{p}/dt - \Lambda q(\,\boldsymbol{E} + \boldsymbol{\beta}\times\boldsymbol{B}\,) = \boldsymbol{A}_S\,\frac{mc}{N^2-1}\Gamma^2 d[\,\lambda\gamma(1+\boldsymbol{A}_S\cdot\boldsymbol{\beta}\,)]/dt.$$

is negligible for $N^{-2}\Gamma^2$ is negligible. Indeed, both $\boldsymbol{Q}$ and $d\,\boldsymbol{Q}/dt$ are negligible for $N^{-2}\Gamma^2 << 1$.

In general, (18) and $\boldsymbol{E} \equiv -\nabla\,\phi^{em} - \partial_t\boldsymbol{A}/c$ yield

$$\boldsymbol{I}^{em}\cdot\boldsymbol{u} = -d(\,\Lambda q\phi^{em}\,)/dt + \Lambda q(\partial_t\phi^{em} - \,\boldsymbol{\beta}\cdot\partial_t\boldsymbol{A}).$$

For atoms at rest, $\partial_t\phi^{em} - \boldsymbol{\beta}\cdot\partial_t\boldsymbol{A}=0$. The numerical success of QED in Lamb shift and magnetic moment phenomena is obtained for zero or very weak static or slowly varying magnetic field. Any static central force field $\boldsymbol{I} = k(r)\,\boldsymbol{r}/r$ yields a total differential exactly: $\boldsymbol{I}\cdot\boldsymbol{u} = -dU(r)/dt$ as long as $\int k(r)dr$ is integrable. Furthermore, as long as $\partial_t\phi^{em} - \boldsymbol{\beta}\cdot\partial_t\boldsymbol{A}=0$ is at least approximately true,

$$\boldsymbol{I}^{em}\cdot\boldsymbol{u} = -dU^{em}(r)/dt = -d(\,\Lambda q\phi^{em}\,)/dt$$

is a total differential with $\gamma d\lambda / dt$ in (19) almost vanishing up to high-energy regions. At the distances from the range of e.m. force $R^f{}_{em} \sim 10^{22+z} cm$ to the 1st e.m. CD $R^c{}_{1em} \equiv 2\sqrt{2}e^2 / \mathcal{H}\gamma_1 m_e c^2 \sim 8\mathcal{H}^{-1}\gamma_1^{-1} \times 10^{-13} cm$ or equivalently at the γ-values ranging from 1 to γ_0, $mc^2 \gamma d\lambda / dt \approx 0$ and its dynamic effect through (19) is negligible. Then $U^{em}(r) = \Lambda q\phi^{em} \approx q\phi^{em}$ (which is the **Coulomb PE** $q\phi_C$ except for the Yukawa-type factor in ϕ^{em}) serves as the e.m. PE. This is why conservation factors put by hand at vertices of Feynman diagrams and **Coulomb PE** could yield numerical success of QED under the condition that it is impossible to get an expression of self-force in classical theory to prove or to be based on the conservation of TE. We do not know whether an expression of self-force can be found in finite-N theory. No satisfactory expression of 4-momentum tensor of gravitational field could be deduced to insure rigorous conservation of TE. If the energy conservation is approximate, its violation will be too small to verify within an earth laboratory. It might cause observable effect in large-scale space in the universe if cumulated for tens of billions of years since the beginning. The author is not sure if **dark energy** phenomenon is caused by accumulation of tiny violation of conservation of TE. (Dark energy has been explained in VIII) However, one thing is clear:

A rigorous expression of TE cannot be deduced in every modern theory of fields and the numerical results everywhere have not been affected by any possible tiny violation.

In both classical and quantum theories, H_I must be PE as long as H is of energy dimension. The fact that e.m. PE in $e^{\pm}$ collisions is numerically Coulomb PE for $R^c{}_{1em} < r < 10^{22+z} cm$ justifies QED-type $H_I = j^\mu A_\mu$ where A_μ may return to Coulomb potential (potential is not a PE when $mc^2 \gamma d\lambda / dt$ term is no longer negligible) if $\boldsymbol{A} \approx \boldsymbol{0}$. Here $R^c{}_{1em} \sim 8 \times 10^{-20} cm$ if $\mathcal{H} \sim 10^2$ and $\gamma_1 \sim 10^5$. It also justifies the further **plane wave** expansion in a Dyson series to yield Feynman rules and numerical success for QED between $R^c{}_{1em} \sim 8 \times 10^{-20} cm$ and $R^f{}_{em} \sim 10^{22+z} cm$. In general, numerical validity of inverse PE between force range $R^f \sim 1/\mu$ and the 1st CD $R^c{}_1 \sim 2\underline{g}^2 / \mathcal{H}\gamma_1 mc^2$ is a necessary condition for $H_I = j^\mu V_\mu$ to be valid numerically within ($R^c{}_1$, R^f).

Once $\boldsymbol{f}^{em} \cdot \boldsymbol{u} = -dU^{em}(r)/dt = -d(\Lambda q\phi^{em})/dt$ is generalized for non-e.m. central forces, new **smooth** PEs similar to (19c) can be deduced with CDs

$$R^c{}_1 \sim 2\underline{g}^2 / \mathcal{H}\gamma_1 mc^2 \text{ and } R^c \sim 2\underline{g}^2 / Nmc^2$$

proportional to coupling strength $\underline{g}^2$. This indicates 1st and deepest **asymptotic freedom** at short distances or large momenta for all attractive and repulsive forces. However, the *deepest* asymptotic freedom takes effect iff the characteristic ratio of interaction

(44) $$R^c : l_{min} = 2\underline{g}^2 / Nmc^2 : \hbar / 2Nmc^2 = 4\underline{g}^2 / \hbar c$$

is larger than 1. This is true only for strong force between quarks and for gravity force within a black hole, if any. This conclusion does not depend on the values of a finite N or on the mass (m_q or m). If a black hole with mass M_{BH} ever exists, then $\underline{g}^2 = \underline{g}^2{}_{BH} = GM_{BH}m$. We find

$$R^c{}_{BH} : l_{min} = 4GM_{BH}m / \hbar c > 1, \; R^c{}_{BH} = 2GM_{BH} / Nc = r_S / N > l_{min}$$

and particles within the black hole would experience the deepest asymptotic freedom at distances shorter than r_S / N. (r_S is the Schwarzschild radius of the black hole.)

Now we use Q-feedback to get a classical theory of strong force in order to deduce strong PE. Exchanges of virtual pions may occur in **double-gluon exchanges**. A pair of virtual gluons with opposite spins is effectively a virtual spin 0 boson that may couple to a pair of a virtual quark and its anti-quark with opposite spins, a virtual π^0, or a pair of $\pi^{\pm}$, and the pion(s) in the middle of two internal lines of double virtual gluons may contribute to pion-force. This possible physical process might also relate to the mass spectrum of hadrons since the double-gluons as spin 0 bosons make the effective coupling ***scalar couplings*** and the corresponding potential energies might appear in Dirac eqn together with quark masses. To interpret quark confinement, the double-gluon diagrams may be of crucial importance (see VI). To deduce the expression of the strong PE, we may temporarily forget the difference between vector and scalar couplings.

Now we apply the simple unified model for all massive vector fields, shown in (31). If quantum 3-body problems (not solvable) can be solved *incompletely* but to show definitely that identical colorless quarks inside a spin 3/2 baryon (such as N^{*-}, N^{*++}, and Ω^-) can be in different quantum states though they have the same spin state, colors must be removed. Here we apply colors. Our convention is to label flavors and colors with ordered pairs of numbers, called flavor and color indices (FCIs). The symbol q_{ij} represents a **quark** of flavor i and color j if both i and j are **positive** integers. It represents an **antiquark** of flavor i and color j if both i and j are **negative** integers. The symbol $q_{s,ij}$ denotes the strong charge carried by q_{ij}. Any quark and its antiquark possess *opposite* FCIs and strong charges. Gluons carry no strong charges. With signed indices in general,

$$q_{s,kl} = -q_{s,-k-l} = s(k)e_s \,, \quad s(k) \equiv k / |k| \,.$$

The Lorentz-force formula (18) ($\Lambda = 1$ in relativity and $\Lambda \neq 1$ in the open theory) corresponds to photon propagator. Its mass term vanishes exactly in Lorentz invariant QED. It can be ignored in massive QED with finite N for all non-IR e.m. phenomena. When we generalize (18) to get a force formula of strong force, it must correspond to an effective massless propagator with a running effective coupling constant. The force formula of classical **gluon force** acting on q_{kl} by a vector gluon field $G_{\mu,k'l'}$ produced by $q_{k'l'}$ is written

$$I^{gl}{}_{(\mu)}(kl,k'l') = s_{klk'l'}k_{eff}q_{s,kl}G_{\mu\tau,k'l'}u^{\tau}/c \,, \tag{45}$$

$$s_{klk'l'} \equiv (2\delta_{kk'}\delta_{ll'} - 1)kk'/|kk'| \,.$$

The coefficient k_{eff} is determined by Q-feedback from LH mechanism. The larger the ratio $\underline{g}_s{}^2/M^2$, the larger the k_{eff}.

Note: If strong force between two identical quarks is also attractive, $2\delta_{ij}\delta_{jl} - 1$ should be replaced by -1. Pauli exclusion principle is solely caused by the property of the wave functions of identical particles, no matter they are elementary or composite, attract, repulse, or do not interact on each other.

The unfamiliar and seemingly complicated coupling style in (45) just means two simple things:

1) The **magnitude** of strong interaction force is **independent of** *colors* and *flavors*. (DS-explanation of FQA fact would not be affected by whether the strong force is flavor/color-dependent.)

2) Strong force between any 2 non-identical quarks q_{kl} and $q_{k'l'}$ is attractive regardless the 2 strong charges $q_{s,kl}$ & $q_{s,k'l'}$ are both e_s, both $-e_s$, or opposite.

Massive Maxwell eqs for gluon fields are

$$\partial^{\mu}G_{\mu\nu,kl} - \mu_{gl}{}^{2}G_{\nu,kl} = -\frac{4\pi}{c}J_{\nu,kl} \,, \tag{46}$$

$$G_{\mu\nu,kl} \equiv \partial_{\mu}G_{\nu,kl} - \partial_{\nu}G_{\mu,kl} \,, \quad \mu_{gl}{}^{2} = M^2c^2\lambda^2/\Lambda\hbar^2 \,.$$

Here $J_{\mu,kl} \equiv \lambda_K{}^{-4}\rho_{kl} * U_{\mu} = \Lambda\rho_{kl}u^{\mu}$, $\rho_{kl} \equiv dq_{s,kl}/dV$.

Now we <u>remove color and flavor indices</u> as we evaluate only the **magnitude** of the **binding/attractive** strong force and deduce the corresponding PE to explain quark confinement. Write

$$\partial^{\mu}G_{\mu\nu} - \mu_{gl}{}^{2}G_{\nu} = -\frac{4\pi}{c}J_{\nu} \,. \tag{47}$$

Suppose a quark inside a nucleon is effectively in a central static massive gluon field

$$G_{\mu} \equiv (\,0,\; i\phi^{gl}\,), \phi^{gl} = b_3q_se^{-\mu_{gl}r}/r$$

as the solution of (47) with a *rest* point-like strong charge b_3q_s, where $b_3 > 0$ is a 3-body factor. Then for a bound quark of strong charge q'_s in such effective central **attracting** field, $sq'_s q_s = -e_s^2$ and (45) with flavor and color indices removed yields

(48a) $$\boldsymbol{f}^{gl} = b_3 k_{eff} e_s^2 \nabla (Y_{gl}/r), Y_{gl} \equiv e^{-\mu_{gl} r}$$

with very short force-range. It is not the observed strong force. Virtual pions between double internal gluon lines and/or quark loops contribute additional terms to and modify propagators, PE, and force. It is very crucial that they extend the range of strong force from gluon-force range to a much longer one (the observed range of strong force) since gluons are much heavier than pions and quark pairs. We write

(48b) $$\boldsymbol{f}^{st} = \boldsymbol{f}^{gl} + \boldsymbol{f}^{pi} = b_3 k_{eff} e_s^2 \nabla [(Y_{gl} + \chi Y_{pi})/r] .$$

The pion/loop parameter $\chi < 1$ is determined by the way the double-gluon and /or u/d-loop diagrams further modify a massless propagator for any given pair of e_s^2 and M. At the present, we do not know exact values of $b_3 k_{eff}$ and of χ. To explain quark confinement, the knowledge of these values is not crucial. For instance, $k_{eff} e_s^2$ is roughly proportional to the ratio e_s^2 / M^2. Quark confinement can be explained for large variety of the values of e_s^2 and M. Weak and e.m. forces make negligible roles in binding quarks. We omit them.

Repeating the deduction process from (19) to (19d) with $\boldsymbol{f}^{em}$ replaced by $\boldsymbol{f}^{st}$ and with model λ (14), we obtain the strong PE $\underline{U}^{st}$, which is just (19c) with replacements

(49) $$\underline{U} \to \underline{U}^{st} , \quad Y \to Y_{st} \equiv Y_{pi} + \chi^{-1} Y_{gl} ,$$

$$\underline{R} \to \underline{R}^{st} \equiv 2\chi b_3 k_{eff} e_s^2 / Nmc^2 \equiv 2\eta e_s^2 / Nmc^2 .$$

Model (14) is of limiting LV only. The deepest e.m. CD $R^c{}_{em}$ is too short and $R^c{}_{st}$ is also too short unless $\eta e_s^2 / e^2$ is large enough. But such large ratio would lead to invalidity of perturbation method. As long as perturbation method is valid, any deviation from inverse PE at the distances between 10^{-14}cm and $R^c{}_{st}$ would require LV to explain. Especially, the graph of strong PE drops down (as distance decreases) quasi-vertically or quasi-linearly to confine quarks and this must be caused by LV at least partially if the ratio e_s^2 / M^2 is not terribly large. Applying model λ (20) to calculate $\gamma d\lambda / dt$ one can get the 1st portions of the corresponding e.m. and strong PEs, which are numerically almost the same as (19c) with the following replacements

(49a) $$\underline{U} \to \underline{U}^{em} , \quad \underline{R} \to \underline{R}^{em} \equiv 2\zeta\zeta' e^2 / \underline{N}mc^2 , \quad Y \to Y_{em} ,$$

(49b) $\underline{U} \to \underline{U}^{st}$, $\underline{R} \to \underline{R}^{st} \equiv 2\eta e_s^{\,2} / \underline{N}mc^2$, $Y \to Y_{st}$.

Here $\underline{N} = N$ within $[1,\ \gamma_0)$ with $(\gamma/\gamma_0)^n \ll 1$, and $\underline{N} = \lambda_0 = \mathcal{H}\gamma_1/\sqrt{2}$ within (γ_0, γ_2) with $(\gamma/\gamma_0)^n \gg 1$. Please note, within $[1,\ \gamma_0)$ with $(\gamma/\gamma_0)^n \ll 1$ the two CDs are too short due to the setting $\underline{N} = N$ and thereby the PEs are inverse ($\propto 1/r$) PEs with high precisions. Within (γ_0, γ_2) with $(\gamma/\gamma_0)^n \gg 1$, the PEs are of the form of (19e) with $\lambda_0 = \mathcal{H}\gamma_1/\sqrt{2}$. These are the crucial portions of the e.m. and strong PEs at distances $r > r_{2em}$ and $r > r_{2st}$ respectively. Here r_{2em} and r_{2st} correspond to γ_2 through conserved TE in e.m. and strong interaction respectively associated with given initial conditions. For central attracting force field, the e.m. and strong PEs are monotone decreasing as r decreases as long as $\lambda\gamma$ and $E = m\lambda\gamma c^2$ is monotone increasing as r decreases. This is due to the conservation of TE. Accordingly, the kinetic energy and γ-value increase as r decreases as long as $\lambda\gamma$ is monotone increasing as γ increases. We may write the correspondence $\gamma_0 \leftrightarrow r_0$, $\gamma_1 \leftrightarrow r_1$, $\gamma_2 \leftrightarrow r_2$ and $\gamma = N \leftrightarrow r = r_N$ and so on. Here $r_0 \sim R^c{}_1 = 2k_{eff}\,|qq'|/\lambda_0 mc^2$ with $\lambda_0 = \gamma_1\mathcal{H}/\sqrt{2}$. Nearby r_0 , the graph of the PE drops down quasi-vertically and after it, the graph becomes quasi-horizontal until r reaches r_2 . After r_2 ($r < r_2$) and before r_N ($r > r_N$), the graph of the PE slops down quasi-inversely. After r_N ($r < r_N$), the graph turns to be quasi-horizontal again.

At current energy scales, the PE and the graph for $r < r_2$ play no important role for the phenomena observed in accelerators. We now focus on the crucial portion of the PEs (19c) with replacements in (49a, b). For e.m. central force between two electrons,

$$R^c{}_1 = R^c{}_{1em} = 2\sqrt{2}e^2 / \gamma_1 \mathcal{H} m_e c^2 .$$

If $\mathcal{H} \sim 10^2$, $\gamma_1 \sim \gamma_0 \sim 10^5$, then $R^c{}_{1em} = 2\sqrt{2}e^2/\gamma_1\mathcal{H}m_e c^2 \sim 8\times 10^{-20}cm$. This explains why the LV shown in (20) has not affected the e.m. phenomena in the high-energy $e^{\pm}$ collisions. For strong central force field,

$$R^c{}_1 = R^c{}_{1st} = 2\sqrt{2}\eta e_s^{\,2} / \gamma_1 \mathcal{H} m_q c^2 .$$

Suppose $\eta e_s^{\,2} \sim 100 e^2$, $m_q = m_u \sim 5MeV$. Then $R^c{}_{1st} \sim 8\times 10^{-19} cm$. For such small quark core confinement, the formula $E_n = \hbar^2\pi^2 n^2 / 8ma^2$ for infinitely deep square PW deduced from Schrödinger eqn clearly shows large kinetic energy and γ-value for bound quarks and hence the formula itself becomes invalid even just for an order-of-magnitude estimation. The 1st CDs and sizes of quark cores are sensitive to the functional form of λ . We do not know the accurate form of λ and we do not know for sure the sizes of quark cores.

Changes in initial conditions cause changes in geometrical shapes of the graphs of the PEs. In current basic theory with $d\lambda / dt \equiv 0$, however, changes in initial conditions only cause translations of the graphs of inverse PEs and the geometrical shapes of the graphs will not change.

Chapter 6 New Dirac Eqn in Massive Vector Fields

The new de Broglie wave function gives the replacements

$$p^{\mu} \to \hat{p}^{\mu} \equiv -i\hbar\partial^{\mu} = -i\hbar\Lambda\partial_{\mu}$$

for <u>free</u> particles. Namely, $p \to \hat{p} \equiv -i\hbar\nabla$ and $E \to \hat{E} \equiv i\hbar\Lambda\partial_t$. Define **high-speed velocity operator** for a ***free*** particle as

(50) $$\hat{u}\,|E> = \frac{c^2}{E}p|E>,\quad \hat{u}^2\,|E> = \frac{c^4}{E^2}p^2\,|E>, \ldots .$$

Here $|E>$ is any energy eigenstate of free particle with energy eigenvalue E equal to the energy of the particle. It is defined on the base of the Hilbert space and therefore works for the entire Hilbert space. Define operator $\hat{\lambda} = f_S(\hat{u})^{-1/2}$ by replacing the classical quantities in the expansion of $\lambda = f_S(u)^{-1/2}$ with the corresponding operators in (50). Then $\hat{\lambda}\psi = \lambda\psi$ holds for free particles.

In the presence of massive vector fields $V_{\mu} = (V, i\phi)$, an **energy eigenvalue** E must be the **conserved TE**: $E = E_p + \underline{U}$ ($E_p = m\lambda\gamma c^2$ is the energy of a particle as the sum of its kinetic and rest energies). One can make replacements of and with classical quantities in the presence of massive e.m. fields

(51) $$p^{\mu} \to p^{\mu} - \Lambda q\underline{A}_{\mu}/c\,,\; u = \frac{c^2}{E}p \to u_{em} \equiv \frac{c^2}{E - \Lambda q\underline{A}_0}(p - \Lambda q\,\underline{A}/c)$$

We call them **classical replacements** (**CRs**). This induces another CR

$$\lambda \equiv f_S(u)^{-1/2} \to \lambda_{em} \equiv f_S(u_{em})^{-1/2}.$$

A **necessary condition** for the CR $p^{\mu} \to p^{\mu} - \Lambda q\underline{A}_{\mu}/c$ to be correct is that $\Lambda q\underline{A}_0 = \underline{U}^{em}$ if ***A*=0**. From (19a) and (19c) with (49a), we see that $\underline{U}^{em}$ returns to the Coulomb PE if ***A*=0**, ϕ^{em} is Coulomb scalar potential and λ is constant or $N \to \infty$. At this moment, the exact form of $\underline{A}_{\mu}$ and the corresponding coupling is unknown. However, one does not need it in explaining atomic energy levels, Lamb shift, and anomalous magnetic moments. For the said phenomena, $\underline{\boldsymbol{A}}=\boldsymbol{A}$ and $\Lambda q\underline{A}_0 \approx \underline{U}^{em} \approx -Ze^2/r$ would yield very accurate numerical results. Potentials (not PEs) depart on-0-mass-shell potentials only near and after force ranges while $d\lambda/dt$ in (19) causes effect at the VSS and beyond, or equivalently, at distances near and shorter than the CDs. For e.m. interactions with a very long force range, slow moving bound electrons in atoms are not sensitive to the departures and we do not need to use $\underline{U}^{em}$, which can be obtained from (19c) with replacement (49a), to figure out the exact form of $\underline{A}_{\mu}$ in the replacement. We may set

$$\underline{U}^{em} \cong -Ze^2 / r$$

as an approximation and have the current CR $p^{\mu} \rightarrow p^{\mu} - \Lambda q A_{\mu} / c$. Then the settings $p^{\mu} \rightarrow \hat{p}^{\mu} \equiv -i\,\hbar\,\partial^{\mu} = -i\hbar\,\Lambda\partial_{\mu}$ complete the transition from classical 4-momentum to its quantum operator.

Energy eigenstates $|\,E >$ in the presence of e.m. fields serve as base vectors in a Hilbert space. For the classical quantities u^{n}_{em} ($n = 1,2,3,\ldots$) in the expansion of $\lambda_{em} \equiv f_S(\,\boldsymbol{u}_{em}\,)^{-1/2}$, we can write the **velocity operators** $\hat{\boldsymbol{u}}^{\,n}{}_{em}$ in terms of the operators $\hat{\boldsymbol{D}} = (\hat{D}_1, \hat{D}_2, \hat{D}_3)$:

(51a) $$\hat{\boldsymbol{u}}^{\,n}{}_{em}\,|\,E > \equiv c^{2n}(E - \Lambda q \phi^{em})^{-n}\;\hat{\boldsymbol{D}}^{\,n}{}_{em}|\,E > \quad (\,n = 1,2,3,\ldots\,)\,,$$

(51b) $$\hat{\boldsymbol{D}}_{em} \equiv -i\,\hbar\,\Lambda\,(\,\hat{\boldsymbol{D}} - iq\,\boldsymbol{A} / \hbar c\,)\,,$$

(51c) $$\hat{D}_i A_j\,|\,E > = A_j \hat{D}_i\,|\,E >\,,\quad \hat{D}_i\,|\,E > = \partial_i\,|\,E >\,,\;(\,i,j = 1,2,3\,)\;.$$

One may use a better notation to indicate $\hat{\boldsymbol{u}}^{\,n}{}_{em}$ in (51a), which should not be understood as the nth power of the operator $\hat{\boldsymbol{u}}_{em}$. Instead, it is the corresponding operator of the *classical* quantity $\boldsymbol{u}^{\,n}{}_{em}$. Expand $\lambda_{em} \equiv f_S(\,\boldsymbol{u}_{em}\,)^{-1/2}$ in terms of powers of $\boldsymbol{u}_{em}$ and then replace $\boldsymbol{u}^{\,n}{}_{em}$ by $\hat{\boldsymbol{u}}^{\,n}{}_{em}$ for each $n = 1,2,3,\ldots$. This defines the operator $\hat{\lambda}_{em} \equiv f_S^{-1/2}(\,\hat{\boldsymbol{u}}_{em}\,)$. The Dirac eqn in classical e.m. field A_{μ} now can be written

(52) $$[\,\gamma^{\mu}(\hat{D}_{\mu} - iqA_{\mu} / \hbar c) + \hat{\mu}_{em}\,]\,\psi = 0\,,\; \hat{\mu}_{em} \equiv mc\hat{\lambda}_{em} / \sqrt{\Lambda}\,\hbar.$$

Expanding ψ in energy eigenvectors and $\hat{\lambda}_{em} \equiv f_S^{-1/2}(\,\hat{\boldsymbol{u}}_{em}\,)$ in $\hat{\boldsymbol{u}}_{em}$ we find that (52) is invariant under U(1) gauge transformations

$$A_{\mu} \rightarrow A_{\mu} + \partial_{\mu}\theta\,,\; \psi \rightarrow \psi\,exp(iq\theta / \hbar c)\,,\; E \rightarrow E - \Lambda q c^{-1}\partial_t\theta\,,$$

no matter what f, f_S, λ, and Λ can be. This gauge invariance is due to the neglect of the departure near the range of e.m. force that is extremely long but finite.

For atoms at rest in S , $u << c$ for bound electrons and $\lambda_{em} \approx 1$ is an approximation with extremely high precision. For instance, if $N \sim 10^{40}$ then $\lambda \sim 1-$

$\frac{1}{2}\times 10^{-82}$ for $\gamma\sim 1.005$. Thus,

(53) $$E_n \approx \Lambda^{-1} mc^2[1-\frac{1}{2}(Z\alpha/n)^2] .$$

Here α remains the well-known fine structure constant.

The **covariant 4-wave vector** is

$$K_\mu \equiv (\boldsymbol{K}, i\omega/c) \equiv p_\mu/\hbar = \Lambda^{-1}p^\mu/\hbar \text{ with } \omega \equiv 2\pi\nu$$

and the new Einstein and de Broglie relations are

(54) $$E = \Lambda h\nu , \quad \boldsymbol{p} = \Lambda \hbar \boldsymbol{K}, \quad \hbar \equiv h/2\pi$$

Since $c\boldsymbol{p} = \boldsymbol{\beta} E$ due to (7), (54) yields the new $\boldsymbol{K}-\boldsymbol{\nu}$ relation for a photon

(54a) $$c|\boldsymbol{K}|/2\pi = \beta_\gamma \nu , \quad \beta_\gamma \equiv u_\gamma /c .$$

It returns to current $\boldsymbol{K}-\nu$ relation if one employs fictitious c-photons with $\beta_\gamma \equiv u_\gamma/c = 1$. Due to (53) and (54), discrete atomic spectra are

(54b) $$\nu = \frac{1}{2} mc^2 h^{-1}\Lambda^{-2} Z^2\alpha^2(n_f^{-2} - n_i^{-2}) .$$

Let ν' be the frequency of a photon emitted from a source **at rest** AND **observed in** S', and ν be the frequency of the ***same*** photon **observed in** S. Let $S'//S$. Consider new parallel transformation containing factor function. The new transformation rule for covariant 4-vectors yields

(54c) $$\nu = f_S(V)^{1/2}\sqrt{(1-V^2/c^2)}\,(1-\beta_\gamma\, \boldsymbol{n}\cdot\boldsymbol{V}/c)^{-1}\,\nu' ,$$

where $\boldsymbol{V} = \boldsymbol{V}_{S'S}$ is the velocity of the light source relative to S. The creation and emission of a **real** photon is a sub quantum physical limiting process that takes place within a vertex of Feynman diagram. At the very beginning of the process, a *timelike* **embryonic photon** with a nonzero mass m_γ is created and accelerated. When its speed reaches the speed allowed by available energy from the source, a real photon is born and emitted. Guided by this physical picture, we now deduce (54c) solely from energy formula in particle mechanics and $K_\mu - p^\mu$ relation without using the transformation rule for the wave-vector K_μ. Let $\boldsymbol{u}$, E and $\boldsymbol{u}'$, E' be the velocities and energies of the **same embryonic photon**, observed in S and S' respectively. Then

$$E/E' = \gamma(\boldsymbol{u})\lambda(\boldsymbol{u})/\gamma(\boldsymbol{u}')\,\lambda'(\boldsymbol{u}'), \; \boldsymbol{u} = \boldsymbol{u}' \hat{+} \boldsymbol{V}_{S'S},$$

$$\lambda'(\boldsymbol{u}') \equiv f_{S'}(\boldsymbol{u}')^{-1/2} .$$

From (5) and (2) we obtain $\Lambda/\Lambda'=1/f_S(V_{S'S})$. Then (54) yields

$$E/E'=\Lambda\nu/\Lambda'\nu'=f_S(V_{S'S})^{-1}\nu/\nu',$$

$$\nu/\nu'=f_S(V_{S'S})E/E'=f_S(V_{S'S})\gamma(\mathbf{u})\lambda(\mathbf{u})/\gamma(\mathbf{u}')\lambda'(\mathbf{u}').$$

Using (6) and (2) we find that

$$\lambda(\mathbf{u})/\ \lambda'(\mathbf{u}')=f_S(V_{S'S})^{-1/2}=f_S(V)^{-1/2}.$$

Applying (8), we get

$$1-\mathbf{u}\cdot\mathbf{V}/c^2=(1-V^2/c^2)/(1-\ \mathbf{u}'\cdot\mathbf{V}/c^2),$$

$$\gamma(\mathbf{u})/\gamma(\mathbf{u}')=\sqrt{1-V^2/c^2}\ /(1-\mathbf{u}\cdot\mathbf{V}/c^2),\ \ \mathbf{V}=\mathbf{V}_{S'S}.$$

Finally, by taking the limit $\mathbf{u}\to c\,\mathbf{n}$ or $\mathbf{u}\to\mathbf{u}_\gamma=u_\gamma\mathbf{n}$ for c-photons in special relativity and timelike photons in finite-N theory respectively, we obtain (54c) with $\beta_\gamma=1$ in special relativity and $\beta_\gamma\equiv u_\gamma/c<1$ in finite-N theory for frame-variant frequency of a photon in the two theories respectively. The said ratio is of no mathematical flaw in finite-N theory for the canceled mass m_γ is nonzero. Thus, the **emission theory** works no matter what the factor functions f_S would be. Of course, light sources are not Galileo-Newton's but Einstein's 'guns'. Mysteriously and amazingly, the energy formula in particle mechanics, $E=m\lambda\gamma c^2$, works perfectly for embryonic photons in the sub-quantum acceleration or limiting processes within vertices of Feynman diagrams.

Let ν_S be the frequency of the photon emitted from a source at rest and observed in S. Even for the same kind of sources and the same kind of quantum level differences (e.g., from the 1st excited state to the ground state) $\nu_S\neq\nu'$ because of the violation of the relativity principle. In the open theory with the violation, (54c) is **not** the formula for **Doppler shift**. For any fixed ordered pair (n_i, n_f) of the quantum numbers, (54b), (4), and (2) tell that

$$\nu'/\nu_S=\Lambda^2/\Lambda'^2=f_S(V_{S'S})^{-2}=f_S(V)^{-2}.$$

Then (54c) yields

(54d) $$\nu=f_S(V)^{-3/2}\sqrt{1-V^2/c^2}\ (1-\ \mathbf{u}_\gamma\cdot\mathbf{V}/c^2)^{-1}\nu_S.$$

This is the new formula for Doppler shift. No LV would be found experimentally through Doppler effect since $f_S(V)^{-3/2}$ is almost constant 1 up to very high speeds V and $u_\gamma = c$ is too accurate for all detectable photons. Numerically, (54d) is almost the same as the formula for Doppler shift in special relativity.

In general, Dirac eqn in a pure massive vector field V_μ is

$$(55) \qquad [\gamma^\mu(\hat{D}_\mu - iq\underline{V}_\mu / \hbar c) + \underline{\hat{\mu}_v}]\psi = 0, \; \hat{\mu}_v \equiv mc\hat{\lambda}_v / \sqrt{\Lambda}\,\hbar.$$

Here $\hat{\lambda}_v$ is defined in the same way as $\hat{\lambda}_{em}$.

The approximation $\Lambda q\underline{V}_0 = \underline{U}^{st} \approx \Lambda q V_0$ is invalid for bound quarks within small quark cores where $\underline{U}^{st}$ severely departs $\Lambda q V_0$ due to the effect caused by $mc^2\gamma d\lambda / dt$ term in the ET (19). Since gluons are very heavy and the range of gluon force is too short, pure vector gluon fields without double-gluon and/or loop-diagrams in LH mechanism may play no role in binding quarks. Then (55) does not apply to the quark dynamics needed to explain the QSI facts, if one sets $\Lambda q\underline{V}_0 = \underline{U}^{st} \approx \Lambda q V_0$. In view of LH mechanism, the double-gluon diagrams in radiative corrections with opposite spins as well as the virtual pions in the middles and/or quark-loop diagrams in radiative corrections feedback into classical strong PE. Thus, $\Lambda q\underline{V}_0 = \underline{U}^{st}$ is given by (19c) with (49b).

The new Dirac eqn (55) and the Lorentz invariant Dirac eqn are of the same type for bound quarks in stable protons if their γ-values and λ-values in the mass term are approximately constants. If $\underline{V}_\mu = (\mathbf{0}, -g^2/r)$ (Coulomb-type PE) and the coupling is vector coupling (the PE is the 4th component of a 4-vector), then the expression of the solution indicates that the condition $k^2 > (g^2/\hbar c)^2$ is a necessary and sufficient condition for the existence of bound states. With $k^2 = 1$ and $g^2 = \hbar c / 137$ we have Bohr radius $r_B \sim 10^{-8} cm$. No bound states with $k^2 = 1$ at shorter-than $137 r_B \sim 10^{-11} cm$ distances. Here k is the eigenvalue of the operator

$$\hat{k} \equiv \gamma_4(\boldsymbol{\sigma}' \cdot \boldsymbol{L} / \hbar + 1), \; \boldsymbol{\sigma}' \equiv \begin{bmatrix} \boldsymbol{\sigma} & \mathbf{0} \\ \mathbf{0} & \boldsymbol{\sigma} \end{bmatrix}.$$

A quark in a proton is of $k = 1$. Then $g^2 / \hbar c < 1$ leads to a fictitious proton with its size larger than $10^{-11} cm$. Therefore, $\underline{V}_\mu = (\mathbf{0}, -g^2/r)$ is incorrect for bound quarks. Coulomb-type or inverse PEs do not work for composite hadrons.

Linear PEs $U = ar$ were used in many potential models of quark dynamics to phenomenologically explain quark confinement and energy spectrum of composite hadrons. It works with Schrödinger eqn. However, when solving Dirac eqn

$$[c\,\boldsymbol{\alpha} \cdot \hat{\boldsymbol{p}} + \beta(mc^2 + U_0' + a'r + b'/r)]\psi = (E - U_0 - ar - b/r)\psi$$

it was found [21] that a linear PE could not lead to bound states unless the scalar linear PE **dominates** the vector linear PE or the linear PE is purely scalar. Namely, $a'^2 > a^2$ is a necessary and sufficient condition for the existence of bound states. Otherwise, the bound states could not exist and this was called '**Klein-Paradox**'. Although composite models of leptons and quarks are hypothetic and their sizes seem to be much less than estimated in [21], some features revealed in many potential models (published in 70's and early 80's) seem to have **unexpected radical implications**. Those potential models are for the strong force biding quarks or for the unknown force binding hypothetic constituent particles of composite leptons and quarks. The full meaning of those implications is not clear. It seems more complicated than the transition from Schrödinger to Dirac eqn.

The eqs of motion in classical particle mechanics since Newton are of the same type, whether or not they contain γ or $\lambda\gamma$. They do not have a feedback from quantum theory. It is unknown whether exchanging scalar and vector force-carriers should lead to different places of the corresponding PEs in Dirac or new Dirac eqs.

One thing has been clear since the solution of Dirac eqn was obtained in 1930s. That is, the solution of Dirac eqn for bound states with $k^2 = 1$ at shorter-than $10^{-11} cm$ distances does not exist if the PEs are pure Coulomb-type (inverse) PEs $U = -g^2 / r$ as the 4th component of a 4-vector. We may call this inverse-size contradiction. It explicitly tells that LV is needed to lead to a departure from pure Coulomb-type PEs at short distances. Vector theory without non-limiting LV will not work, no matter it is massless or massive. We never saw any model, such as QCD, which tries to build a non-phenomenological theory of strong force, ever showing a way to get rid of the inverse-size contradiction.

Remark: New development seen in DS theory reveals crucial importance of MFs while LV may play no significant role in QD. It's crystal clear that MFs cause departure from inverse PEs in SA world.

J. Schwinger [22] thought, "The picture of an infinite sea of negative energy electrons is now best regarded as a historical curiosity, and forgotten." To quote Weinberg [3], "And if the hole theory does not work for bosonic antiparticles, why should we believe it for fermions?" Loop diagrams and vacuum polarizations seem to be the only relic of the concept and picture we actually do not believe. May be, the loops in QED should be replaced by double-photon diagrams with a pair of virtual e^+e^- in the middle, and all the catastrophic results and ghosts including Landau ghosts will disappear. If the strong PE (19c) with (49) is found to be unable to produce bound states as long as it is put in the place where atomic Coulomb PE was in the Dirac eqn, we have an extra-reason for the double-gluon diagrams in the LH mechanism to replace the quark-loops.

The new Dirac eqn in massive gluon fields is unknown at the present for $\underline{V} \neq V$ is unknown. For high-speed quarks, the effect of vector gluon field $\boldsymbol{G}$ is not negligible. However, the strong force formula (45) tells that the force component associated with $\boldsymbol{G}$ is perpendicular to velocity and therefore it would not accelerate escaping bound quarks to be against binding. Even if the strong PE is not solvable strictly, one can still see the reality of confinement.

It is not terrible to see an eqn that cannot be solved strictly. Quantum wave eqs are solvable only for two-particle systems. Even for Hydrogen, once the departure from Coulomb PE at shorter-than proton size distances is considered, it might be no longer solvable. Exact solutions of Einstein field eqs so far have only been found for static spherically symmetric sources, which do not exactly exist in the universe. Even if one uses Newton's laws, only two-body dynamic system is solvable. However, two body systems do not exist in nature without taking approximations. Whether a solvable strong PE can be devised from a particular model λ will be investigated in the future.

Letter E has always been used to denote energy eigenvalue in QM. So we set

$$E_p \equiv mc^2\lambda\gamma = (c^2\boldsymbol{p}^2 + m^2\lambda^2c^4)^{1/2}.$$

If **Dirac replacement**

$$(c^2\boldsymbol{p}^2 + m^2c^4)^{1/2} \to c\ \boldsymbol{\alpha}\cdot\hat{\boldsymbol{p}} + \beta mc^2$$

can be generalized to

$$(c^2\boldsymbol{p}^2 + m^2\lambda^2c^4)^{1/2} \to c\ \boldsymbol{\alpha}\cdot\hat{\boldsymbol{p}} + \beta mc^2\lambda$$

for **free** particles and

$$(c^2\boldsymbol{p}^2 + m_q{}^2\lambda_{st}{}^2c^4)^{1/2} \to c\ \boldsymbol{\alpha}\cdot\hat{\boldsymbol{p}} + \beta m_q c^2\hat{\lambda}_{st}$$

for quarks in strong interactions, and $E = E_p + \underline{U}_{st}$ is the energy-eigenvalue of a bound quark, the Dirac eqn for bound quarks can be written

(55a) $$[\gamma^\mu(\hat{D}_\mu - iq_s\underline{V}_\mu/\hbar c) + \hat{\mu}_{st}]\psi = 0\ ,\ \Lambda q_s\underline{V}_0 = \underline{U}^{st}\ .$$

Here $\hat{\mu}_{st} \equiv m_q c\hat{\lambda}_{st}/\sqrt{\Lambda}\,\hbar$. In particular, taking approximation $\underline{\boldsymbol{V}} = \boldsymbol{0}$, we write

(55b) $$[c\ .\boldsymbol{\alpha}\cdot\hat{\boldsymbol{p}} + \beta m_q\hat{\lambda}_{st}c^2]\psi = (E - \underline{U}_{st})\psi\ .$$

Here $\hat{\lambda}_{st}$ may still be a constant for bound quarks in stable baryons. The high-speed quarks are of such large γ-values that the classical quantity $\lambda(\gamma)$ is not a constant function even as approximation. But bound quarks in stable baryons may be in the s-states and their large γ-values may be constant or quasi-constant.

Chapter 7 Concepts of Quantum Theory of Massive Vector Fields

Violations of discrete symmetries such as **P** and **CP** are related to the different transformation properties of right- and left-handed fermion fields. The violations are independent of the factor function f. Moreover, a finite N does not have an effect on **CPT**. Clearly, dS^2 is invariant under ***non-boost*** transformations (**C**, **P**, **T**, and their combinations). To ensure the c-postulate, the consistence condition holds for the transformation factors written in terms of space-inversed inertial systems. Then it can be shown that $\lambda \equiv f_S(\boldsymbol{u})^{-1/2}$ and $\Lambda = f(V_{SK})$ are invariant under **non-boost** transformations. For example, if S_{inv} is the **space inversion** of S then $dS_{inv}{}^2 = dS^2$. Similarly, $dK_{inv}{}^2 = dK^2$ and $dS^{*2}_{inv} = dS^{*2}$. Then

$$f_{S_{inv}}(\boldsymbol{V}_{K_{inv}S_{inv}}) = f_S(\boldsymbol{V}_{KS}) \text{ and } f_{S_{inv}}(\boldsymbol{V}_{S^*_{inv}S_{inv}}) = f_S(\boldsymbol{V}_{S^*S}).$$

Hence $\Lambda = f(V_{SK}) = f_K(V_{SK}) = 1/f_S(V_{KS})$ and $\lambda \equiv f_S(\boldsymbol{u})^{-1/2}$ ($\boldsymbol{u} \equiv \boldsymbol{V}_{S^*S}$) are invariant under the space inversion, in spite of the fact that the velocity $\boldsymbol{u}$ of any particle changes its sign under the space inversion

$$\boldsymbol{V}_{S^*_{inv}S_{inv}} = -\boldsymbol{V}_{S^*S}, \ \boldsymbol{V}_{K_{inv}S_{inv}} = -\boldsymbol{V}_{KS}.$$

The only difference between finite-N theory and the current theory is that the eqs contain Λ and/or λ, which hide in the current theory in the form of constant 1 factor. For instance, $m \cdot 1 = m$, and we can say that the factor 1 hides there. Therefore, the CPT theorem now has a generalized version:

A quantum theory of fields is CPT invariant if it obeys the c-postulate, regardless of what functional form of f, which produces an entire set of factor functions $\{f_S\}$ via consistence condition, matches the reality.

Now that the CPT is due to the c-postulate, CPT violation requires c-postulate violation. It's difficult to figure it out how to choose spacetime manifold and the transformations so that the c-postulate as well as the CPT will be violated in certain way that seems to be true. But we can infer that even if the CPT violation is one day to be found true due to some reasons we do not know right now AND indeed causing particle-antiparticle symmetry breaking, the violation of lepton and quark number conservation will be still either zero or extremely small. A tiny violation would unlikely to lead to the observed rareness of antimatter if particles and antiparticles were equal in amount at the beginning. A law with severe CPT violation under the extreme condition at the beginning but with almost no violation under current condition in the universe is hardly imaginable. May be, if one believes equal amount of matter and antimatter at the beginning, he/she must believe the existence of abundant antimatter somewhere in the space T times larger than the observed part of the universe. T might be ranging from 2 to 10^{40}. From the current experimental observations and records, we have no further information about the size of the universe and the total amount of matter and antimatter. We are too tiny in the universe and we always have more unknown things than what we have known.

Remark: The above argument contains a ***mistake*** the author made. He should not support the idea of a bizarre initial condition of ***equal numbers*** of ***matter*** particles and ***antimatter*** particles. It's corrected by noticing that nearly all quarks and electrons would have irreversibly annihilated at the beginning of the Big Bang and soon after, if the said bizarre initial condition were true. See page 242 for more details.

In massless CED, it was ***postulated*** ([19] J.D. Jackson) that $L_{\text{int}} = -j^{\mu}A_{\mu}/c$ "*from the form of the electrostatic and magnetostatic energies, or from the charge-particle interaction Lagrangian*". The two premises lead to the same L_{int} and support the H_{int} used in massless QED at any distances. In massive CED, L_{int} is only required to yield the eqn of motion with $\boldsymbol{F}$ determined by the new force formula, while the PE contains the effect of the term $mc^2\gamma d\lambda/dt$ at very large momenta or very short distances. The e.m. PE (19c) with replacement (49a) indicates that it is numerically Coulomb PE between $R^{c}{}_{1em}$ and $R^{f}{}_{em}$. Thus, the space force and its effect (asymptotic freedom and the existence of deep PW at $r < r_{0em} \sim R^{c}{}_{1em}$) can be neglected in massive QED for all e.m. phenomena, taking place at $r >> R^{c}{}_{1em}$. Then we can construct a Lagrangian density that includes IR effect of e.m. YF for massive QED:

$$(56) \qquad L_{em} = -\frac{1}{4}F_{\mu\nu}F^{\mu\nu} - \overline{\psi}(\gamma^{\mu}\partial_{\mu} + \hat{\mu}_{em})\psi + j^{\mu}A_{\mu} \,,$$

where $j^{\mu} \equiv iq\overline{\psi}\gamma^{\mu}\psi$ and $\hat{\mu}_{em} \equiv mc\hat{\lambda}_{em}/\sqrt{\Lambda}\,\hbar \approx mc/\hbar$ is very accurate for the e.m. interactions within the accessed energy scales. The interaction term is numerically valid at all distances $r >> R^{c}{}_{1em}$ and, whenever vector potential $\boldsymbol{A}$ can be neglected, returns to Coulomb PE between $R^{c}{}_{1em}$ and $R^{f}{}_{em}$. The Coulomb PE appears as the 1st order approximation of the e.m. PE in new physics with high precision within $(10R^{c}{}_{1em}, R^{f}{}_{em}/10)$. To get accurate numerical results, we may set $\Lambda = 1$ and $A_S = 0$ for all earth frames, and $\lambda = 1$ for all real particles at accessed energy scales. Once $\Lambda = 1$, the difference between a covariant and contravariant index disappears.

We should feel good about all appropriate approximations. For example, massless QED would be unsolvable if the integration intervals for t were replaced by true finite intervals and gravitation must be considered in Feynman diagrams and rules. The infinite intervals provide with elegant Dyson series. Appropriate approximations in QFT help to shape a solvable theory. No matter how small the photon mass is, neglecting Yukawa-type range factor for e.m. force, however, causes IR catastrophe in radiative corrections and hence is not appropriate.

The significant modification of standard model in a **unified massive vector**

field theory in finite-N framework is the removal of all Abelian and non-Abelian gauge symmetries and gauge-byproducts (Higgs terms and GS coupling terms) associated with the concepts and mass terms shown in Proca and Lee-Yang's Lagrangians. No single experimental fact and the merit of the standard model would be affected by this major change. The simplicity is obvious in the sense that gauge arbitrariness and self-couplings disappears. The complexity is the details of the factor functions that enter the eqs of the laws of physics. All algebraic coupling styles except for Higgs and GS couplings can be preserved. The unified quantum theory of fields in finite-N framework can recover the existence of neutral currents, the masses of weak bosons, the Cabibbo and Weinberg angles, the mass splitting, and many other merits.

As long as Hamiltonian H is of energy dimension, H_I must be a PE. Strong PE (19c) with (49b) can be used to build H_S. To explain quark confinement and asymptotic freedom, we now use (19c) with (49b) as a **potential model** built with Q-feedback to explain fundamental experimental facts of strong interactions without knowing how to build a quasi-Lagrangian or a Lagrangian with a strong PE that would include 3-vector component $\boldsymbol{G}$ of the gluon field G_μ. One thing we do know is that as we neglect $\boldsymbol{G}$, the strong PE (19c) with (49b) serves as interaction term H_I just like one may neglect $\boldsymbol{A}$ and use Coulomb potential energy to calculate radiative corrections and to interpret Lamb shift in QED. Radiative corrections are only able to explain small portions of the experimental values of quantum energy levels in both massless and massive QED for the e.m. coupling is weak and it brings about only tiny effect in higher-order effects. However, in view of LH mechanism, the range of gluon force is too short and the observed strong force is solely due to the double-gluon and/or u/d quark-loop diagrams in radiative corrections. The strong PE in massive QCD now itself is the consequence of radiative correction with double-gluon and/or u/d quark-loop diagrams, obtained from the ET (19) and the Q-feedback with LH mechanism.

Although the new Dirac eqn with the strong PE may not be solvable, it is expected that similar to a deep ***square*** PW, the wider the deep PWs in the strong PE (19c) with (49b), the smaller the energy level differences of bound quarks and anti-quarks in the well. Since the width is roughly of the same order as the 1st strong CD $R^c{}_{1st} = 2\sqrt{2}\eta e_s{}^2 / \hbar\gamma_1 m_q c^2$, the shorter the 1st strong CD, the larger the energy level differences between ground and radially excited states. Small differences between ground and higher-spin baryons are due to the small orbit-spin coupling caused by the quasi-constant portion of the strong PE at the bottom of the strong PW.

Gluons are very heavy in view of LH-model. It explains why radiation of real (not virtual) gluons can never happen in all today's accelerators regardless of the sizes of quark cores, the values of $R^c{}_{1st}$, and the differences between quantum energy-levels of bound quarks. Radiation can happen at accessed energy scale with unchanged flavors only in the form of unstable pairs of quark-antiquark such as pions and J/ψ particles. Thus, neutral pions and pairs of $\pi^+\pi^-$ produced in high energy collisions were produced by double virtual timelike gluons (with opposite spins), while J/ψ particles by single virtual timelike gluons via t-channels. On the other hand, when quarks change their flavors, charged unstable hadrons and/or l-ν_l pairs can be emitted to release the energies. But radiation of real gluons could never happen in accessed energy scales, regardless of the size of quark cores and the adjustable ratio e_s^2 / M^2.

Two attracting quarks have interaction represented by the strong PE in the form of (19c) with (49b). The deep PW at distances shorter than the 1st strong CD can be called **external** and **internal** PW for two quarks within different nucleons in collisions and two quarks within the same nucleon respectively. A quantum particle penetrates a potential hill higher than its kinetic energy through *quantum tunnel* and an intruding quark penetrates the deep external PW through quantum tunnel too. Here we must explain the term 'quantum tunnel':

The well-known example of quantum tunnel is the **penetration** of wave function into a **classically forbidden** region (under the condition of energy conservation, of course). However, we would like to include PW-related phenomena as quantum tunnel phenomena too. It is dangerous for a classical 'child' to play nearby a deep well. Once the 'child' leaves the ground of the edge of the well and is over the well, 'he/she' will fall into the deep until hits the bottom of the well. However, a 'quantum child' with positive ME playing nearby the well is safe. 'He/she' may visit with small probability but will not fall into the deep and the probability to reach the bottom in the visitation is almost zero for a very deep well. We would like to say that the 'child' **passes by** the well through quantum tunnel **without falling** into the region 'he/she' **classically must** go to. To quantum 'child' with positive ME, the existence of an external deep PW nearby is of perturbation effect only.

The probability for a bound quark to be outside the internal PW of finite depth is never zero. This is not the penetration we are talking about for its main region of probability remains inside the PW. It jumps out from the well and becomes a free particle with non-negative ME iff it receives large enough amount of energy. This is why quark 'evaporation' has never been observed. In general, the concept of 'evaporation' of particles bound in deep PW is a misunderstanding of quantum tunnels. Even if the 1st strong CD and the width of the strong PW are approximately equal to the diameter of a proton, any two quarks in high-energy head-on collisions will not leave their own families to form a new bound state as long as they were not bound in the same nucleon before the collision.

Experimental data of asymptotic freedom, the running of effective strong coupling, low through high energy N-N and l-N scatterings and collisions, mass spectra including excited states of hadrons as composite particles, decay rates and lifetimes of hadrons would join forces to determine the values of $e_s^2, M, R^c{}_{1st}$, and a model λ that produces a working strong PE and interaction term in Hamiltonian. For many adjustable values of the pair (e_s^2, M), quarks are always confined in deep PWs. The algebraic coupling styles in the standard model might have no effect on a classical formula for strong force such as (45) and on a strong PE such as (19c) with (49b). Dirac matrices did not affect the Lorentz force formula and the Coulomb PE. At least the qualitative feature of the strong PE and the deep PW will not be affected and the quark confinement remains a reality.

In a small quark core model, $R^c{}_{1st} << R^f{}_s \sim 10^{-13} cm$. Set $a_0 = 1$ for N-N scattering/collision. Then for all $R^c{}_{1st} << r << R^f{}_s$, the strong PE (19c) with (49b) yields an inverse PE with high precision if η is treated as a constant. The LH mechanism shows that η is running smoothly without any divergent pole. Taking account of the effect of vector potential $\boldsymbol{G}$ of gluons, we may write $H_{st} = H_s + H_{ms}$, where the SU(3) symmetry-breaking term $H_{ms} = \beta\lambda_8$ and the SU(3) symmetric term

(57) $$H_s = \sum_q \bar{q} \underline{g}_s \sum_{a=1}^{8} \gamma_\mu G^\mu{}_a \frac{1}{2}\lambda_a q$$

in accordance with the algebraic coupling style in the standard model. The difference is that $\underline{g}_s$ is very large while the gluon fields $G^\mu{}_a$ are very massive and the strong PE is determined by the classical theory with Q-feedback from (57) using LH mechanism. The SU(3) algebraic structure does not change the deep PW of the strong PE produced by model λ (20) or alike and is consistent with quark confinement. This situation is similar to the Dirac matrices that do not change the quantum levels of Schrödinger eqn greatly. The significant change (spin-orbit coupling) there is caused by the replacement of Newton's $E - \boldsymbol{p}$ relation with Einstein's. The replacement demands Dirac matrices for a 1st order differential eqn. Quark confinement now is the reality for large variety of the values of e_s and gluon mass M.

For massive QED, the e.m. CDs are too short for us to consider an e.m. interaction Hamiltonian working at such short distances. In any LH-model of strong interaction, the 1st strong CD may be an accessed distance and the issue to construct interaction Hamiltonian near and within the 1st strong CD and after is serious. At the present, the author does not know how to construct it. However, as we indicated before, it is not needed for the purpose to explain quark confinement and the phenomena that occur at distances longer than the sizes of small quark cores.

Future accelerators may be across states, countries, entire globe, the solar system, even the Milky Way. It is too adventurous to claim that Einstein's formulas will always be valid. As an exact theory with $\lambda(\gamma) \equiv 1$ at any high-energy scales, it has not been proved experimentally and experiment will never prove it in any future due to the MFP of physics. It is impossible to accelerate an electron to such a high speed that its energy would equal Mc^2, where M is the total mass of the observed part of the universe, and to see if Einstein's formulas and eqs are still valid in that energy regions. This situation itself is not an evidence that the foundation has been disproved. We have shown many EDF facts that demand innovation. In particular, massless particles and their by-products have been disproved by overwhelming experimental facts. Beyond the reach of today's accelerators, we have mysterious 'desert' with no direct experimental data about the accurate speeds and accurate γ-values of large-γ particles. Then no comparison of $E = m\gamma c^2$ and $E = m\lambda\gamma c^2$ can be made and tested. What we do know right now is that the limit $N \equiv \lim \lambda\gamma$ as $u \to c$ is finite and all particles are massive. The meaning of the limiting LV for the limit

$\gamma \rightarrow \infty$ (far beyond the reach of accelerators in any future) and of non-limiting LV for the γ-values near the critical γ_0 and within the interval (γ_0, N) has been shown. The experimental evidences are clear. However, we still do not know an accurate form of non-limiting LV from low to high-energy scales within the reach of accelerators. This does not affect the qualitative and half-quantitative explanations to a series of experimental facts.

We have replaced the 2nd postulate by the c-postulate. When we undo the 1st postulate, what else do we have put into the axiom system to replace it, to include it as a special case, and to have it as an approximation? The new eqs of laws of physics are covariant under the new transformations. This can be called the **general principle of covariance**. It itself alone cannot determine the choice of manifold and the coordinate systems in the manifold. It is the axiom system that determines the choice. This principle is the common essence of Galileo-Newton's relativity principle and Einstein's special and general principles of relativity. Surely, any frame-invariance is just a special and simple case of covariance.

Chapter 8 Dark Matter and Dark Energy

8.1 Major Failures in General Relativity and Remedy

Iconic EF-expansion formula (see page 193, eqn 58a) can be rewritten as

$$|dR/dt| = c(const \cdot G \cdot R^{-1} - 1)^{1/2}$$

It is wrong due to the reasons given on pages 8-9.

Non-constant G (Dirac's idea according to Weinberg [6]), Mach's principle with constant G, spontaneous breaking of G-symmetry, and constant G with a running factor are equivalent for the issue of expansion. Such departure from EF-type gravitation and cosmology that claims constant G without any running factor and without any Mach factor as realization of Mach's principle is proved to be necessary and sufficient to solve the problems. Please note, Einstein did not put anything in his field eqs to realize Mach's principle he had so cherished. Now that the four notions, interpretations, and opinions, are equivalent when used to explain the expansion, we will keep using one notion: spontaneous breaking of G-symmetry or non-constant G. All the deduced eqs and formulas will be the same no matter which notion we like the most.

Denser dust particles and molecules in ancient **smaller** universe may also make remote supernovas dimmer than expected. Non-constant G or constant G with a running scalar factor (spontaneous breaking of the G-symmetry), written $G = G_1 b(R)$, with $R = R_1$ at the present time t_1, $G_1 = 6.67 \cdot 10^{-8} cm^3 / g \cdot s^2$, and $b(R_1) = 1$, explains **any** empirical behaviors of **genuine** $R(t)$ and $h(t)$, if G obeys the G-eqn: $G = \pi R(c^2 - R^2 dh/dt)/2M_{total}B$ (see **G-theorem** late). It has nothing to do with whether the supernova results really mean acceleration and dark energy and how much they mean such things. If the dimmer supernovas do not mean acceleration and dark energy, they must mean denser dusts and molecules in the ancient smaller universe. Then discovery of remote dimmer supernovas provides stunning 2nd evidence of the expansion of universe since the only Doppler-interpretation of Hubble's discovery was made long times ago. In constant G notion with Mach's principle, $b(R)$ serves as the Mach factor. The G-theorem also holds in any special case, including the EF-model with constant G (no Mach factor nor running factor), $B \equiv 1$. Then the **price** is that monotone decreasing dR/dt conflicts with supernova results in the sense that it is unlikely for denser dusts and molecules in ancient smaller universe to block light to such degree that remote supernovas appear dimmer than expected in the case of decreasing rate of expansion. Furthermore, in the EF-expansion, **fatally**, faster-than-c and near-c receding leads to c-failure that induces **age-failure** for the expansion had been too fast for very long period of time. For the EF-universe, age$< 2/3H_0$ (see **age-theorem** later). With recent WMAP-h, $2/3H_0 \sim 9.18 \cdot 10^9$ yr. The proved young age of the EF-universe supports an Einstein's [7] lifetime conclusion: "I see no reasonable solutions" (due to the young age). The age is not

$1.377 \cdot 10^{10} yr$ since the formula $age = 1/H_0$ is neither a theoretical nor an empirical formula. Please check the rigorous proof of the age-theorem later.

With T_i^i -terms of **CHSPs** (**cosmic high-speed particles**), field eqs with cosmological ***constant*** (negative pressure or not) are ***inconsistent*** (see **CHSP-theorem** later). Thus, WMAP actually proves that universe is not flat just because it is claimed that WMAP matches theoretical prediction of flat universe accurately. Since inconsistent model departs reality definitely, its accurate agreement with observation stands as a decisive empirical disproof. Let's view from another angle: indeed, if a tiny dot became nearly flat, the expansion would be too fast, producing the c-failure as a fatal conflict with observation.

A sample parabola model for quantity $c^{-2}R^2 dh/dt$ is proposed in case acceleration is real. It is equivalent to a model of non-constant *G* according to the *G*-eqn. The recession speed is **less than** c all the times. Positive curvature condition now only requires positive density according to the new formula for curvature. The new formula returns to the formula in GR whenever one returns to EF-setting (constant *G* without non-constant factors and without CHSPs). Supernovas within 5.01 billion ly appear normal. The relative departure from the *G*-symmetry in a time-period of 500 years is less than $6.50 \cdot 10^{-8}$. An old age (about 20 billion years) is obtained. Some **non-parabola** models with faster expansions at early times can easily **reduce** the age to any value between 11 and 20 billion years (depending on the choice of parameters), without faster-than-c and near-c recessions. We do not have empirical value of the age of the universe and empirical curves of *R(t)* and *h(t)*. Radioactive dating is unable to tell the age of stable hydrogen born before unstable nuclei. Many CBR photons are non-relic, produced in star-era by '**converters**' (planets, bodies, and dusts), that convert part of incident photons' energy into thermo-energy and release it in the form of infrared and microwave photons (see pages 229, 230). The difficulty to get empirical curve of *h(t)* or *h(R)* is explained. However, the age-failure now disappears together with the c-failure in the EFD-type gravitation & cosmology.

The celebrated formula in the **EF-model** can be written

(58a) $$\left(\frac{dR}{dt}\right)^2 = c^2\left(\frac{A}{R} \mp 1\right), \quad A \equiv \frac{8\pi G}{3c^2}\rho R^3 = const.$$

We call *A* the **EF-constant** for *G* is a constant in all EF-type models, w/wo (with or without) LV, w/wo inflation.

Immediately, we see infinite and near infinite speed of expansion at the beginning, terribly faster-than-c receding at early times, and the recession too fast in a very long period of time. Inevitable by-products of such expansion mode are near-c recessions some times between, producing extremely large red shifts never seen. We call this the **c-failure**. Besides, the monotone decreasing rate (58a) fails to explain *acceleration* implied by dimmer supernovas. It is called the **a-failure**.

From (58a) one can easily deduce an $\dot{h} - equation$:

(58b) $$\dot{h} \equiv \frac{dh}{dt} = \frac{c^2}{R^2}\left(\pm 1 - \frac{3A}{2R}\right), \qquad h \equiv \frac{1}{R}\frac{dR}{dt}.$$

When CHSPs are considered with large enhancement of their energy density from the LV factor function, A in (58b) must be replaced by the product AB where B is the CHSP-quantity. See (68) and (74) for details.

Positive dh/dt is sufficient (but not necessary) to yield acceleration. A necessary (not sufficient) condition for $dh/dt > 0$ is a positive curvature. Positive curvature could not lead to acceleration in the EF-model since the rate dR/dt in (58a) is monotone decreasing as R increases in the expansion. Thus, $h \equiv R^{-1}dR/dt$ is also decreasing monotonically. Negative curvature always leads to double failures: c-failure and a-failure, no matter LV is ignored or considered. Recent WMAP observations are interpreted as that total mass (baryonic and dark matter) accounts about 30% of the 'critical density'. If we accept the EF-model and do not mind the said failures, WMAP can only be interpreted as showing negative curvature due to the unique curvature condition deduced in the EF-model. Amazingly, once the a- and c-failures disappear in a new theory of gravitation and cosmology based on the c-principle, the meaning of 'critical density' is dramatically changed. Then the same WMAP results prove a positive curvature. The curvature condition will be rigorously discussed, where the EF-critical density and its full meaning appears as a special case. See (75, a, b) and (83a, b, c, d) for details.

A cosmological model without acceleration might be right for dimmer supernovas might be due to denser dusts and molecules in ancient (smaller) universe. But it is unlikely that the dusts and molecules were so dense five billion years ago that the ancient supernovas could be dimmer than expected even if *dR/dt* was decreasing.

The c-failure is definite. At the very beginning when Einstein and Friedman published their work, (58a) contradicted the c-principle established in SR. The observation of stars' red shifts from Hubble to WMAP, has repeatedly proved the c-failure: No single galaxy has ever been found to have extremely large red shifts, which are the by-products of near-c recessions in the EF-type model w/wo LV, w/wo inflation. This c-failure induces **age-failure** for the EF-universe had expanded too fast for a very long period of time. Thus, it is still young when it has grown to its present size.

Now let us prove how young the EF-universe is. Denote by R_1 and h_1(= H_0) the present values of R and h respectively. We now state and prove an age theorem.

Age-Theorem: The present age t_1 of the EF-universe with positive curvature is absolutely less than $2/3H_0 \equiv 2/3h_1 = 9.18$ billion years, no matter how many galaxies and dark matter will be found in any future and how large the universe is at the present. (Here $H_0 \equiv h_1 \sim 71 km/\sec/Mpc$).

Proof: In the case of *positive* curvature, (58a) in the EF-model yields

$$(58) \qquad \left(\frac{dR}{dt}\right)^2 = c^2\left(\frac{A}{R}-1\right),\ R \le R_{\max} = A \equiv \frac{8\pi G\rho R^3}{3c^2} = const,$$

$$(59) \qquad h^2 = \frac{c^2}{R^2}\left(\frac{A}{R}-1\right), \qquad A \equiv \frac{8\pi G\rho R^3}{3c^2} = const.$$

Obviously, (59) is equivalent to

(59a) $$R(1+\frac{h^2}{c^2}R^2)=A.$$

Now that A is a constant, we have an eqn for present size R_1:

(59a)′ $$R_1(1+\frac{h_1^2}{c^2}R_1^2)=A.$$

From now on we always use subscript 1 to label **present** values. Rewrite (59a) as

(59b) $$u(1+\frac{h^2A^2}{c^2}u^2)=1,\quad u\equiv\frac{R}{R_{\max}}=\frac{R}{A}.$$

$$A=\frac{c}{h}\frac{1}{u}\sqrt{\frac{1}{u}-1}=\frac{c}{h_1}\frac{1}{u_1}\sqrt{\frac{1}{u_1}-1}=const.$$

Then the solution of (58) leads to a t-R relation (**age-size relation**)

(59c) $$t=\frac{\pi A}{2c}-\frac{1}{c}\left(R\sqrt{\frac{A}{R}-1}+A\tan^{-1}\sqrt{\frac{A}{R}-1}\right)$$

$$=\frac{1}{h}\frac{1}{u}\sqrt{\frac{1}{u}-1}\left(\frac{\pi}{2}-u\sqrt{\frac{1}{u}-1}-\tan^{-1}\sqrt{\frac{1}{u}-1}\right),$$

$$u\equiv\frac{R}{R_{\max}}=\frac{R}{A},\qquad A\equiv\frac{8\pi G\rho R^3}{3c^2}=const.$$

It gives an EF-**age formula** to tell the **present age**

(59d) $$t_1=\frac{1}{h_1}\frac{1}{u_1}\sqrt{\frac{1}{u_1}-1}\left(\frac{\pi}{2}-u_1\sqrt{\frac{1}{u_1}-1}-\tan^{-1}\sqrt{\frac{1}{u_1}-1}\right),$$

$$u_1\equiv\frac{R_1}{R_{\max}}=\frac{R_1}{A},\qquad A\equiv\frac{8\pi G\rho R^3}{3c^2}=\frac{8\pi G\rho_1 R_1^3}{3c^2}=const.$$

From (59) we get the 1st **SD-eqn (size-density eqn)** (59e) and a **SD-relation** (59e)′ in general:

(59e) $$R_1=\sqrt{\frac{3c^2}{8\pi G\rho_1-3h_1^2}},$$

(59e)′ $$R=\sqrt{\frac{3c^2}{8\pi G\rho-3h^2}}.$$

With (59e) only, R_1 and ρ_1 determine each other without real determination. A 2nd SD-eqn is needed badly together with (59e) to form a system of eqs to determine them. Without knowing u_1 in (59d), the present age t_1 remains unknown even in the presence of the nice EF-age formula (59d). As the first step, we need to know the present size R_1 badly. We now deduce a desired 2nd SD-eqn.

Let n_m be the statistic average number of the galaxies that can be observed in a sky-area of the same size of a full moon at its average distance from us. It takes about 195 thousand full moons to cover the entire northern and southern galactic hemispheres. Let $k_n \cdot 2 \cdot 10^{11}$ be the total number of luminous stars in our galaxy and $M_o = k_o k_n \cdot 2 \cdot 10^{11} M_{sun}$ be its total luminous mass. The original EF-model without LV did not consider any **cosmic high-speed particles** (**CHSPs**) that are not at rest in the celebrated coordinate system that co-moves with receding galaxies. The average mass of all galaxies **including** all the non-CHSP **dark matter** (within and between galaxies) that **co-moves** with receding galaxies is written

$$\overline{M}_{glx} = k_d k_g k_o k_n \cdot 2 \cdot 10^{11} M_{sun} .$$

It holds rigorously at the price that all **factor-parameters** are of uncertain values. For them, we don't have empirical values but rather rough empirical estimations. Since most galaxies are reported smaller than our galaxy, $k_g < 1$. Because dark matter obviously contributes, we have $k_d > 1$. The value of k_o is determined by the average mass of the stars in our galaxy, largely determined by the percentages of large stars. Once these percentages are obtained from observation, k_o can be determined or estimated. The total mass of all the non-CHSP matter considered in the EF-model with positive curvature and without LV (*total EF-mass*) is

(59f) $$M_{total} = k_m n_m 10^{49} g , \qquad k_m \equiv 1.95 \cdot 2 \cdot 1.99 k_d k_g k_o k_n .$$

In a word, the uncertain value of k_m is related to the uncertain value of the *average mass* of all galaxies with all co-moving dark matter within their 'metro-spaces', while the uncertain value of the product $k_m n_m$ is solely related to the uncertain value of the *total mass* of the universe considered in the EF model.

Since $M_{total} = \rho V_{total} = 2\pi^2 \rho R^3$, (59f) leads to a 2nd SD-eqn

(59g) $$\rho_1 R_1^{\ 3} = \frac{k_m n_m}{2\pi^2} 10^{49} g . \qquad (\rho R^3 = \rho_1 R_1^{\ 3} = const).$$

It induces a formula to estimate the value of the EF-constant A:

(59g)′ $$A \equiv \frac{8\pi G}{3c^2} \rho R^3 = 3.145 k_m n_m 10^{20} cm = const .$$

Obviously, (59a)′ and (59e) are equivalent. Either one of them can be called the 1st SD-eqn. Now, for each given value of the product $k_m n_m$, (59a)′ and (59g) form a consistent system of eqs in two variables R_1 and ρ_1. Using (59g)′ we write the system as

(59h)
$$\begin{cases} R_1(1+\frac{h_1^2}{c^2}R_1^2) = 3.145k_m n_m 10^{20}cm \\ \rho_1 R_1^3 = \frac{k_m n_m}{2\pi^2}10^{49}g \end{cases}$$

Once k_m and n_m are determined by observation, the solution (R_1 , ρ_1) as well as the EF-constant A in (59g)′ will be uniquely determined. Consequently, the important quantity $u_1 = R_1 / A$ will be determined too. It then will immediately determine the present age t_1 via the EF-age formula (59d). Unfortunately, we do not have empirical values but rough empirical estimations of k_m and n_m . It would probably take thousands of years to figure out M_{total} of the universe and the value $k_m n_m = M_{total}10^{-49}g^{-1}$. Theoretical values of R_1 , ρ_1 , A, u_1 , and t_1 are actually unknown and unclear even in the presence of a pile of nice formulas.

The essence of this age theorem we are now proving is that the upper bound of the EF-age t_1 of universe can be surely obtained by treating u_1 in (59d) as a variable and finding out the absolute maximum value of the EF-age function $t_1(u_1)$ in (59d). The upper bound we are proving will confirm Weinberg [6] age-inequality: $t_1 < 1/h_1$, $h_1 \equiv H_0$. The new age-inequality shown in the age-theorem is finer: $t_1 < 2/3h_1$.

Recent **WMAP** *h*-value H_0 (known as present value h_1 in this paper) is $H_0 \equiv h_1 \sim 71km/\sec/Mpc$. Take $h_1 \sim .726\cdot 10^{-10}yr^{-1}$ and $h_1^2 \sim 5.29\cdot 10^{-36}s^{-2}$. However, WMAP *h*-value does not solve the age-failure problem since $1/H_0 \equiv 1/h_1 \sim 13.7\cdot 10^9 yr$ is not the age of the universe. The formula $t_1 = 1/h_1 \equiv 1/H_0$ is neither theoretical nor empirical. Such estimation is too rough.

Set $R_1 \equiv k_1 10^{28}cm$ and rewrite (59h) as

(59h)′
$$\begin{cases} k_1(1+\frac{5.29}{9}k_1^2) = 3.145k_m n_m 10^{-8} \\ \rho_1 = \frac{k_m n_m}{2\pi^2 k_1^3}10^{-35}g/cm^3 \end{cases} \qquad k_1 \equiv \frac{R_1}{10^{28}cm}.$$

About one million galaxies with 19th magnitude and above have been counted in the northern galactic hemisphere. Hence, $n_m \sim\geq 10$. The observation indicates that $k_m \leq 10^4$. If an estimation is $k_m n_m \leq 10^6$, then

$$k_1 \leq 3.14\cdot 10^{-2}, \; R_1 \leq 3.14\cdot 10^{26}cm = 3.32\cdot 10^8 ly .$$

Such small present size of the EF-universe is in conflict with the observed large red shifts.

The ‘*critical density*’ in the EF-model is $\rho_c = 3h_1^2/8\pi G \sim 0.947\cdot 10^{-29}$ g/cm^3, according to recent WMAP-*h*. One may hand-put a value, such as $\rho_1 \sim 2\cdot 10^{-29}g/cm^3$ into the 1st SD-eqn (59e), and get a large size, such as $R_1 \sim$

$1.306 \cdot 10^{10} ly$. This large size does not help to explain the observed large red shifts at all since the age-theorem we are proving reveals a young age of the EF-universe (less than $9.18 \cdot 10^9$ years). Actually, the said hand-put values for the pair (ρ_1, R_1) determine via (59d) that the present age is $t_1 = 7.76 \cdot 10^9 yr$. The larger value $h_1 \sim 0.8 \cdot 10^{-10} yr^{-1}$, measured by Hubble decades ago, yielded a larger size $1.45 \cdot 10^{10} ly$ and an older age $8.2 \cdot 10^9 yr$.

If $R_1 \sim 1.306 \cdot 10^{10} ly$, then $k_1 \equiv R_1 / 10^{28} cm \sim 1.237$. Thus, from (59h)′ we would get a very large value: $k_m n_m \sim 7.47 \cdot 10^7$. This is in disagreement with the present empirical estimation $k_m n_m \leq 10^6$. Due to (59e) and (59h), any density that is less than $2 \cdot 10^{-29} g/cm^3$ but larger than the critical density $0.947 \cdot 10^{-29} g/cm^3$ would lead to a size larger than $1.306 \cdot 10^{10} ly$ and a value of $k_m n_m$ larger than $7.47 \cdot 10^7$. As $\rho \to \rho_c^+$ (right limit), $k_m n_m \to \infty$. From (59g)′, (59h), and (59h)′ we see that as $k_m n_m$ **increases** monotonically, so do A, k_1, $k_m n_m / k_1$, and $u_1^{-1} \equiv A / k_1 10^{28} cm$. Namely, u_1 ***decreases*** monotonically as $k_m n_m$ ***increases***. One can see that

$$\text{(59i)} \qquad \lim_{\Theta\to\infty} R_1 = \infty\,,\ \lim_{\Theta\to\infty} A = \infty\,,\ \lim_{\Theta\to\infty} u_1 \equiv \lim_{\Theta\to\infty} \frac{R_1}{A} = 0\,,\ (\Theta \equiv k_m n_m).$$

Due to (59e) and (59i), the five limits $u_1 \equiv R_1 / R_{max} = R_1 / A \to 0$, $R_1 \to \infty$, $A \to \infty$, $k_m n_m \to \infty$, and $\rho_1 \to (3h_1^2 / 8\pi G)^+ = \rho_{c1}^+ \sim (0.947 \cdot 10^{-29} g/cm^3)^+$ (right limit), are all equivalent. This age theorem will prove that the age function in (59d) monotonically increases as $k_m n_m$ increases. But its absolute maximum is the upper limit $2/3h_1 = 9.18 \cdot 10^9$ years. The upper-bound of the age of the EF-universe is independent of whether the density is close to critical density and how close. It is independent of the size and total mass of the universe, and of how many contributions the dark matter has brought to the value of $k_m n_m$.

If future super-telescopes would discover galaxies dimmer than 19th magnitude 100~1000 times more than what astronomers have counted so far with 19th magnitude and above, most among those new discoveries must be beyond the reach of today's largest telescopes. Then the new discoveries would be associated with deeper spaces more distant than what modern telescopes have ever probed. This definitely requires an old age. Now we continue proving the said young age of the EF-universe and see how any enlarged total mass and size of the universe do not help to yield an older age through the EF-model.

More and more galaxies and/or dark matter might be discovered using future large and/or super-telescopes, or simply the total mass $M_{total} = k_m n_m 10^{49} g$ of the universe may actually be larger and larger than we would expect. If so, we might find that $k_m n_m \sim 10^7$, 10^8, 10^9, and so forth. However, the age-theorem we are proving will tell us that the present EF-universe is absolutely less than $2/3h_1$ =

$9.18 \cdot 10^9$ years old regardless of the values of $k_m n_m$ and how large the universe is at the present.

To allow the limit processes ($\rho_1 \to \rho_{c1}^{+}$, $R_1 \to \infty$, $A \to \infty$, $u_1 \to 0$, and $k_m n_m \to \infty$) to scan and cover any possible future empirical values of the size, total mass, and density and then find out the absolute maximum age, we treat u_1 as the independent variable of the function $t_1(u_1)$ in (59d), where h_1 appears as a fixed constant. One can check from (59d) that

(59j)
$$h_1 u_1^{2} \frac{\sqrt{u_1 - u_1^{2}}}{3/2 - u_1} \frac{dt_1}{du_1} = f_1(u_1) - f_2(u_1),$$

$$f_1(u_1) \equiv \tan^{-1} \sqrt{\frac{1}{u_1} - 1}, \quad f_2(u_1) \equiv \frac{\pi}{2} - \frac{3\sqrt{u_1 - u_1^{2}}}{3 - 2u_1}.$$

The graph of f_1 is below f_2. The numerical calculation shows that $f_1 < f_2$ for any u_1. Thereby, $dt_1 / du_1 < 0$ for any u_1 ($0 < u_1 \equiv R_1 / R_{\max} \leq 1 < 3/2$) and $t_{1\max}$ is not a relative but the absolute maximum, that is obtained by taking the limit $\lim t_1(u_1)$ as $u_1 \to 0$ (equivalently, $k_m n_m \to \infty$).

To get the absolute maximum value of the EF-age function $t_1(u_1)$ in (59d) **analytically**, we rewrite (59j) as

(59k)
$$\frac{\sqrt{u_1 - u_1^{2}}}{3/2 - u_1} h_1 u_1^{2} \frac{dt_1}{du_1} = q(\eta_1) \equiv \tan^{-1} \eta_1 - \frac{\pi}{2} + \frac{3\eta_1}{1 + 3\eta_1^{2}},$$

$$\eta_1 \equiv \sqrt{\frac{1 - u_1}{u_1}} = \sqrt{\frac{1}{u_1} - 1}, \ (\eta_1 \to \infty \text{ as } u_1 \to 0).$$

Now that $0 < u_1 \equiv R_1 / R_{\max} \leq 1 < 3/2$, dt_1 / du_1 and $q(\eta_1)$ have the same sign. One can check

(59l)
$$\frac{dq(\eta_1)}{d\eta_1} = \frac{4}{(1 + \eta_1^{2})(1 + 3\eta_1^{2})} > 0.$$

Namely, $q(\eta_1)$ is monotone increasing and $q_{\max}$ is obtained as $\eta_1 \to \infty$. One can see from (59k) that

(59m)
$$q_{\max} = \lim_{\eta_1 \to \infty} q(\eta_1) = \frac{\pi}{2} - \frac{\pi}{2} + 0 = 0.$$

Therefore, $q(\eta_1) < 0$ for any $\eta_1 < \infty$. Then dt_1 / du_1 is negative for any possible value of u_1. So there is no relative maximum but absolute maximum of $t_1(u_1)$:

(59n) $$t_{1\max} = \lim_{u_1 \to 0} t_1(u_1) = \frac{2}{3} h_1^{-1} \sim 9.18 \cdot 10^9 \, yr \, .$$

The L'Hôpital rule has been used to rigorously calculate the $\infty \cdot 0$-type limit of $t_1(u_1)$ in (59d).

The limit $u_1 \to 0$ in (59n) is equivalent to the limit $k_m n_m \to \infty$ as shown in (59i). The limiting process $k_m n_m \to \infty$ has scanned and covered all possible future empirical values of the factor-parameters in $k_m \equiv 1.95 \cdot 2 \cdot 1.99 k_d k_g k_o k_n$ and $k_m n_m$. The proved upper bound of the young age has nothing to do with how many galaxies will be counted in any future and how many dark matter particles and bodies will be discovered in any future. One can make the factor-parameter k_d as large as needed to match any postulated amount of hypothetic massive stable particles and invisible matter. Anyway, the present age of the EF-universe is absolutely less than $9.18 \cdot 10^9$ years. Precisely, the upper bound of its present age is independent of its size and of the total mass with ANY percentage of dark matter. ■

Not only the observed large red shifts indicate an old age of the universe, but also the trial and error method further reveals the necessity of an old age to quantitatively explain why dimmer supernovas must be about 5 billion ly or farther. God does not make errors. Trial and error method is human's method, used to get a better understanding of the laws of physical world in the absence of some important empirical data.

Why is the expanding EF-universe always so young and the upper bound of its present age is independent of its present size? The reason is as simple as this:

By just looking at (58a), we see that the EF-universe always exploded with infinite speed at the beginning and had expanded with terribly faster-than-c speed for so long. The '*drama*' is: the larger the A, the lager the present size in (59a)′, the larger the product $k_m n_m$ in (59g)′, the larger the total mass in (59f), and the larger the rate dR/dt in (58) at each fixed R. Then the faster expansion speed could lead to a young age with any large present size and any large total EF-mass (with CHSPs not counted).

Although no one (including Einstein) ever defined frame-invariant speed c in curved spacetime and global initial systems do not exist in the universe, the extremely solid **empirical c-principle** remains too accurate and has been verified experimentally nowhere else but in curved spacetime on earth without a single example of violation. The experimental errors are very small. Indeed, the motion of planets only shows very small discrepancies such as an advance of the perihelion of 42.9″ per century for Mercury. One can see how small the effects of the curved spacetime upon the motion of planets are. Theoretically, the metric in the EF-model explicitly shows quasi-Minkowski spacetime at small cosmological scale since long time ago. At large scale, is there any violation of the c-principle? The overwhelming experimental evidences and theoretical results do not support such opinion that we may discard the empirical c-principle due to such irrelevant fact that a coordinate system in curved spacetime is inertial at most at one point and allow the speed of expansion defined and formulated in the EF-model to go from infinite or 'near-infinite' speed at the beginning to zero at the turning point between expansion and contraction, to slow down from terribly faster-than-c receding to the speed c, and then to normal less-than-c receding leaving unreasonable faster-than-c expansion to be a history in the past. A by-product of going from infinite/near-infinite speed to normal less-than-c speed is inevitable **near-c receding**, e.g., $0.99c<v<c$, sometime between, producing **extremely large red shifts**. Among all the galaxies, no single one has ever been seen to have extremely large red shifts. We should respect the merit of non-gravitational physics and the c-principle because global initial coordinate systems and the laws written in terms of them are part of a general theory written in curved spacetime. Without knowing what would happen if GF (gravitational field) approaches zero or can be neglected, local inertial system approaches global inertial system, and curved Einstein spacetime approaches global flat Minkowski spacetime, no eqn in curved spacetime can be obtained. Any generally covariant tensor eqn must tell what would happen when the GF approaches zero and the metric approaches global Minkowski metric. If such limit does not agree with what have been verified in non-gravitational physics tested in curved spacetime on earth, the theory in curved spacetime is immediately ruled out by the nature of the truth. We respect global initial systems in global flat Minkowski spacetime and the laws written in term of them not because they exist, but because they are important and solid, serving as a touchstone to test any general theory in curved spacetime as well as any interpretation or opinion.

Since faster-than-c receding is a common feature of the EF-type models, regardless the LV is ignored or considered, the only way for a theory of gravitation and cosmology to be consistent with the solid empirical c-principle without even changing the looking of the field eqs is a **variable** *G(t)*, to be given in various models. For each given *G(t)*, *R(t)* is to be determined according to Friedman-type non-static solutions while the coefficient υ in the cosmological term is to be determined by a consistency condition proved in the CT-theorem (see later). The formula for υ is part of the solution just like the formula for *R(t)* that is another part of the solution.

Constancy of *G* is symmetry, called ***G*-symmetry**. Along the two opposite directions of time-flow (past and future directions), *G* keeps the same value–the present value. Any non-constant *G* means *G*-symmetry breaking. We'll develop EFD-type gravitation and cosmology with spontaneous breaking of *G*-symmetry to obey the empirical c-principle and simultaneously explain the supernova results with full, less, or no acceleration interpretation, depending on how much the denser dusts and molecules in the ancient smaller universe make remote supernovas dimmer than expected. The explanation, in principle, is of no any uncertainty under the condition that the values of many fundamental quantities of the universe will remain uncertain in the coming centuries. Before we do so, orbiting speeds of stars in the spherical core of our galaxy, on the luminous disc, in the periphery beyond the edge of the major luminous disc, orbiting speeds of galaxies about the centers of clusters and super-clusters should be explained to see the effect of LV in dark-matter phenomena. The v-r curve of stars' orbiting in our galaxy, fast orbiting of lonely stars beyond the luminous disc, the dark, transparent, and resistance-less non-baryonic dark matter seen in 'bullet cluster' must be explained.

8.2 CHSPs and Dark Matter

Any large central mass, such as a super-large black hole, if any, at the center of our galaxy, never causes fast orbiting within the bright spherical core more than thousands light years away from the center nor quasi-constant orbiting on the luminous disc outside the core simply because Newton's theory works very well for the orbiting of planets about the sun as well as the orbiting of stars about the center of their galaxy. For instance, a super-large black hole violates Newton's law significantly only in the places not too far away from it and the departure from Minkowski metric is no longer negligible. Both GR and Newton's theory prove slower and slower orbiting in the places farther and farther away from a central mass. A super large black hole at the center, if any, attracts matter. When bodies and particles nearby fall into the black hole, their speeds are expected to be faster and faster as they get closer and closer to the black hole. Within a galaxy, a typical empirical **v-r curve** is a quasi-horizontal line between the edges of luminous galactic disc and the central bright spherical core plus a line that climbs steeply from small values near the center of galaxy to a large value to meet the quasi-horizontal line at the edge of the spherical core. Of cause, how steep the climbing line is depends on the choice of the units. Let us just call such v-r graph a **Γ-line**. We now prove that this feature is due to the distribution pattern of luminous stars and any baryonic dark matter that has similar pattern of distribution as the luminous stars. The CHSPs with large $\overline{\lambda_d^4}$-enhancement of their energy density play crucial role for stars in the periphery and for the orbiting of galaxies about the centers of clusters and super-clusters. Please refer to (61c) to see the definition of $\overline{\lambda_d^4}$.

Let ρ be the mass density of a homogeneous spherical distribution of matter with radius a. Then

$$\text{(60a)}\qquad \frac{v^2}{r} = \begin{cases} \dfrac{\frac{4}{3}\pi G\rho r^3}{r^2} & r \le a \\[2ex] \dfrac{\frac{4}{3}\pi G\rho a^3}{r^2} & r \ge a \end{cases}, \qquad v = \begin{cases} kr & r \le a \\[2ex] ka\sqrt{a}/\sqrt{r} & r \ge a \end{cases} \qquad k \equiv \sqrt{\frac{4\pi G\rho}{3}}.$$

The 1st piece of the graph of (60a) with $r \le a$ is a line going up linearly and steeply (if k is large enough). This immediately shows that the 1st portion of the empirical Γ-line is attributed to the spherical distribution of luminous stars in the bright core and any dark matter between them and with the similar distribution. The 2nd piece immediately shows that central masses never produce fast orbiting nor quasi-constant orbiting in all places far away from the central mass, where the metric is close to Minkowski metric and Newton's laws of gravitation and of motion work there locally and for stars moving much slower than light.

Spherical distribution of matter makes the eqs solvable in both Newton's and Einstein's theories. It is difficult to discuss exact effects of disc-like distribution, especially with a bunch of luminous bending arms, if any. But one thing is clear that if one neglected the contribution of the massive stars on the disc, the behavior of orbiting would be similar to the 2nd part (outside the sphere) in (60a): The orbiting would slow down as r increases.

Here is the reason for the quasi-horizontal part of the Γ-line. A star on the luminous galactic disc divides the disc into two parts, inner and outer parts. The inner part provides with gravitational force to pull the star inward while the outer part provides negligible effect due to balance. In fact, on the outer part, some masses pull it toward the center while the others pull it away from the center. The farther the star away from the center, the larger the inner part. Consequently, once the force from outer part is neglected and the inner part can be approximately treated as a central mass $M \sim \sigma\pi r^2$ where σ is the mass per unit area on the disc, we have

$$\frac{G\sigma\pi r^2}{r^2} = \frac{v^2}{r} \Rightarrow v = \sqrt{G\sigma\pi}\sqrt{r}\ .$$

We might expect the fast orbiting on the luminous disc if there were no galactic spherical core. For the stars on the luminous disc, the disc distribution of matter causes fast orbiting while the masses in bright spherical core cause slow orbiting as r increases and is larger than the radius of the spherical core. The overall effect would be quasi-constant orbiting speed. Accurate result caused by the masses on the disc may need numerical integration. At the present, we are sure that the empirical v-r curve can be explained, at least half-quantitatively, by the distribution of luminous masses. Baryonic dark matter would agree with the observed v-r curve as long as their distribution is similar to luminous stars or their effect is relatively small within the luminous galaxy.

To explain the fast orbiting of the lonely stars beyond the luminous disc and the dark matter phenomena within a cluster or super-cluster of galaxies, **cosmic high-speed particles (CHSPs)** are needed. They refer to non-soft photons, energetic neutrinos, and high-energy protons and electrons in primary cosmic rays. In new physics with LV, high-speed particles with $\gamma > \gamma_0 \geq 10^5$ are part of the dark matter. Non-soft photons and energetic neutrinos are non-baryonic. CHSPs are so 'hot', i.e., so fast that they are not at rest in the coordinate system co-moving with receding galaxies. Their energy density $\varepsilon_d = \rho_d \lambda_d \gamma_d c^2$ receives large $\overline{\lambda_d^4}$-enhancement. Please see (61c) for the detail. The distribution of 'relic' and 'old' CHSPs radiated long times ago from the sources far away from and around an orbiting star/galaxy/cluster is approximately homogeneous in a space containing the orbit and much larger, unable to produce a central force toward the orbit center, unless there is a physical process to localize them gradually. As long as the process is very slow, the central force needed to cause fast orbiting of the size (radius) of *t ly* is mainly attributed to the CHSPs radiated by the stars within the orbit in the past and in a time period of *t yr*, to the order of magnitude.

Photons and neutrinos radiated from each star are contained in 'out-going' spherical waves centered at the star. As their density in each spherical shell is concerned, the farther the shell from the center, the larger the shell, and the smaller the density. Photons and neutrinos radiated from each galaxy move away from the stars in 'out-going' spherical shells centered at source stars. Outside a galaxy, its radiated CHSPs are moving away from the galaxy in 'out-going' shells whose geometric shape depends on the distribution of the stars in the galaxy. Again, the farther the shell away from the galaxy, the larger the shell, and the smaller the density. Let C_{ri} be the energy-radiation rate per unit time of CHSPs from a source s_i, where $i=1,2,3$ refers to a star, galaxy, and cluster of galaxies respectively. The out-going CHSPs radiated by a spherical source are contained in an expanding sphere centered at the source. Far away from non-spherical sources, we may use, approximately, spheres to calculate the total effective gravity masses of the out-going CHSPs. Let r_i be the distance of a place from the source s_i. It takes the time $t(r_i)=r_i/c$ for the CHSPs radiated from the source to reach the place. The sphere of radius r_i, centered at the source, contains the CHSPs whose total effective gravity mass is

(60b) $$M_d^{eff}(r_i)=\overline{\lambda_d^4}C_{ri}r_i/c^3.$$

At that place, the orbiting speed caused by the CHSPs (as a source of *G*-field) is given by eqn $v(r_i)^2/r_i=GM_d^{eff}(r_i)/r_i^2$ so that the orbiting speed is

(60c) $$v(r_i)=\sqrt{\overline{\lambda_d^4}C_{ri}G/c^3}=const.$$

Suppose there are about $2\cdot 10^{11}$ luminous stars in a galaxy and two hundred galaxies in a cluster. Then $C_{r2}\sim 2\cdot 10^{11}C_{r1}$ and $C_{r3}\sim 200C_{r2}\sim 4\cdot 10^{13}C_{r1}$. For the sun, C_{r1} is too small for the CHSPs radiated from the sun to make a considerable contribution to the orbiting of the solar planets. We know $C_{sun}=2cal/cm^2\cdot\min$ is the solar constant. Then sun's radiation rate is about $C_{rsun}\sim 2.14\cdot 10^{-21}M_{sun}c^2/\sec$. It takes about few minutes to few hundred minutes for the CHSPs radiated from the sun to reach the nearest and farthest solar planets. The effective gravitational mass of the CHSPs that contribute to the orbiting of the solar planets is only about $10^{-19}\sim 10^{-17}M_{sun}$, multiplied by the LV enhancement factor $\overline{\lambda_d^4}$. It is too small to consider even for the LV as large as $\overline{\lambda_d^4}=10^4\sim 10^{10}$. But for a galaxy and a cluster of galaxies, the radiated CHSPs do make nice contributions to the orbiting. For a galaxy with 200 billion stars, we have $C_{r2}=2\cdot 10^{11}k_*C_{rsun}$ (k_* is related to the average rate of radiation). Consider a star on a place $r_2=k_2 10^5 ly$ away from the center of the galaxy. The sphere with radius r_2 contains the out-going CHSPs whose total effective gravitational mass is roughly about

(60c) $$M_d^{eff}\sim\overline{\lambda_d^4}C_{r2}r_2/c^3=1.35k_*k_2\overline{\lambda_d^4}\cdot 10^3M_{sun}.$$

The farther the place beyond the luminous disc, the larger the r_2, the more accurate the eqn (60c). In new physics with LV, λ is almost constant 1 for $\gamma<\gamma_0$ ($\gamma_0\geq 10^5$). In velocity shell space ($\gamma_0,\gamma_0+\delta$), $\lambda=\lambda(\gamma)\sim 10^2$, more or less. The highly adjustable LV factor function λ and enhancement factor $\overline{\lambda_d^4}$ that may range from ten thousand to ten billion, is automatically able to explain the fast orbiting the luminous matter plus baryonic dark matter fail to explain. For example, if $\overline{\lambda_d^4} \sim 5\cdot10^7$ and $k_* \sim 5$ then $M_d^{eff} \sim\ 5\cdot10^{11} M_{sun}$ at the place $2\cdot10^5 ly$ away from the center of the galaxy. This provides with sufficiently large gravity force to make the orbiting faster than anticipated. Even an amateur could easily see the fast orbiting. Due to the uncertain quantity of baryonic dark matter, the value of $\overline{\lambda_d^4}$ cannot be determined accurately by accurate measurement of the orbiting speeds of the stars on the luminous disc and beyond it. However, the agreement with the observation is remarkable. After all, CHSPs with large $\overline{\lambda_d^4}$ -enhancement explain the unexpected fast orbiting of lonely stars far beyond the luminous galactic disc as well as the dark matter phenomena associated with clusters and super-clusters.

Non-soft photons and energetic neutrinos are the major components of CHSPs. They are **dark**, **transparent**, and **resistance-less**.

More than 99.99% of them never reach earth and are absolutely dark to us. The linear superposition of e.m. waves and light rays when they meet and the lack of interactions among photons and neutrinos and between a photon and a neutrino make them transparent, passing through each other without resistance. Two clusters passed through each other in a magnificent collision and formed a new cluster *1E 0657-556* ("**bullet cluster**"). It is an outstanding example for us to see the dark, transparent, and resistance-less CHSPs as verified non-baryonic dark matter filling the spaces btw any 2 among the hundreds of galaxies in the cluster. Baryonic dark matter is contained in two clumps of 'hot' gas. Because baryonic matter interacts with each other, the two clumps encountered resistance during the collision, separated from their original clusters after the collision, and left behind so that the two clusters looked like 2 bullets after they passed through each other. One can see NASA's photo on the web site. No collision btw two galaxies happened in that particular event for they were scattered in a huge space so that no two galaxies in the two clusters got statistical chance to make a head on collision, while non-CHSP dusts and particles did not scatter but filled the space btw galaxies, with large chance to collide when 2 clusters collided.

Photons with frequencies not super low and energetic neutrinos with energy larger than 0.1 MeV are non-baryonic dark matter. They do not interact with e.m. radiation in the sense that they do not or rarely 'collide' with 99.9999% of the photons radiated from stars to change moving direction of light rays via e.m. force, fully consistent with iconic empirical and theoretical facts of superposition. But collectively, too many such non-baryonic DM particles in super-large space of a cluster create powerful gravity forces that cause not only fast orbital velocity of a galaxy but also gravitational lens effect to force radiated photons to change their moving directions during long Journeys in the gigantic space.

Without the $\overline{\lambda_d^4}$ -enhancement in the new physics with LV, CHSPs could only play negligible effect in dark matter/energy phenomena. The solar constant indicates that it takes 1.48 billion years for the total energy of the CHSPs radiated from the sun to equal $M_{sun}c^2 10^{-4}$. It is the large value (large jump) of the LV factor function λ near $\gamma \sim \gamma_0 \geq 10^{5\pm\sigma}$ and the large values of λ after and before γ reaches $0.1N$ provide CHSPs with needed large $\overline{\lambda_d^4}$ -enhancement.

8.3 EFD-Type Gravitation and Cosmology with LV

We do not consider matter-related pressure terms in cosmological problems because not only the EF-model did not include such term but also a 'gas' of photons and neutrinos do not have inner pressure. Non-baryonic dark matter fill the spaces between the galaxies in the 'bullet cluster' 1E 0657-556. It proves that pressure takes place for baryonic matter only. There is no guarantee that $\gamma - \gamma$, $\nu - \nu$, and $\gamma - \nu$ interactions will not occur at extremely high-energy scales and in dense condition. Non-linear optics might work for the dense photons inside the optical fibers used everywhere in communication industry. But it is sure that the linearity of both massless/massive Maxwell eqs is consistent with overwhelming experimental evidences of the well-known linear superpositions of e.m. waves and light beams when they meet. The spaces between stars, galaxies, clusters, and super-clusters do not provide dense condition for photons and neutrinos. Surely, matter related pressure terms are negligible as we study cosmological problems at large scales.

We do not consider a negative pressure term associated with the hypothetic 'vacuum energy'. J. Schwinger [22] and S. Weinberg ([3], p. 14) pointed out that infinite sea of negative energy electrons in Dirac's hole theory is not needed and the existence of Bosons supports their idea. The radiative corrections in quantum theory of fields with virtual particles sometimes are interpreted as a sign of "vacuum energy'. We know that only a very small portion of an empirical value of energy-level attributes to radiative corrections. Classical heavy bodies and we feel no any vacuum energy. If 'vacuum energy' could move stars in a galaxy, it would move any heavy body on earth. No such bizarre phenomenon has ever been seen. Gravity force moves stars in any galaxy and forces galaxies receding through its cosmological effect at large scale as shown in the solution of the field eqs at large scale. This is so natural for we see too many examples of how gravity force moves heavy bodies locally on earth and in solar system. Moreover, a CT theorem will prove a **consistency condition** (72) for the coefficient υ of the $g_\mu^\nu - term$, no matter it is a pure cosmological term, matter-related pressure term, or a combination of them. The CHSP-theorem in this paper will prove that once positive T_i^i-terms of CHSPs enter field eqs with the $g_\mu^\nu - term$ having constant coefficient (zero or nonzero), the field eqs form an inconsistent system of eqs.

Now that the existence of CHSPs is an obvious **physical reality**, we cannot consider any $g_\mu^\nu - term$ that has a constant coefficient. Its 'solutions' inevitably depart reality. One may postulate that all *inflation* with exponential expansion and all faster-than-c and near-c recessions took place before stars were born to radiate. However, once the consistency condition (72) is violated by a constant coefficient in the $g_\mu^\nu - term$ of a model, the solution does not exist at all for the field eqs in that model form an inconsistent system of eqs.

We now calculate the contributions of CHSPs. The four coordinate systems $S_C(x,y,z,ct)$, $S_r(r,\theta,\phi,ct)$, $S_\chi(\chi,\theta,\phi,ct)$, and $S_{\chi\eta}(\chi,\theta,\phi,\eta)$ in the EF-model, are called '**co-moving coordinate systems**' in the literatures. Non-CHSP mater (such as galaxies and all the non-CHSP dark matter) co-moves with receding galaxies and is at rest everywhere ***locally*** in these co-moving coordinate systems. By replacing spatial coordinates in S_C with spherical coordinates, one can establish S_r out of S_C. In view of Einstein-Friedman embedding, $r = R(t)S(\chi)$. Here, $S(\chi) = \sin\chi \,/\, S(\chi) = \sinh\chi$ in positive/negative curvature cases respectively. The cosmic scale factor function $R(t)$ is the radius of the universe in positive curvature case. One gets $S_\chi(\chi,\theta,\phi,ct)$ from $S_r(r,\theta,\phi,ct)$ via the transformation $r = R(t)S(\chi)$. Re-labeling time coordinate t in S_χ with $cdt = R(t)d\eta$ and $\eta = c\int dt/R(t)$ leads to $S_{\chi\eta}(\chi,\theta,\phi,\eta)$. In terms of $S_{\chi\eta}$, the EF-metric is written

(61) $$dS^2 = R^2(\eta)[d\chi^2 + S^2(\chi)(d\theta^2 + \sin^2\theta d\phi^2) - d\eta^2],$$

$$g_{\mu\nu} = g_{\mu\sigma}\delta^\sigma_\nu = 0\ (\mu \neq \nu),\ g_{\eta\eta} = g_{00} = -R^2,\ g_{\chi\chi} = g_{11} = R^2,$$

$$g_{\theta\theta} = g_{22} = R^2S^2(\chi),\quad g_{\phi\phi} = g_{33} = R^2S^2(\chi)\sin^2\theta.$$

In the new physics with LV, $d\tau = dt/\lambda\gamma$, $p^\mu = m\lambda\gamma u^\mu$, $U^\mu = \lambda\gamma u^\mu$ with $u^\mu \equiv dx^\mu/dt$. The kinetic effects of moving rod and clock lead to: $dV = \lambda^3\gamma^{-1}dV^*$ and $\rho^* = \lambda^3\gamma^{-1}\rho$. In a co-moving system, $\vec{u} = \vec{0}$ locally for **non-CHSP** matter of mass density ρ (luminous and non-CHSP dark matter), i.e., $\gamma = 1$, $\lambda = 1$, $\rho^* = \rho$ for galaxies and all the baryonic matter (including electrons) co-moving with them, while $\lambda = \lambda_d$ and $\gamma = \gamma_d$ for **CHSPs** with density $\rho_d \neq \rho_d^*$. In $S_{\chi\eta}$, $x^0 = \eta$ and the notations $u^\mu \equiv dx^\mu/dt$ and $U^\mu \equiv dx^\mu/d\tau = \lambda\gamma u^\mu$ now are changed to

(61a) $$u^\mu \equiv dx^\mu/d\eta\ (u^0 = 1),\qquad U^\mu \equiv dx^\mu/d\tau = \lambda\gamma dx^\mu/dt = \lambda\gamma c R^{-1}u^\mu.$$

One may add subscript d for CHSPs. Then,

(61b) $$T^{\mu\nu} = c^2R^{-2}\,(\rho u^\mu u^\nu + \Sigma\,\rho_d\lambda_d^5\gamma_d u_d^\mu u_d^\nu)$$

$$(u^\mu) = (0,0,0,1),\quad (u_d^\mu) = (u_d^\chi, u_d^\theta, u_d^\phi,\ 1)$$

$$T_\mu^{\ \nu} = g_{\mu\sigma}T^{\sigma\nu} = g_{\mu\mu}T^{\mu\nu} = c^2R^{-2}(\rho g_{\mu\mu}u^\mu u^\nu + \Sigma\,\rho_d\lambda_d^5\gamma_d\,g_{\mu\mu}u_d^\mu u_d^\nu)$$

Matter of density ρ with $(u^\mu) = (0,0,0,1)$ is non-CHSP matter (luminous and dark) and is at rest in the co-moving coordinate systems locally while CHSPs move in all directions. Thus, in terms of $S_{\chi\eta}$

(61c) $$\rho g_{\mu\mu}u^\mu u^\nu = 0\ \ (\mu \neq \nu,\ \mu = \nu = 1,2,3),$$

$$\Sigma\rho_d\lambda_d^5\gamma_d g_{\mu\mu}u_d^\mu u_d^\nu = 0\ (\mu \neq \nu),\ T_\mu^\nu = 0\ (\mu \neq \nu).$$

$$c^2R^{-2}\rho g_{00}u^0u^0 = -\rho c^2,$$

$$c^2R^{-2}\,\Sigma\rho_d\lambda_d^5\gamma_d g_{00}u_d^0u_d^0 = -c^2\Sigma\rho_d\lambda_d^5\gamma_d = -c^2\,\overline{\lambda_d^4}\Sigma\rho_d\lambda_d\gamma_d = -\rho c^2 k_\rho\,,$$

$$\overline{\lambda_d^4} \equiv \Sigma\rho_d\lambda_d^5\gamma_d \,/\, \Sigma\rho_d\lambda_d\gamma_d,\;\; k_\rho \equiv \overline{\lambda_d^4}\varepsilon_d \,/\, \rho c^2 = \overline{\lambda_d^4}\varepsilon_d \,/\, \varepsilon\,,$$

$$\varepsilon_d \equiv \Sigma\rho_d\lambda_d\gamma_d c^2,\;\; \varepsilon \equiv \rho c^2\,.$$

(61d) $$T_i^i = c^2R^{-2}\,\Sigma\,\rho_d\,\lambda_d^5\gamma_d g_{ii}\,(u_d^i)^2 \text{ (no summation over index } i)$$

Statistically, **equipartition** hypothesis in *curved* spacetime yields

(61e) $$T_1^1 = T_2^2 = T_3^3 = \frac{k'}{3}\overline{u_d^2}\Sigma\rho_d\lambda_d^5\gamma_d = \frac{k}{3}\rho c^2 k_\rho,\;\; k \equiv k'\overline{u_d^2}\,/\,c^2\,.$$

Here u_d is usual speed of CHSPs in S_C ($u_d^i=dx_d^i/dt$) while k and k' are statistic parameters with $k \approx 1$, at least to the order of magnitude. Therefore,

(62) $$T_0^0 = -\rho c^2(1+k_\rho),\;\; k_\rho \equiv \overline{\lambda_d^4}\frac{\varepsilon_d}{\varepsilon},\;\; \varepsilon = \rho c^2,\;\; \varepsilon_d = \sum\rho_d\lambda_d\gamma_d c^2\,,$$

$$T_1^1 = T_2^2 = T_3^3 = \frac{k}{3}\rho c^2 k_\rho,\quad T_\mu^\nu = 0\;(\mu \neq \nu)$$

In statistical physics, the law of equipartition of energy holds under the condition of wide distribution of molecules' velocities. The energy-equipartition in curved spacetime (61e) is assumed under the condition of an extremely narrow interval in which the velocity spectrum of CHSPs is squeezed. CHSPs possess an extremely wide spectrum of energy while their speeds are too close to each other, too close to their average speed, and too close to the frame-invariant speed c.

Consider $S_r(r,\theta,\phi,ct)$, $S_\chi(\chi,\theta,\phi,ct)$, and $S_{\chi\eta}(\chi,\theta,\phi,\eta)$. There is a boost between S_r and S_χ because a fixed χ does not correspond to a fixed r due to the expansion. But the relative velocity is surely less than $(1-10^{-9})c$, the LV factor function between them equals 1 with extremely high precision, and $dS_\chi = dS_r$ holds. Numerically renaming coordinate time does not lead to a boost. So $dS_\chi = dS_{\chi\eta}$ holds exactly and the EF metric is still valid.

Due to the large values of the $\overline{\lambda_d^4}$-enhancement factor of the energy density ε_d of the CHSPs, whose distribution is approximately homogeneous at large scale, the well-known **cosmological principle** is valid regardless the distribution of galaxies/stars/dusts is homogeneous or not. Therefore, the original **EF-metric** with *undetermined* cosmic scale function $R(t)$ is still **valid**.

Please note, in the EF-model without LV, the EF-type model with LV (together called EF-type models w/wo LV), and the EFD-type model with LV and non-constant *G*, the metric, originally put into the field eqs before one solves them to determine $R(t)$, looks the same and is given the **same name**, called the **EF-metric**. It actually varies from model to model. One can see the differences explicitly once $R(t)$ is obtained from the solutions in those models respectively to show that the behavior of $R(t)$ varies from model to model.

The sun is approximately homogeneous and spherical. If we consider any departure such as huge **spurts** from the sun, metric would definitely depart from Schwarzschild metric and will be unsolvable. Similarly, if we consider departure from homogeneous distribution such as a **hole** in the distribution of the **CBR** (cosmic background radiation), we would have departure from the cosmological principle and from the EF-metric. One can go to pages 159, 160 to see why CBR holes contain many CBR photons not moving towards our earth and CHSPs, including relic and non-relic CBR photons, obey cosmological principle. We now have a good reason to believe in the cosmological principle: Distribution of CHSPs as dominating major dark matter and sources of gravitational masses at large scale is really quite homogeneous.

From the EF-metric before $R(t)$ is determined by the solution, we get

(63)
$$G_1^1 = G_2^2 = G_3^3 = -\frac{1}{c^2}(h^2 + 2R^{-1}\frac{d^2R}{dt^2} \pm \frac{c^2}{R^2}),$$

$$G_0^0 = -\frac{3}{c^2}(h^2 \pm \frac{c^2}{R^2}), \quad G_\mu{}^\nu = 0 \quad (\mu \neq \nu).$$

Including every its special case (the EF-type model w/wo inflation, w/wo LV, and the EFD-type model w/wo LV), the **field eqs** are written

(64)
$$G_\mu^\nu - \upsilon g_\mu^\nu = \frac{8\pi G}{c^4} T_\mu^\nu,$$

$$G \equiv G_1 b(R), \quad G_1 \sim 6.67 \cdot 10^{-8} cm^3 / g \cdot s^2, \quad b(R_1) = 1.$$

Here υg_μ^ν term is called the g_μ^ν-term, or **cosmological term (CT)**. A pressure term of the form Pg_μ^ν can always be absorbed by the g_μ^ν-term in (64). We write field eqs in terms of the same mathematical form (64), regardless whether a matter-related pressure term Pg_μ^ν is actually included or not. In general, pure cosmological term changes the way how source tensor and a pressure term Pg_μ^ν determine metric and expansion mode. It itself is not associated with any physical object(s) such as a gas of molecules, particles, or a density of hypothetic 'vacuum energy'. A pressure term of the form $PU_\mu U^\nu / c^2$ can be absorbed by $\rho^* U_\mu U^\nu$ term and then ρ^* becomes effective proper mass density.

We will solve (64) in the most general way leading to the most general solution that covers all cases with constant/non-constant *G* and υ, w/wo LV. In particular, the most general form of the consistency condition for the coefficient

υ of the $g_\mu^\nu - term$ shown in (72) applies to every special case. If it is violated in a model, that model is then inconsistent.

With the G-factor $b(R)$ and LV factor λ hidden in G and T_μ^ν respectively, the field eqs in (64) look like the same as in Einstein's GR without LV. They are EFD-type new with LV because not only the source tensor contains the LV factor function λ but also both υ and G do not have any pre-assumed constant value except for the present value of G: $G_1 \sim 6.67 \cdot 10^{-8} cm^3 / g \cdot s^2$.

The Einstein tensor in (63) is homogeneous and isotropic, containing no θ and ϕ. Also, the source tensor in (62) contains no θ and ϕ. However, it demands more for field eqs in (64) to form a consistent system of differential eqs. Later, we will state and prove **CT-theorem** (CT refers to 'cosmological term' as the combination of pure cosmological term and matter-related pressure term of the form Pg_μ^ν). It will be shown that for each given $G \equiv G_1 b(R)$, constant or not, υ is uniquely determined by a **consistency condition** or a formula for υ. Only such υ makes (64), with or without LV, a consistent system of eqs. On the other hand, $G \equiv G_1 b(R)$ is determined by model. We will state and prove a ***G*-theorem** to show that CHSP-quantity $B \equiv 1 + (1 + k/3)\Sigma \rho_d \lambda_d^5 \gamma_d / \rho$, G-quantity $G \equiv G_1 b(R)$, and expansion quantity $c^{-2} R^2 dh/dt$ are all related by a single eqn, called G-eqn. Thereby, it is equivalent to propose a model of $c^{-2} R^2 dh/dt$ in lieu of a model of $G \equiv G_1 b(R)$. A sample parabola model for $c^{-2} R^2 dh/dt$ will be proposed. Non-parabola models will be discussed too. Right now, we study a theorem.

CHSP-Theorem: Once positive T_i^i-terms of CHSPs enter the field eqs (64), with or without LV, the field eqs form an **inconsistent** system of eqs, as long as the coefficient υ in the cosmological term is pre-assumed to be a constant, no matter it is zero or nonzero and whether G is a constant or not.

Pf: The contradictions caused by constant (zero/nonzero) coefficient υ in the cosmological term can be seen clearly in the explicit expressions of the solutions with constant (zero/nonzero) υ. By solving (64), we can show the consequences of constant (zero/nonzero) coefficient υ. Here, we omit the proof. ■

The consistency condition with CHSPs will be written in (72), serving as a formula for υ, which is part of the solution of the field eqs (64). Friedman solution is the solution of (64) with special settings: $G \equiv G_1 b(R) \equiv G_1$, $k_\rho \equiv 0$, and $\upsilon \equiv 0$. The consistency condition (72) is met in the EF-model with above three identities, when one chooses $c_1 = 0$. Here c_1 is one of the two arbitrary constants explicitly contained in a general solution $t = t(R)$ of the 2nd order differential eqn (68).

Due to the effect of CHSPs, we must treat υ as a variable like $R(t)$ and leave it to be determined as part of the solution. In the next, contradictions will be seen in a very easy way once υ in (67) is regarded as a (zero/nonzero) constant. Because contradictions will be seen so clearly and explicitly, one does not need to work on the details in the proof of the CHSP-theorem.

Generally, one may freely write $G \equiv G_1 b(R) \equiv G_0 a(R)$. G_0 is a constant too. The equal value of the products yields the same field eqs. From (62), (63), and (64) we obtain the following eqs in terms of the co-moving coordinate system.

(65) $$h^2 \pm \frac{c^2}{R^2} + \frac{1}{3}\upsilon c^2 = \frac{A}{R^3}(1+k_\rho)c^2 ,$$

$$A \equiv \frac{8\pi G \rho R^3}{3c^2} = A_0 a(R) , \qquad A_0 \equiv \frac{8\pi G_0 \rho R^3}{3c^2} = const .$$

(66) $$h^2 \pm \frac{c^2}{R^2} + \frac{2}{R}\frac{d^2 R}{dt^2} + \upsilon c^2 = -k\frac{A}{R^3}k_\rho c^2 , \quad A \equiv \frac{8\pi G \rho R^3}{3c^2} = A_0 a(R) .$$

The operation (66)–(65) yields

(67) $$\frac{d^2 R}{dt^2} + \frac{Ac^2}{2R^2}[1+(k+1)k_\rho] + \frac{1}{3}\upsilon c^2 R = 0 , \qquad A = A_0 a(R) .$$

If υ were treated as a constant (zero or nonzero), then the solution of (67) became

(67)′ $$t - t_0 = \pm \int \frac{dR}{\sqrt{c_1 - \frac{1}{3}\upsilon c^2 R^2 - A_0 c^2 \int \frac{1+(k+1)k_\rho}{R^2} a(R) dR}} .$$

Its time-derivative then would yield

$$1 = \pm \frac{1}{\sqrt{c_1 - \frac{1}{3}\upsilon c^2 R^2 - A_0 c^2 \int \frac{1+(k+1)k_\rho}{R^2} a(R) dR}} \cdot \frac{dR}{dt} ,$$

(67)″ $$\frac{dR}{dt} = \pm \sqrt{c_1 - \frac{1}{3}\upsilon c^2 R^2 - A_0 c^2 \int \frac{1+(k+1)k_\rho}{R^2} a(R) dR} .$$

Immediately, this would **contradict** what one could get from (65). Note that (67)″ contradicts (65) whenever k_ρ is time-depending and hence R-depending, no matter $a(R)$ is a constant or not. The contradictions are so obvious that one does not even need to know the details in the proofs of the CHSP-theorem.

The operation 3(65)–(66) cancels $\upsilon-$ term (where υ appears as a non-constant variable). We obtain

(68) $$\frac{d^2 R}{dt^2} - \frac{1}{R}\left(\frac{dR}{dt}\right)^2 + \frac{3ABc^2}{2R^2} \mp \frac{c^2}{R} = 0 ,$$

$$A = A_0 a(R) , \quad B \equiv 1 + \frac{k+3}{3}k_\rho , \quad k_\rho \equiv \overline{\lambda_d^4}\frac{\varepsilon_d}{\varepsilon} .$$

We call B the **CHSP-quantity**. It equals 1 in the EF-model that ignores CHSPs with the setting $\varepsilon_d = 0$. It is very close to 1 if the energy density ε_d of CHSPs is considered without any enhancement such as the $\overline{\lambda_d^4}$ -enhancement in the new physics with LV. (Without LV, $\lambda \equiv 1$ & $\overline{\lambda_d^4} \equiv 1$.) Due to radiation and the 2nd law of thermodynamics, B is monotone increasing. One can freely write $B = B(t) = B(R(t))$. It is large and plays an important role in cosmology when large $\overline{\lambda_d^4}$ -enhancement for CHSPs is taken into account.

The general solution of (68) is written

(69)
$$t - t_0 = \pm \int \frac{dR}{\sqrt{c_1 R^2 + 3A_0 c^2 R^2 I(R) \mp c^2}},$$

$$I(R) \equiv -\int_{nc} \frac{Ba(R)}{R^4} dR, \quad A_0 \equiv \frac{8\pi G_0 \rho R^3}{3c^2} = const.$$

Here c_1 and t_0 are arbitrary constants, to be determined by initial conditions. The symbol $\int_{nc}$ means indefinite integral without any constant term (arbitrary or fixed). If $I(R)$ were defined as a regular indefinite integral, any constant c_2 it contained could be absorbed into c_1 ($c_1 + 3A_0 c^2 c_2 \to c_1$) for A_0 is a constant. The 1st sign pair $\pm$ in (69) corresponds to expansion and contraction respectively while the 2nd sign pair $\mp$ corresponds to positive and negative curvature respectively. Label

(69a)
$$C(R) \equiv \frac{c_1 R^2}{c^2} + 3A_0 R^2 I(R), \quad I(R) \equiv -\int_{nc} \frac{Ba(R)}{R^4} dR.$$

Using (69a) we rewrite (69) as

(69b)
$$t - t_0 = \pm \frac{1}{c} \int \frac{dR}{\sqrt{C(R) \mp 1}}.$$

Taking time-derivative of (69b), we find

(70)
$$\frac{dR}{dt} = \pm c\sqrt{C(R) \mp 1}, \quad C(R) \equiv \frac{c_1 R^2}{c^2} - 3A_0 R^2 \int_{nc} \frac{Ba(R)}{R^4} dR,$$

(71)
$$h \equiv \frac{1}{R}\frac{dR}{dt} = \pm \frac{c}{R}\sqrt{C(R) \mp 1}.$$

From (65), (71), and (69a) we get a **consistency condition**:

(72)
$$\upsilon = -\frac{3c_1}{c^2} - 9A_0 I(R) + 3A_0(1 + k_\rho)\frac{a(R)}{R^3}, \quad I(R) \equiv -\int_{nc} \frac{Ba(R)}{R^4} dR.$$

Because of (72), the following CT-theorem is evident.

CT-Theorem: The field eqs in (64), with or without LV, with constant or non-constant *G*, form a consistent system of eqs if and only if the consistency condition (72) is satisfied.

Proof: In the presence of CHSPs, the general and consistent way we solve (64) by treating υ as a non-constant variable and reaching the condition (72) proves this theorem. ■

Please note, the CHSP-theorem proves a non-constant υ in the presence of CHSPs. It does not tell what it is and what if CHSPs are ignored. The CT-theorem provides us with a general condition of consistency. It even tells why & how the EF-type model is consistent with or without inflation, with or without a constant negative pressure term:

If we ignore CHSPs by setting $k_\rho \equiv 0$, then $B \equiv 1$. If we further assume constant *G* by setting $a(R) \equiv a(R_1) = 1$ and $G \equiv G_0 = G_1 \sim 6.67 \cdot 10^{-8} cm^3 / g \cdot s^2$, then $I(R) = 1/3R^3$. Thus, υ in (72) equals the constant $-3c_1/c^2$. In the EF-model without inflation, c_1 is set to equal zero and hence υ vanishes. Once Einstein saw Friedman's solution, he was impressed by its consistency and the agreement with Hubble's observation. He said that it was his biggest blunder to have introduced a cosmological term. However, we can see that introducing cosmological term itself is not a mistake, as long as consistency condition is deduced and satisfied (CT-theorem). It must be introduced as a non-constant function if CHSPs are not ignored (CHSP-theorem). If they are ignored and *G* is taken as a constant, it can still be introduced as a non-zero constant as it is in the model of inflation. In such special case with $\upsilon = -3c_1/c^2$, $C(R) \equiv c_1 R^2/c^2 + A/R$. Then, (70) becomes

(70)′ $$\frac{dR}{dt} = \pm c\sqrt{\frac{A}{R} \mp 1 + \frac{c_1}{c^2}R^2}\ .$$

When Friedman chose his special solution with the special setting $c_1 = 0$, $R_{\max} = A$ in positive curvature case, the rate of expansion became monotonically decreasing, and c-failure and age-failure were doomed. When one set $c_1 > 0$, υ became negative. Especially, once $c_1 > c^2(1 - A/R)/R^2$ holds for any R (e.g., if $c_1 > 4c^2/27A^2$), expansion will last forever even in positive curvature case. If so, the expansion would be almost exponential (almost constant h and $R \sim R_0 e^{ht}$) at all the places where $|A - R|/R^3 << c_1/c^2$ (such as at $R \approx A$ and at all $R >> 3\sqrt{3}A/2$). In particular, when c_1 is thought to be very large, the universe could start to expand exponentially at very early cosmological time so that $R^3 >> Ac^2/c_1$ could be true even for very small values of R at very early times. If it were the case, one might then expect that the universe was a singularity or a tiny dot at the beginning and now it should be nearly flat like a sheet and nearly infinite. However, such near-infinite and fast-than-c speeds of expansion have by-products: near-c recessions and extremely large red shifts. No single galaxy is found to have extremely large red shifts. The fatal conflict with observation is due to ignoring the reality of CHSPs and of the empirical c-principle. Such rapid expansion models make the age failure even worse.

Because the existence of CHSPs and non-existence of extremely large red shifts are physical reality, we are not supposed to reach consistency by ignoring them. Besides, a positive constant c_1 in (72) gives a negative constant portion in the cosmological term, which is effectively equivalent to a negative pressure term. However, we must emphasize that the constant c_1 originally appears in the solution (69) and is solely determined by and related to initial condition needed to write down any special solution of a 2nd order differential eqn. By its very nature, it has nothing to do with a hypothetic concept of 'vacuum energy'.

Using the fact that $dh/dt = d(R^{-1}dR/dt) = R^{-1}d^2R/dt^2 - R^{-2}(dR/dt)^2$, we find

(73) $$R^{-1}\frac{d^2R}{dt^2} = \frac{dh}{dt} + h^2 .$$

From (73), (67), (72), and (71) we obtain the following $\dot{h}-equation$.

(74) $$\dot{h} \equiv \frac{dh}{dt} = \frac{c^2}{R^2}\Delta^{\pm}(R), \qquad \Delta^{\pm}(R) \equiv \pm 1 - \frac{3A_0B}{2R}a(R) = \frac{R^2}{c^2}\frac{dh}{dt}.$$

From (74) one can see immediately that a **necessary** condition for $dh/dt > 0$ to ever possibly happen is a **positive** curvature with

(74a) $$\frac{dh}{dt} = \frac{c^2}{R^2}\Delta(R), \qquad \Delta(R) = \Delta^{+}(R) \equiv 1 - \frac{3A_0B}{2R}a(R) = \frac{R^2}{c^2}\frac{dh}{dt}.$$

Please note, one can also get (74) by taking the derivative

$$2hdh/dt = dh^2/dt = dh^2/dRdR/dt = Rhdh^2/dR,$$

and then using (71) and the fact that (69a) yields

$$dC(R)/dR = 2c_1R/c^2 + 6A_0RI(R) - 3A_0Ba(R)/R^2 .$$

From (71) it follows that

(75) $$\frac{c^2}{R^2} = \pm[\frac{c^2}{R^2}C(R) - h^2],$$

This reveals the NSC (**necessary** and **sufficient** condition) to know the sign of the curvature:

(75a) $$\frac{c^2}{R^2}C(R)\begin{cases} > h^2 & \textit{positive} \\ < h^2 & \textit{negative} \end{cases} \text{curvature.}$$

Using (69a) we rewrite (75a) as

(75b) $$\rho I(R)\begin{cases} > H(R) & positive \\ < H(R) & negative \end{cases} \text{curvature}, \quad H(R) \equiv \frac{h^2 - c_1}{8\pi G_0}\frac{1}{R^3}.$$

If one assumes constant G by setting $a(R) \equiv a(R_1) = 1$, ignores CHSPs by setting $k_\rho \equiv 0$, and sets $c_1 = 0$, then $B \equiv 1$, $I(R) = 1/3R^3$, $C(R) = A/R$. Thus, υ in (72) vanishes and the eqs (62), (65), (66), (67), (68), (69), (69b), (70), (71), (74), and (75a, b) will return to their counterparts in the EF-model without CHSPs and without LV. On the other hand, if one assumes ***constant*** G but considers **CHSPs** and **LV**, then the eqs will return to their counterparts in the EF-model with LV, where a temporary replacement $A_0 I(R) \rightarrow AI(R)$ yields a new $I(R)$ function:

(75c) $$I(R) = -\int_{nc} BdR/R^4 = B/3R^3 - \int_{nc} \dot{B}dR/3hR^4, \quad \dot{B} \equiv dB/dt.$$

The integral in (75c) is difficult to evaluate. Before stars were born to trigger a new radiation mode, we may roughly assume that $\dot{B} \sim C_u/R^3$, where C_u is a constant. That is, the rate is proportional to the density of charged particles (inversely proportional to R^3. At very early times, $V(\pi) = hR\pi = \pi dR/dt \approx c$, $h \sim c/\pi R$, and

(75d) $$I(R) = \frac{B}{3R^3} - \int_{nc} \frac{\dot{B}dR}{hR^4} \sim \frac{B}{3R^3} + \frac{\pi C_u}{5cR^5} \Rightarrow$$

$$\frac{dR}{dt} = \pm c\sqrt{\frac{AB}{R} \mp 1 + \frac{c_1}{c^2}R^2 + \frac{A\pi C_u}{5cR^5}}.$$

If we ignore CHSPs, then $B \equiv 1$ (see (68)), $\dot{B} \equiv 0$, $C_u \equiv 0$, and (75d) returns to (70)′. In general, the rate in (75d) features faster-than-c and near-c recessions, with or without CHSPs and with or without LV.

C-failure is common in all EF-type models with constant *G*, whether or not the CHSPs and/or LV is considered. The only way to avoid c-failure is to allow *G*-symmetry to be broken spontaneously.

Negative curvature always leads to double failures: c-failure and a-failure, no matter LV is ignored or considered. The general curvature condition (75b) leads to EF-critical density formula under the EF-settings and the induced expansion mode. We will soon see that (75b) imposes no restriction on mass density under our new expansion modes, called parabola and non-parabola models. Namely, the celebrated critical density in EF cosmology appears as a special case of the new formula, exactly obtained from the new formula if and only if the EF-settings (e.g., constant *G* without spontaneous symmetry breaking factor, running factor, or Mach factor) are enforced. Thus, WMAP observation results are always irrelevant to curvature issue. We will **only** consider models of **positive** curvature. From now on, whenever we use a formula, deduced for both positive and negative curvature cases, its version for negative curvature will be automatically excluded unless we specify.

The EFD-type gravitation and cosmology is supposed to be able to tell what is the non-constant function $a(R)$ so that its present value $a_1 \equiv a(R_1)$, $G_0 = G/a(R) = G_1/a(R_1)$, and $G = G_0 a(R)$ will all become known on the most fundamental level with $G_1 \sim 6.67 \cdot 10^{-8} cm^3/g \cdot s^2$. We now prove a theorem.

***G*-Theorem**: Suppose the field eqs are of the form of (64) and the EF-metric with undetermined *R*(*t*) is right at large scale. For any smooth function *R*(*t*) that matches reality and therefore both *R*(*t*) and its induced Hubble coefficient *h*(*t*) will pass all future empirical tests, a properly chosen gravitational coupling coefficient *G*=*G*(*t*)=*G*(*R*(*t*)) automatically/immediately explains such behaviors of *R*(*t*) and of *h*(*t*) via the EFD-type gravitation and cosmology.

Proof: From (74a) we get

(76a)
$$a(R) = \frac{2R}{3A_0 B}\left(1 - \frac{R^2}{c^2}\frac{dh}{dt}\right).$$

From (7a, b), (64)′, and (59f) we see some basic constants:

(76b)
$$G_0 = \frac{G}{a(R)} = \frac{G_1}{a(R_1)} = \frac{6.67 \cdot 10^{-8} cm^3/g \cdot s^2}{a(R_1)}, \qquad \rho R^3 = \frac{M_{total}}{2\pi^2},$$

$$A_0 \equiv \frac{8\pi G_0 \rho R^3}{3c^2} = \frac{4G_0 M_{total}}{3\pi c^2}.$$

Then (76a, b) lead to

(76c)
$$G = \frac{\pi c^2}{2M_{total}B} R\left(1 - \frac{R^2}{c^2}\frac{dh}{dt}\right), \qquad B \equiv 1 + \frac{k+3}{3}k_\rho = 1 + \frac{k+3}{3}\overline{\lambda_d^4}\frac{\varepsilon_d}{\varepsilon},$$

$$B \equiv 1 + \frac{k+3}{3}\overline{\lambda_d^4}\frac{\varepsilon_d}{\varepsilon} = 1 + \frac{k+3}{3}\overline{\lambda_d^4}\frac{E_{d\,total}}{M_{total}c^2},$$

$$M_{total}B = M_{total} + M^{eff}_{d\,total}, \quad M^{eff}_{d\,total} \equiv \frac{K+3}{3}\overline{\lambda_d^4} E_{d\,total}/c^2 \quad (K{=}k).$$

Obviously, (76c) is equivalent to

(76d)
$$G = \frac{\pi c^2}{2}\frac{R\left(1 - \frac{R^2}{c^2}\frac{dh}{dt}\right)}{M_{total} + M^{eff}_{d\,total}}.$$

Either one of (76c, d) is called the ***G*-eqn**. Surely, (76d) is equivalent to

(76e)
$$\frac{R^2}{c^2}\frac{dh}{dt} = 1 - \frac{1}{\pi c^2}\frac{2G(M_{total} + M^{eff}_{d\,total})}{R}.$$

Radiations from charged particles and the 2nd law of thermodynamics make both $M_{d\,total}^{eff}$ and B monotone increasing, just like R in the expansion. The G-eqn holds not only in general case with non-constant G but also in every special case with **constant** G:

Case 1

If one sets $\varepsilon_d \equiv 0$, $G \equiv G_1 \sim 6.67 \cdot 10^{-8} cm^3 / g \cdot s^2$, then $B \equiv 1$, $M_{d\,total}^{eff} \equiv 0$, and the G-eqn (76c, d) becomes equivalent to (58b) in positive curvature case. One can use (59f) to verify this equivalence. The bad thing of this special case is that the constant G makes A a constant and dR/dt is monotone decreasing for all $R \leq R_{max} = A$. Thus, it does not agree with the acceleration interpretation of the supernova results. Fatally, it causes c-failure and age failure. Indeed, one can check that (58b) directly yields

$$d^2R/dt^2 - R^{-1}(dR/dt)^2 + c^2R^{-1}(\mp 1 + 3A/2R) = 0.$$

Its solution is shown in (58a), obviously monotone decreasing. Moreover, the recession speed that can be obtained from (58a) leads to c-failure, caused by completely abandoning the empirical c-principle without even noticing that this firm principle was completely and manually violated. The expansion in a model with c-failure had been too fast for long period of time, inducing age-failure.

The age-failure and the c-failure will never be solved by any future WMAP measurement. The two failures are the consequences of inner contradiction in the EF-type models, which have no remedy unless we obey empirical c-principle by allowing G-symmetry broken spontaneously or by allowing G running.

Case 2

If one sets $G \equiv G_1 \sim 6.67 \cdot 10^{-8} cm^3 / g \cdot s^2$ and considers CHSPs and LV, then the G-eqn (76c, d) becomes equivalent to

(76f) $$\frac{dh}{dt} = \frac{c^2}{R^2}\left(1 - \frac{3AB}{2R}\right).$$

Eqn (76f) can be seen in an EF-type model with LV (author's unpublished work). The uncertain values of the CHSP function $B = B(t) = B(R(t))$ and its uncertain influence on the solution $R(t)$ make events of acceleration uncertain, though such events seem to have barely occurred according to the EF-type model with LV and without spontaneous breaking of G-symmetry. Again, ***constant*** G makes A a constant. Moreover, (76f) directly yields

$$d^2R/dt^2 - R^{-1}(dR/dt)^2 + c^2R^{-1}(\mp 1 + 3AB/2R) = 0.$$

Its solution gives

(76g) $$\left(\frac{dR}{dt}\right)^2 = c^2\left[\frac{A(B+B_h)}{R} \mp 1\right] + c_1R^2$$

$$B_h \equiv -R^3 \int (h^{-1} R^{-4}\, dB/dt)\, dR\,, \qquad B \equiv 1 + \frac{k+3}{3}\overline{\lambda_d^4}\frac{\varepsilon_d}{\varepsilon}\,.$$

If we set c_1=0 and CHSPs are ignored with the setting $\varepsilon_d \equiv 0$, then $B \equiv 1$, $B_h \equiv 0$, and (76g) returns to (58a) in the EF-model without LV. As we consider LV seriously, still, B and B_h are bounded from above and below. This repeats the story in the EF-model without LV: The receding speeds determined by (76g) also go from infinity or near-infinite to the observed less-than-c speeds. Again, it leads to near-c recessions with extremely large red shifts nobody ever observed.

The above two special cases show that ***constant*** G is not consistent with the empirical **c-principle**, fatally conflicting with the observation showing that large red shifts are normally large without exception. Moreover, it is unable (without LV) or difficult (with LV) to explain acceleration supernova results may imply. The near-c recession, as a by-product of the violation of empirical c-principle, is disproved by overwhelming evidences: Observations in past 80 years, covering deep space and in all directions, find no single example of extremely large red shift.

Now a subtle but crucial meaning of (76e, d) can be seen:

Whenever the true values of $k_m n_m$, k_*, $E_{d\,total}$, $\overline{\lambda_d^4}$, M_{total}, and $M^{eff}_{d\,total}$ that match the reality (but unknown to us) could not lead to the observed behavior of h with constant G in (76e), a non-constant G, that is given by the G-eqn (76d) or (76c), where *R(t)* and *dh/dt* match the supernova results, automatically and immediately explains such observed acceleration on the platform of EFD-type gravitation with LV and with the spontaneous G-symmetry breaking. The G-eqn is an explicit and quantitative expression of the spontaneous breaking of G-symmetry. If the acceleration interpretation of dimmer supernovas is wrong, the genuine *R(t)* and *h(t)* must not lead to acceleration. Even in this case, the spontaneously broken G-symmetry immediately explains such genuine *R(t)* and *h(t)* as long as G obey the G-eqn where *R(t)* and *dh/dt* are genuine. The explanation power of the EFD-type gravitation and cosmology is of no uncertainty.

In a basic theory, G should be determined on most fundamental level, not phenomenologically by *R(t)*, *dh/dt*, and *B(t)* or $E^{eff}_{d\,total}(t)$ via the G-eqn. However, the G-eqn establishes a solid relation between G and *R(t)* with its induced quantities such as $c^{-2}R^2\dot{h}$ ($\dot{h} \equiv dh/dt$). Practically, we do not have empirical curves for *R(t)*, $c^{-2}R^2\dot{h}$, $R(1-c^{-2}R^2\dot{h})$, and *B(t)=B(R(t))* so that we can use such empirical curves to determine G theoretically via the G-eqn.

In the **presence** of a nice empirical curve of black body radiation and two theoretical curves of Wien and Rayleigh & Jeans, Planck was able to deduce his **genius** formula of black-body radiation. Now in the **absence** of an empirical curve of the function $c^{-2}R^2\dot{h}$, no genius formula for $c^{-2}R^2\dot{h}$ but a **sample model** will be proposed, enlightened by supernova results and subjected to severe velocity regulation to obey the c-principle everywhere and all the times. We are nailed down to a sample model of $c^{-2}R^2\dot{h}$ because it determines $Ba(R)$ via

(76a) and thereby $I(R)$ can be obtained via the definition in (69). Then *C(R)* that is defined in (69a) can be determined too. Finally, dR/dt and h will be determined via (70) and (71). It is Planck's genius sense of the physical reality of quanta (not his genius formula of black-body radiation) that has influenced the modern physics so deeply and largely.

8.4 Parabola and Non-Parabola Models

Although the spontaneous *G*-symmetry breaking is powerful enough to explain any behavior of the expansion of universe, a necessary condition for such explanation to be acceptable is to simultaneously explain the fact that *G* has been a constant within experimental errors since Tycho and Kepler. It is called **TKG fact**. To do so, specific quantitative sample models are proposed. Firstly, we apply acceleration interpretation of the dimmer supernovas and require *dh/dt* be positive in certain period of cosmological time. If such large departure from the behavior of the EF-universe can be in agreement with the TKG fact, a mild departure without positive *dh/dt* would be easier to agree with the TKG fact.

Consider a parabola function that does lead to positive *dh/dt* on an interval:

(77) $$\Delta(R) \equiv 1 - \frac{3A_0 Ba(R)}{2R} = \frac{R^2}{c^2}\frac{dh}{dt} = \beta[r^2 - (u-u_0)^2],$$

$$u \equiv R/R_{\max}, \quad \beta > 0, \quad 0 < u_0 < 1, \quad 0 < r < 1, \quad 0 < u_0 - r < u_0 + r < 1.$$

Obviously, *dh/dt>0* within the interval $u_0 - r < u < u_0 + r$. We know that h is a function of R hence of u. Two local extrema of h are reached at $u = u_0 \pm r$:

(77)′ $$h_{\min} = h(u_0 - r) \text{ and } h_{\max} = h(u_0 + r).$$

Note: Δ(R) is dimensionless so are the parameters r, u_0, and β. Solving (77) for $Ba(R)$ and then calculating $I(R)$ and $C(R)$ defined in (69a), we find

(77a) $$I(R) = \frac{1}{3A_0R^2}[1 + \beta(u_0{}^2 - r^2 - 4u_0u - 2u^2 \ln u)],$$

(77b) $$C(R) = \frac{c_1}{c^2}R^2 + 1 + \beta(u_0{}^2 - r^2 - 4u_0u - 2u^2 \ln u).$$

In the expansion, R and u are monotone increasing. Their values can serve as **cosmological times**. At each cosmological time R or u and in the interval $0 \le \chi \le \pi$,

(77c) $$V_{\max} = |h|\, d_{\max} = |h|\, R\pi = \pi\left|\frac{dR}{dt}\right|, \qquad 0 \le \chi \le \pi.$$

In positive curvature case, (77c), (70) and (77b) yield

(77d) $$V_{max} = \pi c\sqrt{\frac{c_1}{c^2}R^2 + \beta(u_0^2 - r^2 - 4u_0u - 2u^2\ln u)}\ .$$

A **fine** IC (*initial condition*) that obeys the empirical c-principle is

(77e) $$V_{\max} \to k_c c\ (k_c < 1)\ \text{as}\ R \to R_{min} > 0,\ u \to u_{min} \equiv R_{min}/R_{max} > 0.$$

For a galaxy at rest at χ in the co-moving system S_χ, its recession speed is $v = v(\chi) = \dot{d} = \dot{R}\chi = hR\chi$, according to Einstein embedding. The formula $v = \dot{d}$ in flat Euclidean space has been directly generalized and now works in curved spacetime where the distance *d* is no longer the length of a straight line (segment) but the length of a shortest line that is curved. Remote galaxies with $\chi < \pi$ we see are always their ancient images. If we search for their images at even earlier times, we may look in opposite direction with $\chi \to 2\pi - \chi > \pi$. Looking at them this way, the distance the light rays have traveled to reach us and the receding speed increases. Moreover, light rays may make one or more round-the-universe tour in addition, covering a longer distance such as $d = R(2\pi + \chi)$, $R(4\pi + \chi)$, and so forth. Faster-than-c and near-c receding seems to be inevitable in the expanding universe with positive curvature and Einstein embedding. However, the age of the universe and the age of stars are not and will not be old enough and therefore human being may be doomed to never have chance to see such faster-than-c and near-c receding. One thing is clear at this moment: No single galaxy is seen to have extremely large red shifts. Namely, near-c receding has never been seen. This is an empirical foundation for us to extend the solid empirical c-principle verified in curved spacetime on earth and around $\chi = 0$ (within labs, our solar system and the Milky Way) to large scale around $\pi \le \chi_c \le 3\pi/2$. Any positive $R_{\min}$ or $u_{\min}$ removes the singularity at the beginning. To **simplify** calculations, however, we take them to be zeros and set $\chi_c = \pi$ and $k_c = 1$:

(77e)′ $$V_{\max} \to c\ \text{as}\ R \to R_{min} = 0,\ u \to u_{min} \equiv R_{min}/R_{max} = 0.$$

With such **simplified** IC (initial condition) and the extended empirical c-principle at large scale, (77d) yields

(77f) $$\beta = \frac{\pi^{-2}}{u_0^2 - r^2}.$$

The **oscillation** condition (**OC**) is

(77g) $$V = 0\ \text{at}\ u = 1,\ \text{i.e., at}\ R = R_{max}.$$

Then (77d, f, g) lead to

(77h) $$c_1 = \frac{c^2}{\pi^2 R_{max}^2}\frac{4u_0 - u_0^2 + r^2}{u_0^2 - r^2}.$$

Please note, (68) is a 2nd order differential eqn. The two arbitrary constant t_0 and c_1 in the solution (69) are supposed to be determined by **standard** initial condition (**SIC**)

(77i)
$$\begin{cases} R(0) = R_{\min} \\ \left.\frac{dR}{dt}\right|_{t=0} = \dot{R}(0) \end{cases} \quad \left(\begin{array}{c} \dot{R}(0) = c/\pi \quad \text{by simplified IC and} \\ \text{extended empirical c - principle} \end{array} \right).$$

While β is supposed to be determined by the parabola model (77). Apparently, things did not happen that way. It was the oscillation condition (77g) that together with (77d, f) determined c_1 written in (77h), while it was the simplified initial condition $V_{\max} = \pi \mid dR/dt \mid = c$ at the beginning $t = 0$ with $R_{\min} = 0$ that determined β via (77f). Such strange things happened because a c_1 -term disappeared in (77f) due to the approximation $R_{\min} = 0$. In general, the SIC (77i) and (77c, d) lead to

(77j)
$$\frac{c_1}{c^2} R_{\min}^2 + \beta(u_0^2 - r^2 - 4u_0 u_{\min} - 2u_{\min}^2 \ln u_{\min}) = \pi^{-2}.$$

This was supposed to be used to determine c_1. But the simplified initial condition $R_{\min} = 0$ ($u_{\min} = 0$) made it impossible to use (77j) to determine c_1. Instead, we got β written in (77f). Then we had to use another condition, the OC in (77g), together with (77d, f) to determine c_1. This unusual thing is a by-product of the singularity that corresponds to $R_{\min} = 0$. Once we return to the fine initial condition (77e) and (77i) with $R_{\min} > 0$ and $u_{\min} > 0$, that removes the singularity at the beginning, things immediately turn to be normal. To our surprise, the normal procedure to determine c_1 and β justifies (77f) and (77h) with high precession. In fact, (77j) with $R_{\min} > 0$ immediately yields

(77k)
$$c_1 = \frac{c^2}{\pi^2 R_{\min}^2}[1 - \pi^2 \beta(u_0^2 - r^2 - 4u_0 u_{\min} - 2u_{\min}^2 \ln u_{\min})].$$

Now we use parabola model (77) to determine β. Here is the subtlety: the model (77) is written in terms of the variable u ($u \equiv R/R_{\max}$) that contains a parameter $R_{\max}$. Four parameters (β, u_0, r, and $R_{\max}$) are contained in the parabola model (77). However, only three parameters are needed to fix its shape and position: two for the position of its vertex and one for its span. On the other hand, we do need all the four parameters to fix the parabola model (77). One of the four parameters has to be determined by the physical meaning of the parameters in the parabola model. Since 'max' so far is only a label of the parameter $R_{\max}$, we must enforce OC (77g) so that $R_{\max}$ can be endowed with desired physical meaning. The essence of this enforcement with the OC (77g) is to determine the parabola model with clear physical meaning. As a consequence, we find immediately that β cannot be an independent parameter in any oscillation model but dependent, determined by other three parameter (u_0, r, R_{max}) and the initial value R_{min}, i.e., β is determined by u_{min} ($u_{min} \equiv R_{min} / R_{max}$), u_0, and r. Indeed, (77g) and (77d) yield

(77l) $$\frac{c_1}{c^2} R_{max}^2 + \beta(u_0^2 - r^2 - 4u_0) = 0 .$$

Then (77l) and (77k) lead to

(77m) $$\beta = \frac{\pi^{-2}}{(u_0^2 - r^2)(1 - u_{min}^2) - 4u_0 u_{min}(1 - u_{min}) - 2u_{min}^2 \ln u_{min}} .$$

Using (77m), we rewrite (77k) as

(77n) $$c_1 = \frac{c^2}{\pi^2 R_{max}^2} \frac{4u_0 - u_0^2 + r^2}{(u_0^2 - r^2)(1 - u_{min}^2) - 4u_0 u_{min}(1 - u_{min}) - 2u_{min}^2 \ln u_{min}} .$$

Though the fine initial condition $u_{min} > 0$ ($R_{min} > 0$) removes the singularity at the beginning, it may be too small to calculate numerically. Then (77m) and (77n) are too close to (77f) and (77h) respectively. Frankly, we do not know the IC. Physical laws in terms of differential eqs never determine initial conditions. To simplify the calculations, however, we will use (77f) and (77h), which now have been justified by (77m, n) under the assumption of small IC: $u_{min} << 1$.

Using (77f, h) we now rewrite (77b)

(78a) $$C(R) = 1 + \pi^{-2} + \frac{\pi^{-2}}{u_0^2 - r^2} [(4u_0 - u_0^2 + r^2)u^2 - 4u_0 u - 2u^2 \ln u] , \quad u \equiv \frac{R}{R_{max}} .$$

Infinitely many choices of the pair (u_0, r) with $0 < u_0 - r < u_0 < u_0 + r < 1$ will either produce faster-than-c recession or a nonsense $(dR/dt)^2 < 0$ at certain cosmological time(s), while infinitely many other choices of (u_0 , r) with $0 < u_0 - r < u_0 < u_0 + r < 1$ will make all recession-speeds within $0 \le \chi \le \pi$ no more than c at every cosmological time. A sample parabola model that automatically satisfies such strict velocity regulation is given by the single initial condition (77e)′ and such a choice as

(78b) $$u_0 = 0.356 , \quad r = 0.13 .$$

Using (77c) and (70), we write in positive curvature case

(78c) $$V_{max} = ck(u) , \qquad k(u) \equiv \pi\sqrt{C(R) - 1} .$$

Then (78a,c) yield

(78d) $$k(u) \equiv \frac{1}{\sqrt{u_0^2 - r^2}} \sqrt{(4u_0 - u_0^2 + r^2)u^2 - 4u_0 u - 2u^2 \ln u + u_0^2 - r^2}\ .$$

The sample choice (78b) and (78d) lead to

(78e) $$k(u) \equiv \frac{1}{\sqrt{0.109836}} \sqrt{1.314164 \cdot u^2 - 1.424 \cdot u - 2u^2 \ln u + 0.109836}$$

Now from (70), (78c), and (78d, e) we write the rate of expansion/contraction with positive curvature

(78) $$\frac{dR}{dt} = \pm \frac{c}{\pi} k(u)$$

$$= \pm \frac{c}{\pi} \frac{1}{\sqrt{u_0^2 - r^2}} \sqrt{(4u_0 - u_0^2 + r^2)u^2 - 4u_0 u - 2u^2 \ln u + u_0^2 - r^2}$$

$$= \pm \frac{c}{\pi} \frac{1}{\sqrt{0.109836}} \sqrt{1.314164 \cdot u^2 - 1.424 \cdot u - 2u^2 \ln u + 0.109836}\ .$$

Due to (78c), the velocity regulation (**VR**) $0 \le V_{max} < c$ is equivalent to $0 \le k(u)^2 < 1$. One can check that (78e) obeys the VR at every cosmological time $0 < u_{min} \le u \le 1$.

We now deduce some important formulas. Using (78c) we rewrite (71) as

(79) $$h = \pm \frac{c}{\pi R} k(u) = \pm \frac{c}{\pi R_{max}} \frac{k(u)}{u}$$

In (79) the sign pair $\pm$ is for expansion and contraction respectively. From (79), (77)′, and (78b, e) we get two relative extrema and two absolute extrema (in the expansion period):

(79a) $$h_{min} = \frac{c}{\pi R_{max}} \frac{k(.226)}{.226}, \qquad h_{max} = \frac{c}{\pi R_{max}} \frac{k(.486)}{.486}, \qquad \frac{h_{max}}{h_{min}} = 1.454363,$$

$$h_{absmin} = \frac{c}{\pi R_{max}} \frac{k(1)}{1} = 0, \qquad h_{absmax} = \frac{c}{\pi R_{max}} \frac{k(u_{min})}{u_{min}} >> 1.$$

Here the absolute maximum is very large but finite due to the fine (non-zero) initial value $o < u_{min} << 1$. One can easily use numerical calculation or a graphing calculator to get a nice *u-h* graph, where one unit on the vertical h-axis has a value of $c / \pi R_{max}$.

At present time $t = t_1$, $h = h_1 \sim 0.726 \cdot 10^{-10} yr^{-1}$, $u = u_1 \equiv R_1 / R_{max}$, and $R = R_1$. Then from (79) it follows

(79b)
$$R_{\max} = \frac{c}{\pi h_1}\frac{k(u_1)}{u_1} = 4.384\frac{k(u_1)}{u_1}10^9 ly,$$

(79c)
$$R_1 = u_1 R_{\max} = \frac{c}{\pi h_1}k(u_1) = 4.384\cdot k(u_1)\cdot 10^9 ly,$$

$$d_{\max 1} = R_1\pi = \frac{c}{h_1}k(u_1) = 1.377\cdot k(u_1)\cdot 10^{10} ly.$$

Now we must explain any red shifts that may be larger than $d_{\max 1}$. An important **common** feature of finite EF- and EFD-universe with positive curvature (w/wo LV) is the awesome **Einstein embedding**. That is, any t-snapshot of such universe is a 3-dim surface of a 4-dim sphere embedded in a 4-dim Euclidean space. It is a corner stone of the EF-metric and its induced theories of cosmology. It provides with a magnificent 4-dim bird's-eye view that may play important role in future study of cosmology. Red shift is directly related to recession speed of light source by its physical nature, called Doppler-effect interpretation of stars' red shifts. The expanding 4-dim sphere shows a vision that nicely relates recession speed v to distance d via the celebrated formula $v=hd$, where h is a constant in space (not in time). Thus, distance may serve as a measure of red shift. The observed red shift of a light ray sent to us at a past cosmological time, if larger than the maximum distance in (79c), must be explained through an unusual explanation (UE), which claims that such large red shifts are caused by the following reason if Doppler red shifts dominate gravitational red shifts.

UE: The light rays with such large red shifts came from the light sources that were and are on the **backside** of the observer. This is called **BI** (backside interpretation). It claims and stresses the large distances the light rays with large red shifts had traveled. The long journey of such light rays can be visualized in the 4-dim bird's-eye view, usually refers to cosmological effect of gravitational lens. Because of the age-theorem, the EF-universe is too young to make such interpretation valid. This BI works in any model with an old enough age such as the EFD-type model we are developing. But it does not work in the EF-model with young age $t_1 < 9.18\cdot 10^9 yr$ and with the remotest light source receding with less-than-c speed only after cosmological time

$$u = u_c(\pi) = 1/(1+\pi^{-2}) = 0.90800033165.$$

We must realize that if gravitational red shifts dominate Doppler red shifts, the light source is not that far away from us as its red shifts apparently mean. Some galaxies and quasars have very large red shifts that might need UE if Doppler red shifts dominate. However, it is possible that they are not that far and their sizes are small enough for gravitational red shifts to dominate. If this is true, the above UE is not needed. Due to the Schwarzschild metric, the formula of gravitational red shifts is $|\Delta\nu|/\nu = GM/c^2r$. On the surface of the sun, it is about $2\cdot 10^{-6}$. For a quasar with $M_q = 10^{11-s}M_{sun}$ and $r_q = 10^q r_{sun}$, we have

$$|\Delta\nu|/\nu = 10^{11-s-q} GM_{sun}/c^2 r_{sun} \sim 2\cdot 10^{5-s-q}.$$

As long as $5.30103 < s+q < 5.47712$, we have $2/3 < |\Delta\nu|/\nu < 1$ $(2 < Z < \infty)$. It is difficult to measure the mass and the size of a quasar without knowing its distance. It is equally difficult to know the distance without knowing the nature of the red shifts. The nature of quasars' red shifts and how far they are remain unclear. If they are really far and 'old', the old age of the EFD-type universe allows the UE of their large red shifts.

The Milky Way contains about $2\cdot 10^{11}$ stars. Suppose a huge group of babe-stars of total mass $10^{11} M_{sun}$ were concentrated in a sphere whose diameter was no less than .6 ly. This group of babe-stars was not a super black hole. It radiated large amount of energy like a galaxy. This may be a clue to the energy source problem of quasars if they are really as remote as their large red shifts imply. This hypothesis needs discovery of 'toddler' galaxies between quasars and normal galaxies to support. Elliptical galaxies seem to be between 'young' spiral galaxies and quasars. The so-called "young' galaxies are those that sent their images to us not long time ago while elliptical galaxies and quasars we see today are the images of their remote past. But if we identify them by their ages counted from the beginning of star era, the Milky Way we see is one of the oldest galaxies while quasars may be among the youngest we know today as long as their 'birthdays' are close to the average and to each other. The author does not know for sure whether spiral galaxies should be 'younger' or 'older' than elliptical galaxies. One thing is sure due to basic logic and reasoning: The father the galaxy, the older the image we see, and the younger the galaxy at the time it sent out the light rays we see today provided galaxies were born at approximately the same cosmological time.

If a remote galaxy we see today is the image of its past physical existence, for instance, 5 billion years ago, the light rays we see have traveled a distance of 5 billion ly. Where is the galaxy now? In the formula *v=hd*, *h* should be of the past value 5 billion years ago and *d* should be of the **past value** of the **distance** between the past physical existence of the earth and of the galaxy 5 billion years ago. If we know the past and present values for a galaxy at rest at χ in the co-moving system S_χ, we have

(79d) $$d = v/h = R\chi\,,\ d_1 = v_1/h_1 = R_1\chi\,,\ d/d_1 = R/R_1\,.$$

Consequently, a large past value of *h* does not necessarily mean large recession speed *v* because *R* was smaller in the past and so was *d*.

Please note: The distance d_γ a light ray we see today has traveled is not d nor d_1. The 4-dim bird's-eye view and the geometry in 4-dim Euclidean space can be used to calculate d_γ. The light rays were moving along the 3-dim surface of an expanding 4-dim sphere. This visualizes the motion of light rays. One can see that d_γ is always close to d_1 than to d, especially when d is much less than d_1. Approximately, one may use $d_1(\chi) = R_1\chi$ to represent the distance d_γ the light ray has traveled from the light source at rest at χ in the co-moving coordinate system. Then we have

The law of (apparent) brightness: Supernovas appear normal if and only if $k(u)$, not h, is quasi-constant in time. A remote supernova appears dimmer than expected if and only if $k(u)$, not h, was smaller in the corresponding remote past cosmological time u.

Proof: From (79) it follows

$$\text{(79e)} \qquad v(\chi) = hd(\chi) = hR\chi = \frac{c}{\pi}k(u)\chi\,,$$

$$\frac{\Delta d_\gamma}{\Delta v(\chi)} \approx \frac{\Delta d_1(\chi)}{\Delta v(\chi)} = \frac{R_1\Delta\chi}{\pi^{-1}ck(u)\Delta\chi} = \frac{\pi R_1}{c}k(u)^{-1}, \quad d_1(\chi) = R_1\chi\,.$$

For Doppler shifts, the quantity of red shift $Z \equiv \Delta\lambda/\lambda$ is directly related to receding speed v. The larger the $v(\chi)$ and $\Delta v(\chi)$ respectively, the larger the $Z(\chi)$ and $\Delta Z(\chi)$ respectively. Besides, Δd_γ is directly related to the change in brightness. Thus, we call the above rate a **RS-D rate** (*redshift-distance rate*), directly related to the **RS-B rate** (*redshift-brightness rate*). ■

Compare what we get from (60a) in the original EF-model:

$$\text{(79f)} \qquad v(\chi) = hd(\chi) = hR\chi = c\chi\sqrt{1/u-1}\,,$$

$$\frac{\Delta d_\gamma}{\Delta v(\chi)} \approx \frac{\Delta d_1(\chi)}{\Delta v(\chi)} = \frac{\pi R_1}{c}\frac{\pi^{-1}}{\sqrt{1/u-1}}, \qquad d_1(\chi) = R_1\chi\,.$$

From (78e) we find that $1/k(u) > 1/\pi\sqrt{1/u-1}$. This is only a comparison between two theories. We will focus on the explanation of dimmer supernovas using (79e), which is the very comparison between the theory and the observation.

Once the severe violation of the c-principle is gone in the chosen parabola model, an old age (about 20 billion years) of the present universe will be proved. In fact, (69b) and (78c) lead to the **present age**

$$\text{(80a)} \qquad t_1 = \int_a^b \frac{dR}{k(u)} = \frac{\pi R_{\max}}{c}\int_s^w \frac{du}{k(u)}. \quad a = R_{\min},\ b = R_1,\ s = u_{\min},\ w = u_1\,.$$

In general, the **u-t relation** is given by

$$\text{(80b)} \qquad t = \frac{\pi R_{\max}}{c}\int_s^u \frac{d\xi}{k(\xi)}, \qquad s = u_{\min}\,.$$

It is not integrable analytically, according to (78d, e). But the graph of $1/k(u)$ makes it very easy to get numerical results of the integration in (80b). To know the times in (80a, b), extrema of h in (79a), and h in each cosmological time u, we must know $R_{\max}$. To know $R_{\max}$ and R_1 via (79b) and (79c), we must know

u_1. Any direct measurement of the size of the universe requires last Columbus-type voyage. We do not know how long mankind has to wait for this to come true and whether human societies will overcome their problems to survive until that day comes. A round-universe trip would take $2\pi R$ years to finish (*R* is the radius in ly). Mankind will be unable to catch up the expanding/contracting universe. It seems that there is no direct experimental way to measure the size of the universe. We now are searching for any **clue** to the value of u_1.

From (78e) and (79) one can get a graph of *h=h(u)*. The unit in *h*-axis is $c/\pi R_{\max}$, which is an unknown quantity at this moment. Between u=0.226 and u=0.486 the graph climbs up. But between u=0.4 and u=0.58 it is **quasi-horizontal**. The numerical integration of (80b) indicates that it would take about 5 billion years for the size of the universe to grow from u=0.4 to u=0.58 if we assume $u_1 = 0.58$. In other words, the supernovas no more than 5 billion ly away should appear normal while those farther than 5 billion ly away appear dimmer than expected. If observation confirms this situation, $u_1 \sim 0.58$ would be a reasonable estimation. However, there is a ***flaw*** in the above reasoning. Indeed, $v(\chi) = hd(\chi)$ leads to

(80c) $$\frac{\Delta d(\chi)}{\Delta v(\chi)} = 1/h.$$

Due to (80c), a quasi-constant h means a quasi-constant rate $\Delta d(\chi)/\Delta v(\chi)$. But the distance in (80c) is different from the distance $d_1(\chi)$ in the rate one can see in (79e). It is $d_1(\chi)$, not $d(\chi)$, that is approximately equal to the distance d_γ the light ray has traveled to reach the earth. Thus, it is the rate $\Delta d_1(\chi)/\Delta v(\chi)$ in (79e), not the rate $\Delta d(\chi)/\Delta v(\chi)$ in (80c), that is related to the rate of change in brightness of remote supernovas. Due to (79e), the following **law of (apparent) brightness** becomes clear.

Supernovas appear normal if and only if $k(u)$, not h, is quasi-constant in time. A remote supernova appears dimmer than expected if and only if $k(u)$, not h, was smaller in the corresponding remote past cosmological time *u*.

This law indicates that a smaller *h* in the remote past and a positive rate $\dot{h} \equiv dh/dt$ in the remote past is not necessary for explaining acceleration. In the parabola model (77), *dh/dt>0* within the interval $u_0 - r < u < u_0 + r$. Positive rate $\dot{h} \equiv dh/dt$ in the remote past is sufficient to explain dimmer supernovas. Other models will be studied in the future. Enlightened by the law of brightness, trial and error method helps to get a reasonable estimation. One can see (81c) and the table after (81d) for the reason why u_1 is estimated as follows:

(80d) $$u_1 \sim 0.82$$

Eqs (80d), (78e), (79a, b, c), and (80a) provide us with the estimations of following important quantities:

(81) $R_{\max} = 4.911 \cdot 10^9 ly$, $R_1 = 4.027 \cdot 10^9 ly$, $h_{\min} = 0.769 \cdot 10^{-10} yr^{-1}$,

$$h_{\max} = 0.917 \cdot 10^{-10} yr^{-1}, \quad h_1 = \frac{c}{\pi R_{\max}} \frac{k(u_1)}{u_1} = .726 \cdot 10^{-10} yr^{-1},$$

$$t_1 = 2.08 \cdot 10^{10} yr, \quad d_{\max 1} = \pi R_1 = 1.265 \cdot 10^{10} ly.$$

Here $h_{\min} > h_1$ for it is a local minimum. The remotest galaxy with $d = d_{\max 1} =$ $1.265 \cdot 10^{10} ly$ is located in the place opposite to us in the 4-dim bird's-eye view. Any Doppler red shifts larger than this must be explained with UEs (**unusual explanations**) shown in the **UE**, which can be found between (79c) and (79d). The proved old age in (81) and all less-than-c recession validate the UE.

Recent WMAP observations are interpreted as that total mass (baryons and dark matter) accounts about 30% of the 'critical density'. DM takes 22% and ordinary matter 4%. If this is approximately true, then

(81a) $$\rho_1 \sim 0.26\rho_c \sim 2.46 \cdot 10^{-30} g / cm^3$$

is the present density where dark matter is included. We will soon prove rigorously that the spontaneous breaking of *G*-symmetry leads to such a new curvature-condition for the sign of curvature that the meaning of the 'critical' density in the EF-model is completely changed. The positive curvature condition now only requires a positive density of non-CHSP matter.

If the total mass in (59f) includes such dark matter that is considered by some interpreters of the WMAP results, then R_1 in (81), (81a), and (59g) lead to

(81b) $$k_m n_m \sim 2.69 \cdot 10^5.$$

In the original EF-model, the rate in (79f) is in **fatal conflict** with the observed dimmer supernovas. That rate is monotone decreasing as u decreases. In the contrary, (79e) with the sample parabola model (78e) yields such a rate $\Delta d_1(\chi) / \Delta v(\chi)$ that it is significantly increasing as u decreases approximately from u=0.52. Let u=0.52 be the cosmological time, t_{sn} years ago from now. Then (80b, d), (81), and (78e) lead to

(81c) $$t_{sn} = \frac{\pi R_{\max}}{c} \int_{.52}^{.82} k(u)^{-1} du \approx 4.93 \cdot 10^9 yr.$$

Due to (78e), the function $k(u)^{-1}$ is not integrable. But it is easy to do numerical integration. We will see a table telling that *k(u)*, hence the rate $\Delta d_\gamma(\chi) / \Delta v(\chi)$ in (79e) is quasi-constant in the past $4.99 \cdot 10^9$ years. Before that, $1 / k(u)$ and hence the rate $\Delta d_\gamma(\chi) / \Delta v(\chi)$ was significantly larger, causing supernovas dimmer than expected. Namely, supernovas appear dimmer than expected if they are the images of their ancient physical existence approximately more than five billion years ago. This agrees with the observation. To get a rate-u table we write

(81d) $$\Delta d_\gamma(\chi)/\Delta v(\chi) \approx \Delta d_1(\chi)/\Delta v(\chi) = \psi r(u),$$

$$\psi \equiv \pi R_1/c, \quad r(u) = \begin{cases} 1/k(u) & (EFD) \\ 1/\pi\sqrt{1/u-1} & (EF) \end{cases}.$$

With (78e) based on the choice (78b) of the parameter pair, we obtain:

Table 1 of r(u) ($u_0 = 0.356$, $r = 0.13$)

u	0	0.1	0.2	0.3	0.4	0.5	0.51	**0.52**	0.55	0.6	0.7	0.8	**0.82**	0.9
r(u)(EDF)	1	2.031	4.156	2.496	1.585	1.227	1.2047	1.184	1.131	1.068	1.015	1.062	1.089	1.316
r(u)(EF)	0	0.11	0.16	0.21	0.26	0.32	0.325	0.33	0.35	0.39	0.49	0.64	0.73	1.39

Now let us consider another sample parabola model with $(u_0, \ r) = (0.5, \ 0)$ different from (78b). We repeat the same process since (78b) and obtain:

Table 2 of r(u) ($u_0 = 0.5$, $r = 0$)

u	0	0.1	0.2	0.3	0.4	**0.5**	0.55	0.6	0.7	0.8	**0.84**	0.9
r(u)(EDF)	1	1.484	2.264	3.213	3.282	2.710	2.467	2.287	2.093	2.120	2.217	2.559
r(u)(EF)	0	0.11	0.16	0.21	0.26	0.32	0.35	0.39	0.49	0.64	0.73	1.39

The EF-model allows near-infinite and faster-than-c speeds of recession. If we allow initial speed of recession to be faster-than-c such as $v_0 \equiv v(\pi)|_{t=0} \sim 1.54c$, then the age of the EFD-universe becomes 13 billion years. Faster-than-c initial speeds of recession in EFD-type models could easily and trivially lead to any ages younger than 20 billion years. Then we may have to change the sample choice (78b), even to consider non-parabola models, in order to match the supernova results. One can see that there is no any problem for a properly selected EFD-type model to match the facts such as the non-existence of a single galaxy having extremely large red shifts, the supernova results, and an age younger than 20 billion years (13, 14, or 15 billion years). In this paper, we will not discuss the details due to the following reason.

The ***age*** of the universe may not be determined by the ***CBR***. CBR contains too many non-relic microwave photons. At day times, planets absorb photons such as infrared photons and photons with higher frequencies. Only a small portion is reflected with the same or nearly the same frequencies. Others are absorbed and the photon-energies become thermal energies. At night times, part of the thermal energies is radiated out from planets in the forms of infrared and microwave photons. Planets, celestial bodies, dusts, and molecules in the spaces btw stars are '**converters**' that convert a portion of non-relic photons with frequencies higher than microwave photons into lower frequency photons and/or microwave photons. The moon's radius is about $1.738 \cdot 10^3 km$. If it were broken into pieces of the same size $0.01738cm$, the total surface area would be 10^{10} larger than the moon. The conversion effect would become much larger. Each time a beam of non-relic incident photons meet a 'converter', a portion of their energies is transformed into photons with lower frequencies. This process has been going on for more than ten billion years since the birth of stars and galaxies. Now that CBR photons contain many non-relic photons, the observation of CBR is unable to tell the age of the universe. Note, even tiniest dusts contain large numbers of molecules close to each other. The concepts of temperature and thermal energy in thermodynamics apply to them. Hydrogen gases (H_2) may occupy larger spaces than dusts. When an incident photon with energy higher than a microwave photon hits a molecule H_2 in space, other molecules are far away. There is another way to convert. The photon energy can be changed to a 'thermal energy' in a generalized meaning: The incident photon may be absorbed with its energy becoming kinetic energies and mechanical energies of the two atoms in the single H_2 molecule. The 2 atoms are bound. Once the incident photon with energy larger than a microwave photon makes a single H_2 molecule, instead of an atomic electron, excited, the excited state may return to ground state not by a single de-excitation but gradually by a series of de-excitations releasing many microwave photons. So hydrogen gases in space are 'convertors' too. There are many unbound protons and electrons in spaces btw stars. As they move in the magnetic fields in spaces btw stars, the motions are not inertial and they must radiate. Even if their accelerations and speeds are only able to produce radio waves, the situations may change when incident photons with proper energies hit them. Unbound protons and electrons may be part of CBR source matter and part of the 'convertors' too. All such convertors serve as baryonic CBR source matter in spaces radiating non-relic microwave photons. Only those relic and non-relic microwave photons that arrive at the earth can be detected and photographed. They are called CBR.

Radioactive dating is able to detect the ages of unstable chemical elements. But it is disable to tell the ages of stable hydrogen atoms. We do not have any empirical value of the age of the universe. We only have solid empirical lower bound of the age of universe, that has proved once for all that no solution in the models for cosmology is reasonable, exactly as Einstein's lifetime conclusion.

Now we show the TKG fact. From (76c) and (77) we obtain

(82) $$G = \frac{\pi c^2}{2M_{total}} \frac{R}{B} \Omega, \quad \Omega \equiv 1 + \beta(u_0^2 - r^2) + \beta(u^2 - 2u_0 u).$$

With the sample choice (78b), we have

(82a) $$\Omega \equiv 1 + \pi^{-2} + \frac{\pi^{-2}}{0.109836}(u^2 - 0.712 \cdot u).$$

Thus, (82) yields

(82b) $$\frac{G}{G_1} = \frac{R}{R_1} \frac{B_1}{B} \frac{\Omega}{\Omega_1}.$$

There are three major eras in the cosmology: 1. Big Bang era. At the beginning, the initial condition could be zero or small number of CHSPs. On later times, relic CHSPs were produced irreversibly. The quantities with subscript b are for relic CHSPs as birth marks of the universe. 2. Ante-star era when the universe cooled down with radiation negligible in cosmological problems. In this era, only relic CHSPs made considerable contribution to k_ρ ($k_\rho = k_b = const$ in this era). 3. Star era when sufficiently large number of stars were born. Quantities with subscript r are for non-relic CHSPs radiated by stars.

Now we calculate the ratio B_1 / B. Using the CHSP-function $B = B(t) = B(R(t)) \equiv 1 + (k+1)k_\rho / 3$ ($k_\rho \equiv \overline{\lambda_d^4}\frac{\varepsilon_d}{\varepsilon}$) defined in (76c), we write

(82c) $$\frac{\varepsilon_d}{\varepsilon} = \frac{\varepsilon_b + \varepsilon_r}{\varepsilon} = \frac{\varepsilon_b V + \varepsilon_r V}{\varepsilon V} = \frac{E_b + E_r}{Mc^2}, \quad \frac{E_r}{Mc^2} = \frac{M_L}{M}\frac{E_r}{M_L c^2} \equiv k_L \frac{E_r}{M_L c^2}.$$

The solar constant is $C_{sun} = 2cal / cm^2 \cdot \min$. Let $C_r = k_* C_{sun}$ be such an average rate of radiation for all luminous stars that

(82d) $$\frac{E_r}{M_L c^2} = \frac{4\pi R_{SE}^2 k_* C_{sun} t_*}{M_{sun} c^2} = 6.76 k_* 10^{-14} yr^{-1} \cdot t_* = 2.14 k_* 10^{-21} \sec^{-1} \cdot t_*$$

holds rigorously for some unknown value of k_*. When star-radiation in the entire universe becomes stable, k_* becomes a constant. At the star-radiation-time t_* = $k_t 10^{10} yr$ ($k_t = 0$ and $t_* = 0$ corresponds to the beginning of the star era), the CHSP-quantity B defined in (68) can be written using (82d)

(82e) $$B = B(t) = B_b + \frac{k+3}{3} 6.76 k_L k_* k_t \overline{\lambda_d^4} 10^{-4},$$

$$B_b \equiv 1 + \frac{k+3}{3}\overline{\lambda_d^4}\frac{\varepsilon_b}{\varepsilon}, \quad k_t \equiv t_* / (10^{10} yr).$$

Now we find

(82f) $$\frac{B_1}{B} = \frac{B_b + 6.76 \cdot k_0 k_{L1} k_{*1} k_{t1} \overline{\lambda_d^4} \cdot 10^{-4}}{B_b + 6.76 \cdot k_0 k_L k_* k_t \overline{\lambda_d^4} \cdot 10^{-4}}. \quad k_0 \equiv \frac{k+3}{3} \sim 4/3.$$

Let $k_t = k_{t1} + \delta$, $\delta = \pm 500 \cdot 10^{-10} = \pm 5 \cdot 10^{-8}$. Then $t = k_t \cdot 10^{10} yr = k_{t1} \cdot 10^{10} yr \pm 500 yr = t_1 \pm 500 yr$. In a period of $500 yr$, $k_L \sim k_{L1}$ and $k_* \sim k_{*1}$. Then

(82f)′ $$\frac{B_1}{B} = 1 \mp \frac{5}{\xi_1 + k_t} \cdot 10^{-8}, \quad \xi_1 \equiv \frac{1 + k_0 \overline{\lambda_d^4}\varepsilon_b / \varepsilon}{6.76 k_0 k_{L1} k_{*1} \overline{\lambda_d^4} 10^{-4}}.$$

In a <u>cosmologically short period of time such as 500 years,</u>

(82g) $$\frac{u}{u_1}=\frac{R}{R_1}=\frac{R_1\pm\Delta R}{R_1}=1\pm\frac{1}{R_1}\frac{dR}{dt}\bigg|_{t_1}\cdot\Delta t=1\pm h_1\cdot 500yr=1\pm 4\cdot 10^{-8}.$$

$$u=(1\pm 4\cdot 10^{-8})u_1.$$

Taking $u_1 \sim 0.84$ and using (78b) and (82a, g), we get a 1st order approximation:

(82h) $$\frac{\Omega}{\Omega_1}\sim 1\pm 2.50\cdot 10^{-8}.$$

Using (82f)′ and (82b, g, h), we find

(82i) $$\frac{G}{G_1}=1\pm(6.50-\frac{5}{k_t+\xi_1})\cdot 10^{-8},$$

$$\frac{|G-G_1|}{G_1}=\frac{|\Delta G|}{G_1}=|6.50-\frac{5}{k_t+\xi_1}|\cdot 10^{-8}.$$

With largest possible errors, star era began $5\cdot 10^9 \sim 2\cdot 10^{10}$ years ago so that $.5\le k_t \approx k_{t1}\le 2$. We do not and will not have empirical value of k_{*1} in the coming century. We only have empirical estimations. This average rate of radiation mainly depends on the percentages of largest and very large stars. We safely estimate that $10^{-4}\le k_{*1}\le 10^5$. Without LV, $\overline{\lambda_d^4}=1$. With intentionally enlarged errors, $10^{-4}\le\overline{\lambda_d^4}\le 10^{10}$, $10^{-2}\le k_{L1}\le 1-10^{-2}$ and $10^{-8}\le\varepsilon_b/\varepsilon\le 10^4$. Thus, the definition of ξ_1 in (82f)′ and the above estimations with intentionally enlarged largest possible errors lead to

(82i)′ $$0.5<k_t+\xi_1\le 10^{16}.$$

Now (82i) and (82i)′ lead to

(82j) $$\frac{|\Delta G|}{G_1}<6.50\cdot 10^{-8},\qquad |G-G_1|=|\Delta G|<4.34\cdot 10^{-15}cm^3/g\cdot s^2.$$

We have intentionally enlarge the errors in the estimations not because we want safest estimations but because we want to show that the inherent uncertainty of the values of those important quantities of the entire universe is mysteriously absorbed and does not cause any uncertain effect on the certainty of acceleration once the *G*-symmetry is allowed to be spontaneously broken in the EFD-type theory of gravitation and cosmology. The amount of the departure from the *G*-symmetry is so small that it fully agrees with the TKG fact although the values of the universe quantities are uncertain.

The past 3 billion years were crucial for **life** to occur. In this period of time, one can check that $|\Delta G|/G_1 < 0.35$. A better estimation indicates that 3 billion years ago,

$$R/R_1 \sim 1 - h_1 \cdot 3 \cdot 10^9 yr = 0.76,\ \Omega/\Omega_1 \sim 0.88,\ B_1/B \sim 1.26,\ G \sim 0.84 G_1.$$

Does the existence of life require G to be a constant? No, it does not.

A place nearby a black hole was too harsh for life to occur. A place on earth was nice and life occurred about two billion years ago. But G was the same in the two places. Life is sensitive to departure from normal gravity forces if and only if the departure is very large. Electromagnetic force dominated gravity force in the process when molecules formed DNA, RNA, and cells. Bacteria live in the space (inside the orbiting devices) and in the centrifuges. The very big inertial force in a high-speed centrifuge is effectively a very heavy gravity force. Why should we postulate that their oldest ancestors were so sensitive to a change in G-value that they would have died out or never occurred as long as the G-value was 20% larger or smaller than its present value? In general, let l_i be a physical quantity crucial for life to occur and survive and c_i be a fundamental constant in physics. The eqs of fundamental laws of physics indicate that the derivative $\partial l_i / \partial c_i$ is not big. There is no sign of such critical difference in two c_i-values that one created life and the other forbad life. Even if there is critical difference, it is unknown which one did create life about two billion years ago. It may take three hundred years to develop a quantum and chemical theory about the birth of a living DNA and a most primitive life and then to see what is the range of c_i-values that would allow life to occur in the history of nature. But one thing has become crystal clear that if G were a constant then both solutions of the field eqs with and without LV would feature a similar expansion mode: the speed of recession was infinite at $R=0$, terribly faster-than-c when R was very small, and had been faster than c for very long time. The corresponding theoretical results of constant G are near-c recession speeds and extremely large red shifts, which contradict the overwhelming facts of observation.

To see whether the condition for **positive curvature** in (75b) is met, we need to use (77a, f), (79), (78d, h) and the definition of A_0 given in (64)′ and (76b). We find

(83a) $$\rho I(R) = \frac{c^2}{8\pi^3 G_0 R^5}[\pi^2 + \theta(u)], \qquad \frac{h^2 - c_1}{8\pi G_0}\frac{1}{R^3} = \frac{c^2}{8\pi^3 G_0 R^5}\theta(u),$$

$$\theta(u) \equiv 1 - \frac{4u_0 u + 2u^2 \ln u}{u_0^2 - r^2}, \qquad I(R) = \frac{c^2}{8\pi^3 G_0 \rho R^5}[\pi^2 + \theta(u)].$$

Since $\theta(u) < \pi^2 + \theta(u)$, the condition for positive curvature shown in (75b) is obviously satisfied because of (83a), no matter what are u_0, r, and u. The curvature is positive at every cosmological time u and for any choice of (u_0, r). The concept of critical density is no longer needed.

One may insist on a notion of critical density. We now show that the meaning of 'critical' density is dramatically changed. From (75) and (69a), or, from (75b), one may write

(83b)
$$\rho_c = \frac{h^2 - c_1}{8\pi G_0 I(R) R^3} \,.$$

Using (83a, b) we find

(83c)
$$\rho_c = \rho \frac{\theta(u)}{\pi^2 + \theta(u)} = \rho\left(1 - \frac{\pi^2}{\pi^2 + \theta(u)}\right), \qquad \theta(u) \equiv 1 - \frac{4u_0 u + 2u^2 \ln u}{u_0^2 - r^2} \,.$$

In the sample parabola-model with the choice (78b), $\theta(u)$ and $\pi^2 + \theta(u)$ vanish at u_{c1} and u_{c2} respectively ($0 < u_{c1} < u_{c2} < 1$). Solving $\theta(u) = 0$ and $\pi^2 + \theta(u) = 0$ numerically, we can get (u_{c1}, u_{c2}) as accurately as one wants. We may describe the following sign graph:

For $0 < u < u_{c1}$, both $\theta(u)$ and $\pi^2 + \theta(u)$ are positive and (83c) means $\rho_c < \rho$ as long as $\rho > 0$.

For $u_{c1} < u < u_{c2}$, $\theta(u) < 0$ while $\pi^2 + \theta(u) > 0$. Then $\rho_c < 0$ and $\rho_c < \rho$, as long as $\rho > 0$.

For $u_{c2} < u < 0$, both $\theta(u)$ and $\pi^2 + \theta(u)$ are negative and (83c) means $\rho_c > \rho$, as long as $\rho > 0$.

Please pay close attention on the last strange condition $\rho_c > \rho$. Why should such bizarre condition lead to positive curvature? The reason is as simple as this: Once $\pi^2 + \theta(u)$ is also negative, $I(R)$ in (83a) is **negative** as long as $\rho > 0$. Hence, the inequalities in (75b) must be **reversed** whenever we divide both sides by $I(R)$ to see what will happen to ρ . We said, once both $\theta(u)$ and $\pi^2 + \theta(u)$ are negative, (83c) means $\rho_c > \rho$, as long as $\rho > 0$. This is simply because (83c) leads to the following if $\pi^2 + \theta(u)$ is negative

(83d)
$$\rho_c = \rho\left(1 + \frac{\pi^2}{|\pi^2 + \theta(u)|}\right) \text{ when } \pi^2 + \theta(u) < 0.$$

Now that the meaning of ρ_c is complicated and is no longer ‘critical’ any more, we would like to use the original conditions for the signs of curvature shown in (75) and (75a, b). We have proved that the curvature is really positive at every cosmological time u and for any choice of (u_0 , r). The parabola-model based on positive curvature case is **consistent**.

The concept of critical density has played a really critical role in the EF-type gravitation and cosmology w/wo LV, where spontaneous breaking of G-symmetry is not considered. With the powerful c-principle, that demands non-constant G, supported by non-existence of extremely large red shifts and the old age the observed largest red shifts reveal, the concept of critical density that had put restriction on the density ρ of all the matter co-moving with receding galaxies now puts no restriction on it.

At the beginning, there was no light, $\varepsilon_d = 0$, and therefore, $B = 1$. Then from (82), (77f), and (78b) it follows that

(84) $$G|_{t=0} = \frac{\pi c^2 R_{max}}{2M_{total}}(1+\pi^{-2})u_{min}.$$

Using (79b) with $u_1 = 0.84$, (78e), and (59f), we find that

(84a) $$G|_{t=0} = \frac{9.273}{k_m n_m} u_{min} \cdot 10^{-1} cm^3 / g \cdot s^2 .$$

The initial value $u_{min} \equiv R_{min} / R_{max}$ ($R_{min} = R|_{t=0}$) is unknown. Differential eqs in mathematics and in physics, from Newton's to Einstein's, cannot determine initial conditions. A law of physics, if written in terms of a differential eqn, can only be used to exclude those initial conditions that violate the law. But the laws of physics and their eqs are unable to determine initial conditions. This is why Newton contributed the initial conditions of the motion of planets to God's first push. One thing is clear that if $u_{min} \equiv R_{min} / R_{max}$ is really extremely small, the value of G at the beginning can be much less than its present value. This 'flexibility' in the spontaneous breaking of *G*-symmetry seems to be an amazing 'arrangement' that could make all celestial bodies and particles have obeyed the solid counterexample-less empirical c-principle since the beginning.

If the concept of Schwarzschild radius were valid for the universe and $r_s = 2GM_{total} / c^2$ were the Schwarzschild radius of the universe, we would find from (84a) that

$$r_s|_{t=0} = (\pi + \pi^{-1})R_{min} > R_{min}, \quad R_{min} = u_{min} R_{max}.$$

Then the universe would seem to begin as the largest ever black hole. However, we would like to point out that the celebrated formula $r_s = 2GM / c^2$ is valid for and only for an isolated local mass M. Outside the local mass M there are **huge** spaces with **weak** fields, so that the metric can be written with small quantities of weak fields omitted and then the **approximate** formula for the metric could be deduced and used to yield the famous formula of Schwarzschild radius. This can be clearly seen in the deduction of the formula. Namely, GR explicitly tells us that we do not have a formula for Schwarzschild radius for universe mass M_{total}, nor for the total mass of a group of massive particles in early dense matter condition. Should we think that the universe was a super-black hole or a singularity in the beginning?

An interesting thing is the *G*-values at early times of star era. It relates largely and directly to uncertain values of the ratio B_1 / B via (82b). Compare the present time around $u_1 \sim 0.84$ and an early time around $u \sim 0.2$, about 14 billion years ago according to the *u-t* relation (80b), for instance, which is near the beginning of star era with $k_t \sim 0$ in (82e). Then, $R / R_1 = u / u_1 \sim 0.2381$ and

(82a) shows that $\Omega/\Omega_1 \sim 0.83870$. Large LV (with large $\overline{\lambda_d^4}$ -enhancement for CHSPs) and small/zero LV correspond to $\overline{\lambda_d^4}\varepsilon_b/\varepsilon >> 1$ and $\overline{\lambda_d^4}\varepsilon_b/\varepsilon << 1$ respectively. From (82e, b, f) one can check

(85) $$B_b \equiv 1+(k+3)\overline{\lambda_d^4}\varepsilon_b/3\varepsilon \sim \begin{cases}(k+3)\overline{\lambda_d^4}\varepsilon_b/3\varepsilon & \overline{\lambda_d^4}\varepsilon_b/\varepsilon >> 1\\ 1 & \overline{\lambda_d^4}\varepsilon_b/\varepsilon << 1\end{cases}$$

$$\frac{G}{G_1} = 0.1997\frac{B_1}{B} \sim \begin{cases}0.1997(1+6.76k_{L1}k_{*1}k_{t1}\dfrac{\varepsilon}{\varepsilon_b}10^{-4}) & \overline{\lambda_d^4}\varepsilon_b/\varepsilon >> 1\\ 0.1997 & \overline{\lambda_d^4}\varepsilon_b/\varepsilon << 1\end{cases}$$

In case of large LV, write

(86) $$G/G_1 \sim 0.1997(1+6.76k_\varepsilon), \qquad k_\varepsilon \equiv k_{*1}10^{-4}\varepsilon/\varepsilon_b,$$

which can be large whenever k_ε is large. We know that $\varepsilon/\varepsilon_b$ is indeed large. For instance, if $\varepsilon \sim 10^{-29} g/cm^3$ and $\varepsilon_b \sim \varepsilon_{MW} \sim 10^{-34} g/cm^3$, then $\varepsilon/\varepsilon_b \sim 10^5$. If $\varepsilon \sim 10^{-31} g/cm^3$, $\varepsilon_b \sim 10^{-34} g/cm^3$, then $\varepsilon/\varepsilon_b \sim 10^3$. The stars with their rates of radiation larger or much larger than the sun may rise the average rate of radiation so that $k_{*1} \sim 10$, or higher. Besides, in a period of five to fifteen billion years, the frequencies of many non-relic photons have shifted from higher to lower sections due to the conversion process described before (82) (see pages 229, 230). This non-stop process might make star-radiation to contribute to the observed cosmic background radiation (CBR) a lot. If the CBR contains only a small portion of relic photons, the ratio $\varepsilon/\varepsilon_b$ might be larger than we expect. It is possible that in early years of star era, G might be five to fifty times larger than its present value $G_1 \sim 6.67\cdot 10^{-8} cm^3/g\cdot s^2$. With small/zero LV, in the contrary, $G/G_1 \sim 0.1997$ is small.

Eqs (72), (77h), (77a), (76b), (82), (77f), (78b), and (82a) can be used to yield the following formula for υ and estimate its value.

(87) $$\upsilon = \frac{3}{u^2 R_{\max}^2}[\frac{2}{3}(1+k_\rho)\frac{\Omega}{B} - 1 - \pi^{-2}(1-u^2+\frac{4u_0u^2-4u_0u-2u^2\ln u}{u_0^2-r^2})].$$

In the past five billion years, $0.48 < u \le 0.84$. One can use (82a) and (81) to see the value of Ω and $R_{\max}$. Obviously, both υ and υc^2 have been extremely small since 5 billion years ago. This immediately justifies the solution for the local problem ($\rho = \rho_{sun}$ and $\rho = 0$ inside and outside the sun respectively) that led to the stunning successes of GR in the three major tests. Now in the platform of EFD-type gravitation and cosmology, the solution is Schwarzschild metric plus $\upsilon = 0$. The background values of ρ and υ are neglected to lead to the solution. We omit them in the sense that the perturbation effects of remote stars are omitted. It was this solution that led to the stunning successes of the GR when it was used to explain the motion of the perihelion of Mercury and to predict the gravitational red shifts. It was this solution together with $d\tau = 0$ along the path of the photon that led to the stunning success when it was used to explain the bending of light rays passing by the sun. In new physics with LV, $d\tau = 0$ is too accurate for massive and timelike photons for the mass of photons is too small. The stunning success of GR in the classical three major tests is completely covered by EF-type gravitation with LV and EFD-type gravitation with LV.

The secrets in the universe at largest as well as smallest scales are more than what we know. The behavior of the quantity $c^{-2}R^2 dh/dt$ does not have to be a parabola with positive values in an interval. A general meaning of (78) is seen to be a direct proportion between $|dR/dt|$ and $k(u)$. Thus, the RS-B ratio in (79e) indicates a NSC (necessary and sufficient condition) for ancient (remote) supernovas to be dimmer than expected. That is, the rate of expansion $|dR/dt|$ in remote past was less than its quasi-constant value approximately during the past five billion years. An increasing Hubble constant in the remote past is sufficient, but not necessary for an acceleration to occur. Even if it is a parabola, we may use computer to change the choice (78b) and get the corresponding graphs and the values of fundamental quantities of the universe and compare the consequences. We have more to work, to observe, and to think.

Besides acceleration, denser dust particles and molecules in the ancient universe of smaller size might also make supernovas dimmer than expected. The *G*-theorem proved in this paper shows that the behaviors of genuine functions *R(t)* and *h(t)* can be explained if *G* satisfies (76c, d). This has nothing to do with whether dimmer supernovas really mean acceleration and how much they do mean acceleration. The spontaneous breaking of *G*-symmetry works as we use it to explain dark energy phenomenon. The age failure and c-failure problems are solved. But the exact age of the universe remains unknown. Theoretically, non-parabola models, with different geometric shape of the function $1/k(u)$ at cosmological times from the beginning to 5 billion years ago, can easily and trivially provide with different ages ranging from 20 to 11 billion years. Since radioactive dating cannot tell the ages of stable hydrogen atoms and CBR contains many non-relic microwave photons, the empirical value of the age of the universe will remain unavailable in the 21 century.

References

[1] S. Coleman and S. L. Glashow, Phys. Lett. B **405** (1997) 249. Phys. Rev. D59, 116008, 1999.
[2] T. D. Lee and C. N. Yang, Phys. Rev. **128**, 885 (1962).
[3] S. Weinberg, "The Quantum Theory of Fields", Volume I, Cambridge University Press, 1995, p.212.
[4] G. Sterman, "An Introduction to Quantum Field Theory", Cambridge University Press, 1993.
[5] H. A. Bethe, Phys. Rev. **72**, 339 (1947).
[6] S. Weinberg, "Gravitation and Cosmology: Principles and Applications of the General Theory of Relativity", Wiley, New York, 1972.
[7] A. Einstein, "The Meaning of Relativity", fifth ed. Princeton University Press, 1955, p.164, p.129.
[8] P. A. M. Dirac, in "The Past Decade in Particle Theory", E. C. Sudarshan and Y. Ne'eman (eds.), Gordon and Breach, London, 1973.
[9] R. P. Feynman, "Quantum Electrodynamics", Addison-Wesley Publishing Company, 1961, p.150.
[10] T. D. Lee, "Particle Physics and Introduction to Field theory", Harwood Academic Publishers, 1988, 825.
[11] J.D. Bjorken and S.D. Drell, "Relativistic Quantum Mechanics", McGraw-Hill, New York, 1964, p.4.
[12] J.D. Bjorken and S.D. Drell, "Relativistic Quantum Fields", McGraw-Hill, New York, 1965, p.154.
[13] M. Gell-Mann, "The Quark and the Juagar: Adventures in the Simple and the Complex", New York, NY: W. H. Freeman and Co., 1994.
[14] S. Weinberg, "The Quantum Theory of Fields', Volume II, Cambridge University Press, 1996, p.153.
[15] S. Weinberg, "Phys. Lett.", **102 B,** 401 1981, 401.
[16] W. Rindler, "Essential Relativity", Springer-Verlag, 1977.
[17] K. Mikami and A. Weinstein, "Publications of the Research Institute for Mathematical Science", Kyoto University, Vol. 24, No. 1, 1988.
[18] B. O'Neill, "Semi-Riemannian Geometry with Applications to Relativity", Academic Press, New York, 1983.
[19] J. D. Jackson, "Classical Electrodynamics", 2nd Edition, John Wiley & Sons, 1975, 508.
[20] E. G. Harris, "Introduction to Modern Theoretical Physics", Volume 2, John Wiley & Sons, New York, 1975, p.639.
[21] Q. Sun, "Potential Model and Mass Spectrum of Leptons-Quarks", Journal of Chongqing University, 1983, No.2, 114-126. (Chinese with English summary.)
[22] J. Schwinger, "A Report on Quantum Electrodynamics," in The Physicist's Conception of Nature (Reidel, Dordrecht, 1973): p.415.

Appendix

Deeper Structure

Abstract

In spite of unsolvable or unsolved **MB** (**many body**) and **MFs** (**magneton forces**) binding systems and with less number of elementary particles, **DS** (**deeper structure**) and MFs join forces to explain all **SA** (**subatomic**) facts qualitatively and half-quantitatively, including tough facts such as the ironic **FQA** (**free quark absence**) fact that no single free or unbound quark has ever been found among all the **MSPs** (**middle stage particles**) and **FSPs** (**final state particles**) in all recorded cosmic shower events and **HE** (high energy) collision events, nucleons' anomalous **MMs** (**magnetic moments**), bizarre uniqueness of stable free protons (bizarre unique stable d-quarks in unbound protons), bizarre unique long lives of neutrons decaying about 10^8 times slower than any slowest decaying hadrons other than nucleons, chemical elements on stability island containing neutrons (even more than protons), all hadrons other than nucleons with much shorter lifetimes than neutrons, beta-particles' bizarre continuous energy spectrum and preferential spin direction bizarrely parallel to the direction of the **EMF** (**external magnetic field**), bizarre different lifetimes of K_L^0 and K_S^0 that have the same DS content and the same quark content and may decay into the same set of FSPs, ... DS even exactly proves why β^+-decay is forbidden if the available energy is less than $2m_ec^2$ while there is no such restriction for e^--capture and why u-quarks in free nucleons are all stable while a u-quark in Δ^{++}, for example, could change to a d-quark so that a proton could appear among the FSPs in a decay mode of Δ^{++}.

Formulas for MF between 2 spinning charges are deduced from CED (classical electrodynamics). MF almost vanishes at atomic distances, but plays a role and even dominates corresponding Coulombian force at SA short distances. Non-existence of S-states for DS electrons bound in non-spherosymmetric MF fields in SA small spaces, their **OSOD** (**opposite spin-orbital directions**), and then their small net MMs remove the oldest obstacle for DS electron content to agree with nucleons' MMs that are much smaller than electron's spin MM.

The proof of a misconception of composite particles' '*spins*' pointing out that composite particles do not spin for constituent particles of a composite particle do not spin about the same axis nor with the same angular velocity & the proof of the empirical disability to measure the **OAM** (**orbital angular momentum**) of a FSP or an **ISP** (**initial state particle**) with its linear momentum line deviating from the mass center of a decayed particle or of a colliding 2-particle system by an amount about or less than $10^{-11}cm$ give simple explanations of the '***spin crises***', some other facts, and **FBS** (**fermion/boson statistics**): Any composite particle that contains odd/even number of elementary fermions is a fermion/boson. The FBS of *elementary* particles depends on their spins while the FBS of *composite* particles is totally independent of and totally irrelevant to any OAM and the **TAM** (**total angular momentum**) of all its constituent elementary fermions.

Large number of examples have shown unusual ability of DS with MFs and **4TDSP** (**4 types of DS-processes**) to qualitatively and half-quantitatively explain all empirical facts in particle physics: mass spectrum, MMs, hundreds of types of MSPs, thousands of decay modes, lifetimes, branch ratios, and all kinds of SA reactions.

I. Introduction

Legendary solutions of quantum eqs in Coulombian force field and in any inverse potential field tell the atomic binding size and Bohr radius, proving that other type of forces must exist to bind light electrons and neutrinos in SA small spaces. MFs between spinning charges of the same kind do the jobs. Formulas of MFs are deduced in II.

Decisive evidence that neutrino has a weak charge and weak MM is its preferential spin direction in the EMF seen in beta-decay. Its empirical zero value of electric charge is too accurate to allow such behavior to be caused by an **e.m. (electromagnetic)** force. A charged particle has its electric charge accompanied by a weak charge and every magnetic field is then accompanied by a weak magnetic-like field. Such charge-extension is capable to explain many empirical facts about neutrinos. SDL (see details later). Neutrinos are too light to cause considerable effects on the motion of heavy quarks and DS electrons. DS-neutrinos can be ignored when we discuss mass spectrum and other issues about the motion of heavy quarks and DS electrons. We have upper bounds but no empirical values of the masses of neutrinos. DS of muons and taus explain all their decay modes and their masses. DS of neutrinos is not discussed. The reader may feel free to think about DS of muon/tau neutrinos.

Among all quarks/antiquarks, only $u/\bar{u}$ are elementary. A d-quark contains one u-quark and two leptons $e^-\bar{\nu}_e$, called d-leptons or **DSLs** (**DS leptons**), ***in ground states***. A neutron is unstable for its 2 d-electrons, forbidden to be in the same state by Pauli's exclusion principle and crowding in SA small space, undergo **ENM** (**eagle nest mechanism**): the one bound closer to u-quark core slowly pushes out the one bound less tight (SDL). A free proton is stable for its d-electron does not have another electron nearby to push it out. A proton bound in some nuclei can be unstable if it absorbs d-leptons released in a nuclear transmutation, such as seen in shell-electron capture events, and changes to a neutron with other leptons released in the same nuclear transmutation and emitted. Stable elements on the stability island have enough protons nearby to absorb d-electrons released by decaying bound neutrons. Some stable elements can have more neutrons than protons for neutrons do not release beta-particles at the same time so that an equal number of protons nearby is not necessary to relocate the emitted β-particles.

An s-quark has the same DS content as a d-quark, with its d-leptons in the 1st excitation. A b-quark has the same DS content as a d-quark but its d-leptons are in higher excitation and the u-quark in a b-quark may be in excitation too. A c-quark contains 1 u-quark and 2 pairs of DS leptons: $e^-\bar{\nu}_e$ and $e^+\nu_e$ in excitations. A t-quark has the same DS content as a c-quark but in higher excitation that may include excitation of its u-quark. Please note, a u-quark with positive charge can stick an electron and a positron at the same time because both MFs between u-quark and electron and between u-quark and positron can be attractive as long as electron and positron are in attraction zones in the non-spherosymmetric MF field of a u-quark and MF attraction dominates Coulombian repulsion between a u-quark and a DS-positron in a SA small space. Some details can be seen later.

Such SA excitation reality revealed by DS immediately explains short and very short lifetimes of all the baryons other than nucleons: Nucleons are the only members in baryon-family that do not contain any excited SA constituent particle and SA de-excitations always take place quickly or very quickly. Moreover, DS tells that every meson contains one **HMC (hadronic meson core)** $u\bar{u}$, so does every anti-meson. Thus, all mesons live short without a single exception for any $u\bar{u}$-annihilation takes place quickly, even in pions with all their SA constituent particles in ground states.

For each resultant force of an MF & a Coulombian force, the **TME** (**total mechanical energy**) theorem proves positive, zero, and negative TME at the distances shorter than, equal to, and longer than **0TMED** (**0-TME distance**) respectively.

An example of large positive TME is a **LMC (leptonic meson core)** e^+e^-. A muon is formed by a LMC and other DS leptons: $\mu^+ = e^+e^-e^+\nu_e\bar{\nu}_\mu$. Such structure and heavy LMC with large positive TME explains heaviness of muons and agrees with a **QT (quantum tunnel)** decay mode $\mu^+(e^+e^-e^+\nu_e\bar{\nu}_\mu) \rightarrow e^+e^-e^+\nu_e\bar{\nu}_\mu$. All the decay modes of muons are explained. The light positronium e^+e^- is due to its atomic size with negligible MF in the **MPR** (**most probable region**). The lifetime inequalities among different states of positronium e^+e^- and muon are miraculously explained using DS and MF's properties.

DS immediately shows that every baryon contains one **BC** (**baryon core**) uuu while every anti-baryon contains one anti-BC $\bar{u}\bar{u}\bar{u}$. Each BC, anti-BC, and HMC can be called a **q-core**. Excitations of q-cores may be involved in t, b, and c quarks. Excitations of many-body systems with more than one kinds of forces involved are complicated with unknown details. BC and anti-BC, in ground or an excited state, are bound systems. This is one of the reasons causing **BNC** (**baryon number conservation**) under the condition that hundreds of kinds of MSPs seen in **DI** (**deep inelastic**) collisions and thousands of decay modes are explained with reduced elementary particles in DS and only 4 types of DS processes (**4TDSP**): DS excitations/de-excitations, **PAP** (**particle-antiparticle pair**) productions/annihilations, **PQTR** (**particle quantum tunnel relocating**) (in most cases with SA particle(s) nearby in the same SA small space to capture relocating SA elementary particle(s)), SA photon radiations (in most cases followed by SA photoelectric encounters with DS electrons nearby to be hit). Every original particle-production before the birth of MSPs is a **PAP** production leading to non-conservation of meson number and BNC. HMC $u\bar{u}$ can be created in collision or SA de-excitation and can be annihilated when a meson decays. Simple DS explanation reveals why meson number does not conserve. DS-explanations of all kinds of experimental facts are discussed in III.

Note: PAPs refer to $u\bar{u}\,/\,e^+e^-\,/\,\nu_e\bar{\nu}_e\,/\ldots/\,p\bar{p}\,/\,n\bar{n}\,/\,\mu^+\mu^-\,/\ldots$, not $u\bar{d}\,/\,c\bar{s}\,/\ldots$. Any original PAP production in decay and collision events is the production of an elementary PAP such as $u\bar{u}\,/\,e^+e^-\,/\,\nu_e\bar{\nu}_e\,/\ldots$, not $p\bar{p}\,/\,n\bar{n}\,/\,\mu^+\mu^-\,/\ldots$. Most elementary PAP productions are followed immediately by PQTR leading to the births of hundreds of types of MSPs. SDL

Remark: As initial particles form *matter*, the FSPs form *matter* too, though all PAP productions produce equal number of elementary particles and antiparticles. This is because PQTR takes place in the environment of matter that absorbs relocating particles and/or antiparticles. Matter'll remain matter and antimatter will remain antimatter after absorbing QT-relocating particles and antiparticles with equal number. Quick PAP annihilations such as annihilations of HMC and LMC remind us of a reasonable **IC (initial condition)** in the beginning of the Big Bang: Matter and its constituent particles such as u-quarks and electrons were much more than antimatter and its constituent particles such as anti-u-quarks and positrons, just like what we can see every day on earth and in the universe. If the IC were an equal number of particles and antiparticles, they would all or nearly all annihilate in the super dense condition of the beginning and very dense condition of early universe. Today, the observed abundance of matter and much less antimatter disprove such equal number initial condition.

II. Dynamics with MFs

CED gives two well-known formulas about MF: the formula of magnetic field $\vec{B}_1$ produced by *spin* MM $\vec{\mu}_1$, called source MM, and the formula of magnetic force $\vec{f}_{12}$ exerted on spin MM $\vec{\mu}_2$ by magnetic field $\vec{B}_1$:

(1) $$\vec{B}_1 = \frac{3\vec{\rho}(\vec{\rho}\cdot\vec{\mu}_1)-\vec{\mu}_1}{r^3}, \qquad \vec{f}_{12} = \vec{\nabla}(\vec{B}_1\cdot\vec{\mu}_2) \quad (\vec{\rho}\equiv\frac{\vec{r}}{|\vec{r}\,|}\equiv\frac{\vec{r}}{r})$$

__Remark__: Another component of *magnetic force*, called **magnetic orbital force**, can be included if $\vec{\mu}_1$ & $\vec{\mu}_2$ include orbital MMs produced by SA local orbital motions of charges of the same kind. It's negligible if and only if the speeds of local orbital motions aren't near c, i.e., if $\beta \equiv u/c \ll 1$.

Choose spherical coordinates (r,θ,ϕ) with $\vec{\mu}_1 = |\vec{\mu}_1|\vec{k}$ in positive zenith-direction and at the origin. For parallel & anti-parallel MMs ($\vec{\mu}_1 \times \vec{\mu}_2 = \vec{0}$),

(2) $$\vec{f}_{12} = \vec{\nabla}(\vec{B}_1\cdot\vec{\mu}_2) = \vec{\mu}_1\cdot\vec{\mu}_2(\frac{3-9\cos^2\theta}{r^4}\vec{\rho} - \frac{6\sin\theta\cos\theta}{r^4}\vec{\theta})$$

It's well-known that

(3) $$\hbar/2 = 1.93\cdot10^{-11}cm\cdot m_e c = 0.986\cdot10^{-11}cm\cdot Mev/c$$

Quantum connection $\mu \equiv |\vec{\mu}| = |q|\hbar/2mc$ (radiative correction is omitted) and (3) lead to

(4) $$\vec{\mu}_1\cdot\vec{\mu}_2 == sign(\vec{\mu}_1\cdot\vec{\mu}_2)\cdot 3.727463\cdot10^{-22}cm^2\cdot|q_1 q_2|\frac{m_e m_e}{m_1 m_2}$$

From (2) we see that *radial* component of MF vanishes at $\theta = \theta_{zero1}$ and $\theta = \theta_{zero2}$. Here, $\theta_{zero1} \equiv \cos^{-1} 1/\sqrt{3}$ and $\theta_{zero2} \equiv \pi - \cos^{-1} 1/\sqrt{3}$, about 54.7^0 and 125.3^0 respectively. We call them **0R-MF angles**. Region **R0** contains all points where θ is equal or close to 54.7^0 or 125.3^0. The *radial* component of MF btw (between) 2 magnetons of the same kind is turned off or greatly suppressed if they are in R0 at each other with parallel or antiparallel MMs.

Non-radial component of MF vanishes at $\theta = 0, \pi/2, \pi$. Within any sphere centered at source MM, we define **A-region** and **R-region** where MF is attractive and repulsive respectively if **RD** (**radial domination**) condition $|\vec{f} \cdot \vec{\rho}| \geq 2.5 |\vec{f} \cdot \vec{\theta}|$ is satisfied. RD takes place in two 3-dim regions: region **R1** (at source magneton) with *elevation* angle $\theta' \equiv \pi/2 - \theta$ btw -10.5^0 and 10.5^0 and region **R2** with $0 \leq \theta \leq 20.4^0$ and $159.6^0 \leq \theta \leq 180^0$. By 'at source magneton' we mean: it's the center of centrosymmetry of R1 or R2. R1 looks like a disc with thickness increasing with radius. R2 is formed by two circular cones whose vertices coincide at the origin with zenith axis as their axis of symmetry. For **parallel** MMs, **R1**/R2 are **repulsive**/attractive regions respectively. For **anti-parallel** MMs, **R1** and R2 are **attractive** and repulsive regions respectively. We will see that R0, R1, and R2 are three important θ-**ZL**s (**zone-locations**).

Each plane containing the zenith axis intersects R1 and R2 with section looking like a **fan** that has 4 **vanes**. When rotating about the zenith axis, they sweep over space to form R1 and R2. The 4-vane figure is useful for us to see how d-electrons in ground and excited states move in SA spaces and how DS-symmetry and DS-asymmetry may determine mesons' lifetimes qualitatively and half-quantitatively. We may call R1 and R2 **H-vane** (horizontal vane) and **V-vane** (vertical vane) at source magneton respectively.

Before Dirac eqn with MF-binding is solved, we don't have quantitative clues about ground state region. But it must be a net-attraction region, where resultant force is attractive whether or not repulsive force exists. We assume that R1 is the ground state region that is the MPR of a MF-bound d-electron. We found that such assumption agrees with experimental facts. See details in III.

Non-spherosymmetric MF field disproves the existence of S-state of d-electron once for all though MB systems are unsolvable. As a d-electron moves in R1 at source MM, its orbital and spin angular momenta must be in opposite directions in order for both spin and orbital electric currents to be attracted towards the source MM. Such OSOD can be easily seen as we apply hand-rules without computations. OSOD gives small net MM of a d-electron and small MM of a nucleon that contains d-electron.

Take *approximations* $\cos\theta = 0$ and 1 in in R1 and R2 respectively. From (2) we get *approximate* MF-formulas in R1 and R2:

(5)
$$\vec{f}_1 = \frac{3\vec{\mu}_1 \cdot \vec{\mu}_2}{r^4} \vec{n}_1 \qquad (\vec{n}_1 \equiv \vec{\rho}\,|_{\theta=\pi/2}),$$

$$\vec{f}_2 = -\frac{6\vec{\mu}_1 \cdot \vec{\mu}_2}{r^4} \vec{n}_2 \,(\vec{n}_2 \equiv \vec{\rho}\,|_{\theta=0}), \quad \vec{\rho} \equiv \frac{\vec{r}}{|\vec{r}|} \equiv \frac{\vec{r}}{r}$$

For parallel or anti-parallel MMs $\vec{\mu}_1$ and $\vec{\mu}_2$ of the same kind, formula of **MF** with position vector from $\vec{\mu}_1$ to $\vec{\mu}_2$ ($\vec{r} \equiv |\vec{r}|\vec{n} \equiv r\vec{n}$) perpendicular ($\theta = \pi/2$) to or collinear ($\theta = 0, \pi$) with MM-Line is obtained from (2)

(5a)
$$\vec{f}_1 = \frac{3\vec{\mu}_1 \cdot \vec{\mu}_2}{r^4}\vec{n} = sign(\vec{\mu}_1 \cdot \vec{\mu}_2) \cdot 1.24 \frac{|q_1 q_2|}{r^2}\left(\frac{3 \cdot 10^{-11} cm}{r}\right)^2 \frac{m_e}{m_1}\frac{m_e}{m_2}\vec{n}$$

$$\vec{f}_2 = -\frac{6\vec{\mu}_1 \cdot \vec{\mu}_2}{r^4}\vec{n} = -sign(\vec{\mu}_1 \cdot \vec{\mu}_2) \cdot 2.48 \frac{|q_1 q_2|}{r^2}\left(\frac{3 \cdot 10^{-11} cm}{r}\right)^2 \frac{m_e}{m_1}\frac{m_e}{m_2}\vec{n}$$

Note: Extension of **electric charge** to **St** (strong) and **Wk** (weak) **charges** leads to St and Wk MMs. Force btw two MMs of the same kind is e.m., St, or Wk **MF**.

Condition for domination of e.m., St, and Wk MFs over Coulomb-counterparts at certain distances against the **magneton distance (MD)** r_m, can be seen in **MT-C (magneton-Coulomb) ratio**

(6)
$$k_{MT-C} \equiv \frac{|\vec{f}_i|}{|q_1 q_2 / r^2|} = \frac{r_{mi}^2}{r^2},$$

$$r_{mi} \equiv 3.341\sqrt{1+\delta_{i2}} \cdot 10^{-11}\sqrt{\frac{m_e}{m_1}\frac{m_e}{m_2}}cm, \quad (i = 1,2.)$$

$$r_m \sim 3 \cdot 10^{-11}\sqrt{\frac{m_e}{m_1}\frac{m_e}{m_2}}cm$$

For a proton and an atomic electron, the MD $r_m \sim 7 \cdot 10^{-13} cm$. At atomic distances $r \sim 10^{-8} cm$, the MT-C ratio is about $4.9 \cdot 10^{-9}$ so that by ignoring MF one could get good result in atomic physics and QM (quantum mechanics) of hydrogen atom. For a u-quark, a BC, or a HMC $u\bar{u}$ to bind a DS electron, the MD in (6) is roughly written $r_m \sim 10^{-13} cm$, where the MFs lead to:

TME-Theorem:

Part I. Circular motion of classical **1B (1 body)** in every *Coulomb*-type central force field with **IPE (inverse potential energy)** always leads to ***negative*** TME, true for all circular binding sizes $0 < r < \infty$. Equal-mass **2B (2 body)** circular motion with ***Coulomb***-type binding force alone also have ***negative*** TME, true for all circular binding sizes $0 < r < \infty$.

Pf : Omitted. 1B-case was proved in Newton's time. Equal-mass 2B-case is a special case of part III.

Part II. 1-body circular motion in any central force field **alone**, where force $\vec{f} = -a_{MF}\vec{r}/r^5$ is of ***MF***-type (in the sense that it's inversely proportional to r^4) and is attractive ($a_{MF} > 0$), has ***positive*** TME for all circular binding sizes. Its equal-mass 2B version also has ***positive*** TME, true for all circular binding sizes.

Pf: Omitted. Part II is a special case of part III.

An **MT-C resultant force** is the vector sum of a MF-type force and a Coulomb-type force.

Part III. 1-body circular motion in a central field of a MT-C resultant force $\vec{f} = -(k_C / r^2 + a_{MF} / r^4)\vec{r} / r$ has ***positive*** TME for all circular binding sizes smaller than 0TMED r_{0TME} where TME equals 0 and ***negative*** TME for all circular binding sizes larger than 0TMED r_{0TME}. Equal-mass **2B circular motion** with above resultant interaction force has ***positive*** TME rigorously proved to be **true** at **all** circular binding sizes smaller than r_{0TME} while it has ***negative*** TME, true at **all** circular binding sizes larger than r_{0TME}. At unique **critical distance** r_c $(r_c > r_{0TME})$, TME has a negative minimum value. The bizarre positive TME for such special bound state has been rigorously proved to be true for any positive values of k_C and a_{MF} and at ANY radius r (as circular binding size) smaller than 0TMED.

Note: MT-forces in (2) depend on both r and θ. So far TME theorem is proved for special circular motion on zenith plane $\theta = \pi / 2$. Non-radial component of MF outside zenith plane has no influence on classical motion if such classical motion takes place on zenith plane. While it must have influence on quantum motion. Such influence on quantum motion is unknown since quantum eqs in MF field are unsolved and may be unsolvable.

Proof: Since **LV** (**Lorentz Violation**) is omitted for $1 < \gamma < 10^3$, γ is constant in every uniform circular motion and the eqn of motion in **1B uniform circular motion** is

(7) $$f = k_C / r^2 + a_{MF} / r^4 = m\gamma u^2 / r, \qquad \beta^2 \gamma = fr / mc^2,$$

$$\beta^2 = k\gamma^{-1}, \qquad \gamma = \frac{k + \sqrt{k^2 + 4}}{2}, \qquad k \equiv fr / mc^2 = \frac{1}{mc^2}\left(\frac{k_C}{r} + \frac{a_{MF}}{r^3}\right).$$

$$U = \int f dr = -k_C / r - a_{MF} / 3r^3 = -kmc^2 + 2a_{MF} / 3r, \quad E_k = m\gamma c^2 - mc^2$$

$$T_{ME} = E_k + U = mc^2 \frac{\sqrt{k^2 + 4} - k - 2}{2} + \frac{2a_{MF}}{3r^3}$$

Immediately, $a_{MF} = 0$ leads to TME T_{ME} being negative for any $k > 0$ and any binding size $r < \infty$. This is 1B-case in Part I. While $k_C = 0$ with $a_{MF} / r^3 = kmc^2$ and positive TME T_{ME} true for any $k > 0$ and any binding size $r < \infty$. It's 1B-case in Part II.

In general, both k_C and a_{MF} are positive. One can check

$$(8)\quad T_{ME} = 0 \Leftrightarrow 3mc^2kr^3 = a_{MF}(\sqrt{k^2+4}+k+2) \Leftrightarrow \sqrt{k^2+4}+k+2 = 3+\frac{3k_C}{a_{MF}}r^2$$

$$\Leftrightarrow \sqrt{k^2+4}+k = 1+\frac{3k_C}{a_{MF}}r^2$$

Above eqn is not solvable. But last eqn proves: There is 1 and only 1 solution $r = r_{0TME}$. To reach such conclusion under the condition of insolvability one needs to look at LHS and RHS of last eqn in (8): LHS is monotone decreasing from super large positive value to almost 2 as r increases from near-0 to long distances while RHS increases from 1 to super large positive values as r increases from near-0 to long distances. Two graphs must intersect at 1 and only 1 point. Its r-coordinate is just the 0TMED r_{0TME}. Moreover, TME is positive at any r shorter than r_{0TME} and negative at any r longer than r_{0TME}.

Existence of critical distance r_c at which TME takes the minimum value is seen below

$$(9)\qquad dT_{ME}/dr = 0 \leftrightarrow 4a_{MF} = [1-k/(k^2+4)^{1/2}](k_C r^2 + 3a_{MF}) \leftrightarrow$$

$$1-(1+4/k^2)^{-1/2} = 4(k_C r^2/a_{MF}+3)^{-1}$$

Again, last eqn in (9) is not solvable but proves: There is 1 and only 1 solution $r = r_c$. Obviously, $r_{0TME} < r_c$. More rigorous proof of 1B-case is omitted. Rigorous proof for equal-mass 2B-case is shown below. One can easily repeat rigorous proof for 1B-case.

Equal-mass 2B-systems are important for HMC & LMC are such systems. Consider special classical solutions: Uniform circular motion of radius r. Distance btw two equal-mass particles bound by attractive force is $2r$. Write

$$(10)\quad f = k_C/4r^2 + a_{MF}/16r^4 = m\gamma u^2/r,\ \beta^2 = k\gamma^{-1},\ \gamma = \frac{k+\sqrt{k^2+4}}{2},$$

$$k \equiv fr/mc^2 = \frac{1}{mc^2}\left(\frac{k_C}{4r}+\frac{a_{MF}}{16r^3}\right),\quad E_k = 2m\gamma c^2 - 2mc^2,$$

$$U = \int 2f dr = -k_C/2r - a_{MF}/24r^3 = -2kmc^2 + a_{MF}/12r^3$$

$$T_{ME} = E_k + U = mc^2(\sqrt{k^2+4}-k-2)+\frac{a_{MF}}{12r^3}$$

Again, $T_{ME} = 0$ is an unsolvable eqn. Value of r_{0TME} is algebraically unknown.

Existence and uniqueness of solution, however, can be rigorously proved:

(11) $$T_{ME}/mc^2=\sqrt{k^2+4}-k-2+\frac{a_{MF}}{12mc^2r^3}=\frac{a_{MF}}{12mc^2r^3}-\frac{4k}{k+2+\sqrt{k^2+4}}$$

$$=\frac{a_{MF}(k+2+\sqrt{k^2+4})-4k\cdot 12mc^2r^3}{12mc^2r^3\cdot(k+2+\sqrt{k^2+4})}=\frac{A-B}{D}$$

$$A=k+2+(k^2+4)^{1/2},\ B=\frac{12k_C}{a_{MF}}r^2+3,\ D=\frac{12mc^2r^3\cdot(k+2+\sqrt{k^2+4})}{a_{MF}}.$$

Note: $\frac{4mc^2kr^3}{a_{MF}}=\frac{k_C}{a_{MF}}r^2+\frac{1}{4}$, true for the k -function defined in (10).

As r increases, A is monotone decreasing from super large positive values near $r=0$ to nearly 4 at very long distances while B is monotone increasing from 3 at $r=0$ to super large values at very long distances. Hence curves of functions A and B must have 1 and only 1 intersection point, simultaneously proving that TME vanishes at r_{0TME}, positive for all $r<r_{0TME}$, and negative for all $r>r_{0TME}$.

Existence/uniqueness of the **critical distance** r_c can be proved rigorously:

(12) $$4r^4dT_{ME}/dr=[1-k/(k^2+4)^{1/2}]\,(k_C r^2+3a_{MF}/4)-a_{MF}=$$

$$(1-k/\sqrt{k^2+4})(a_{MF}/4)[4k_Cr^2/a_{MF}-(k^2+1+k\sqrt{k^2+4})]=D(A-B),$$

$$A\equiv 4k_Cr^2/a_{MF},\quad B\equiv k^2+1+k\sqrt{k^2+4},\quad D\equiv(1-k/\sqrt{k^2+4})(a_{MF}/4).$$

As r increases, B is monotone decreasing from super large value near $r=0$ to near 1 at very long distances while A is monotone increasing from 0 at $r=0$ to super large values at very long distances. Two graphs of functions A and B have exactly 1 intersection point. Its r-coordinate is denoted by r_c. It simultaneously proves: TME is monotone decreasing in interval $(0,r_c)$, monotone increasing in interval (r_c,∞), and takes the minimum value, obviously negative, at r_c.

Eqs $T_{ME}=0$ and $dT_{ME}/dt=0$ are unsolvable. Values of r_{0TME} and r_c are algebraically unknown. But once k_C and a_{MF} are given one can easily get numerical values of r_{0TME} and r_c as accurately as one wishes.

For HMC $u\bar{u}$ in π^0, MF is suppressed in a special quantum state called **OSTP (orbit-spin-tilting-precession)** the 0TMED is shifted to a much shorter one. Then $u\bar{u}$ in π^0 may be bound in the lowest possible region with very negative TME that naturally explains low mass of π^0 under heavy u-mass. (See

OSTP and π^0-mass eqn later) Experimental evidences of such TME-r curve are listed below:

1. The ***V-shape*** of the curve nearby r_c may create a bunch of bound states nearby r_c, called **NCB-states** (**n**ear-**c**ritical **b**ound **states**) with small level-differences. Mass spectrum of hadrons shows states with small mass differences.

2. Examples of negative TME are BC and HMC. The **ME** (**mechanical energy**) of each u-quark and the d-electron in proton is negative at all times. Negative TME and all-time negative MEs shut QT decay channel, necessary in explaining the FQA and stable unbound protons. (Surely, FQA fact under HE collisions needs something else to explain. SDL)

3. An example of positive TME is LMC e^+e^- in muon or tau, born to within the distances shorter than r_{0TME}.

4. The ME of the quasi-bound d-electron in any neutron is continuous, time varying (see MB theorem and MB postulate and experimental evidences later), and positive sometimes. When its ME becomes positive, it may trigger the QT escaping and the beta decay may occur.

Since quantum eqs in a non-spherosymmetric MF field are not solved and may be unsolvable, the only thing we can do right now is to use **CI** (**classical imitation**) to penetrate the dense 'fog' that covers the secrets of SA world, like we did in TME theorem. **PCQM** (**perfect classical-quantum match**) encourages CI. One may skip the following content about **DBR** (**Dirac-Bohr radius**) and PCQM if they have been read in the Highlights.

Solution of Dirac eqn gives ground state energy and wave function for atomic electron in hydrogen. Average values of $1/r$ and **PE** (**potential energy**) $U=-e^2/r$ as well as DBR can all be deduced from the ground state wave function. The ground state energy and the ground state average PE $\overline{U}$ determine the average **KE** (**kinetic energy**) and the average γ-value $\overline{\gamma}$. But we may also use circular motion CI and see ground state electron bound in hydrogen as a classical particle moving along a circle in classical Coulomb field. By solving classical eqn of motion in relativistic mechanics, we obtain infinitely many classical circular motion solutions associated with corresponding **ICs** (**initial conditions**). Classical theory itself is unable to get quantized orbital sizes and identify which one among infinitely many circular motions best represents the quantum motion of the atomic electron in hydrogen and in ground state. But if we choose classical circular motion with the same size as DBR, corresponding γ value is exactly the same as $\overline{\gamma}$ we got from relativistic QM.

< General DBR >

Consider Dirac eqn for any particle with mass $m=k_m m_e$ ($0<k_m<\infty$) and in any Coulomb-type force field with IPE

$$U=-k_C e^2/r=-k_C\alpha\hbar c/r=-\alpha_C\hbar c/r$$

$$(0 < k_C < 137\ , 0 < \alpha_C < 1,\ \alpha = 1/137.035999679)$$

Solution of such Dirac eqn gives **ground** state wave function and energy, written as

(13) $$\psi_0 = \begin{pmatrix} b_0 a^s r^s e^{-ar} \chi_0 / r \\ d_0 a^s r^s e^{-ar} \chi_0 / r \end{pmatrix}, \quad E_0 = mc^2 \sqrt{1-\alpha_C^2}\ ,$$

$$a = \frac{\sqrt{m^2c^4 - E_0^2}}{\hbar c} = \frac{mc^2\alpha_C}{\hbar c} = \frac{mc^2 k_C e^2}{\hbar^2 c^2} = \frac{mk_C e^2}{\hbar^2} = \frac{1}{r_b^{Sch}}\ ,$$

$$r_b^{Sch} = \frac{\hbar^2}{mk_C e^2} = \frac{r_b}{k_m k_C} \quad (r_b \equiv \frac{\hbar^2}{m_e e^2} = 0.529141 \cdot 10^{-8} cm\ ).$$

Note: The celebrated Bohr radius $r_b \equiv \hbar^2 / m_e e^2$ is the special **SBR (Schrödinger-Bohr radius)** r_b^{Sch} for atomic electron with special settings: $m = m_e$, $k_C = 1$, and $\alpha_C = \alpha \approx 1/137$.

Ground state in a spherosymmetric potential field is S-state, indicating constant χ_0 . Then normalization gives

(13a) $$1 = (b_0^2 + d_0^2)\chi_0^2 a^{2s} \int_0^\infty 4\pi e^{-2ar} r^{2s} dr = \frac{4\pi s \Gamma(2s)}{2^{2s} a}(b_0^2 + d_0^2)\chi_0^2\ , \quad s \equiv \sqrt{1-\alpha_C^2}$$

$$\frac{1}{r_b^{Dirac}} = \overline{\left(\frac{1}{r}\right)}_{Dirac} = (b_0^2 + d_0^2)\chi_0^2 a^{2s} \int_0^\infty 4\pi e^{-2ar} r^{2s-1} dr =$$

$$= \frac{4\pi\Gamma(2s)(b_0^2 + d_0^2)\chi_0^2}{2^{2s}} = \frac{4\pi s\Gamma(2s)(b_0^2 + d_0^2)\chi_0^2}{2^{2s} a} \cdot \frac{a}{s} = \frac{a}{s}$$

With $a = 1/r_b^{Sch}$ and $r_b^{Sch} = r_b / k_m k_C$, obtained in (13), we get

(14) $$r_b^{Dirac} = \frac{s}{a} = s r_b^{Sch} = \sqrt{1-\alpha_C^2}\, r_b^{Sch} = \frac{\sqrt{1-k_C^2\alpha^2}}{k_m k_C} r_b$$

One can easily check that this DBR is also the most probable radius.

For ***atomic*** electron in hydrogen, $m = m_e$, $k_m = 1$, $k_C = 1$, $\alpha_C = \alpha$. Then

(14a) $$r_b^{Sch} = \frac{\hbar^2}{mk_C e^2} = \frac{\hbar^2}{m_e e^2} = r_b\ , \qquad r_b^{Dirac} = \sqrt{1-\alpha^2}\, r_b\ .$$

For ***atomic*** electron in hydrogen, DBR r_b^{Dirac} is almost the same as Bohr radius r_b because α^2 is too small. But from general formula $r_b^{Dirac} = \sqrt{1-\alpha_C^2} r_b^{Sch}$ in (14) we see that for HMC $u\bar{u}$ in π^0, St force with $\alpha_C^2 = \alpha_s^2 >> \alpha^2$ makes DBR r_b^{Dirac} in (14) significantly less or much less than $r_b^{Sch} = r_b / k_m k_C$. Please note, this SBR for bound u-quark itself is much less than Bohr radius r_b for the bound atomic electron in hydrogen due to St force with $k_C >> 1$ and heavy u-quark with $k_m >> 1$. Now we show the details of an example of PCQM.

< Example of **Perfect Match** btw **QM** and **CI** >

Solution of Dirac eqn for electron in Coulomb field with PE $U = -e^2 / r$ yields ground state energy

(15) $$E_0 = m_e c^2 \sqrt{1-\alpha^2} = m_e c^2 \gamma + U = m_e c^2 \bar{\gamma} + \bar{U}$$

Ground state wave function yields average value of $1/r$ and average PE $\bar{U}$:

(16) $$\frac{1}{r_b^{Dirac}} = \overline{\left(\frac{1}{r}\right)}_{Dirac} = \frac{1}{\sqrt{1-\alpha^2} r_b^{Sch}} = \frac{1}{\sqrt{1-\alpha^2} r_b}$$

$$\bar{U} = -e^2 / r_b^{Dirac} = -\frac{e^2}{\sqrt{1-\alpha^2}} \cdot \frac{m_e e^2}{\hbar^2} = -\frac{\alpha^2}{\sqrt{1-\alpha^2}} m_e c^2$$

(15) and (16) Lead to

(16a) $$\bar{\gamma} = \sqrt{1-\alpha^2} + \alpha^2 / \sqrt{1-\alpha^2} = 1 / \sqrt{1-\alpha^2}$$

We now prove that $\bar{\gamma}$ in (16a) is exactly the same as what we can obtain from classical mechanics for circular motion CI. For $\gamma < 10^3$ we assume $\lambda = 1$ and $d\lambda / dt = 0$. Note quantum ground state energy in (15) is also obtained under 0-LV approximation condition $\lambda = 1$. Uniform circular motion's speed is constant, with $d\gamma / dt = 0$ and $\vec{u} \cdot d\vec{u} / dt = 0$. Centripetal acceleration's written as $\vec{a} = d\vec{u} / dt = -r^{-2} u^2 \vec{r}$, $a = u^2 / r$. One can check

(17) $$f = e^2 / r^2 = m_e \gamma u^2 / r \Rightarrow \beta^2 = k \gamma^{-1},$$

$$k \equiv e^2 / m_e c^2 r \Rightarrow \gamma = (\sqrt{k^2 + 4} + k) / 2$$

Setting $r = r_b^{Dirac} = \sqrt{1-\alpha^2} r_b$ yields $k = \alpha^2 / \sqrt{1-\alpha^2}$, $\gamma = 1 / \sqrt{1-\alpha^2}$. Such γ deduced from classical mechanics plus setting $r = r_b^{Dirac}$ is exactly the same as $\bar{\gamma}$ in (16a) drawn from (15) and (16) that are obtained via QM. Such PCQM is ***natural*** for QM uses exactly the ***same*** formulas of energy, momentum, and $E - p$ relation as used in classical mechanics for classical heavy bodies.

Remark: MB problem and non-spherosymmetric MF field make SA dynamics unsolved and/or unsolvable. But CI and TME theorem help to probe the SA world. At the same time, DS and MFs are found to be able to explain nearly everything of SA physics in the absence of quantitative solutions. We'll see how DS and MFs explain facts after facts. The author has not found anything in SA world which DS and MFs are unable to explain qualitatively or half-quantitatively.

The unsolved quantum eqs in awful θ-dependent MF fields remind us of an ignored issue about OAM & TAM of the constituent particles in any composite particle as well as **spin/TAM misconception** about composite particles. Also, the **empirical disability to measure OAM** of any FSP and **TAM** of all the FSPs after decay of any unstable particle is pointed out below.

< Spin/TAM **Misconception** and Empirical **Disability** to Measure OAM/TAM >

Composite particles do not spin for constituent elementary particles do not spin about the same axis and do not spin with the same angular velocity. They do not have the same charge-mass ratio. Elementary fermions with the same spin ½ even do not spin with the same angular velocity if their masses are different. The J-label of a ***composite*** particle should be the **TAM** as the vector sum of spins and orbital angular momenta of all its ***constituent*** elementary particles in the **CMF** (**C**enter of **M**ass **F**rame). For any ***unstable particle*** decaying into the final state particles, its correct spin-label (if believed to spin) or J-label must be **TAM** of all the ***final state particles***, measured in the CMF.

TAM of all the constituent particles of any composite particle is *empirically* ***immeasurable*** for OAM of any constituent particle is *empirically* ***immeasurable***. TAM of FSPs is *empirically* ***immeasurable*** since tiny departures of momentum-lines of FSPs from their mass-center yield large orbital angular momenta for high-speed FSPs but tiny departures are *empirically* ***immeasurable***. ***TAM*** of all constituent particles of a composite particle is *theoretically* ***unknown*** because their **OAM** numbers are theoretically unknown due to unsolvable MB-problems and unsolvable or unsolved 2B-systems with **mag-forces** involved. Here mag-forces refer to the forces btw moving charges of the same kind. S-state with 0-orbital number $l = 0$ does not exist in SA world due to non-spherosymmetric MFs. The value of l, hence the value of J, is theoretically unknown and empirically immeasurable. Thus, the spin/J-values for composite particles in **SM** (Standard **M**odel) are only **labels**. Even TAM of constituent particles may not equal SM spin/J-value of the composite particle they form while exact empirical value of the sum of spin-vectors of all final-state particles does not scientifically prove any value of total OAM nor TAM.

Spin/TAM misconception/disability in SA physics refer to two simple facts:

Fact 1. ***Composite Particles Do Not Spin***.

Fact 2. **OAM/TAM** of each/all FSPs and of each/all Constituent Particles is ***Empirically Immeasurable*** and ***Theoretically Unknown***.

Remark: In classical physics and quantum atomic physics, the OAM and TAM for any 2B or MB system bound by an inverse-square-law force, such as the solar system or an atom, are *theoretically known* due to the solvability of the classical and quantum systems bound by the forces that obey the inverse-square-law and with negligible small perturbations from other members in each such MB system and *empirically measurable* (directly via astronomical observations or indirectly via spectrum measurement because the well-known theoretical solutions in QM provide with reliable quantitative connections btw empirical atomic spectra and many quantum states with different quantum theoretical orbital numbers).

Example 1

In the rest frame of any decaying particle, even very small deviation of a final state particle's linear momentum line from accurate **MCtr-MCtr** (**M**ass **C**en**t**e**r** to **M**ass **C**en**t**e**r**) line would invalidate a TAM-measurement where the sum of only their spin-vectors is treated as TAM with OAM uncounted. For a FSP, MCtr-MCtr line means that the 1st MCtr is its MCtr and the 2nd MCtr is the MCtr of all the final state particles. Using the reduced Planck constant in (3), we can easily calculate and see that a flying out 1 MeV electron's OAM is $\hbar/2$ if the said deviation is $1.147 \cdot 10^{-11} cm$. Once future tech enables experimenters to discover such deviation, they will reveal lots of basic facts of SA structure via directly measuring fundamental physical properties of final state particles. The OAM of any FSP has never been measured since the beginning of particle physics. Even worse than that, we do not know how many centuries mankind must wait to see empirical disability replaced by ability.

Now that the TAM of all the constituent particles in each SA composite particle, as correct J-label, is unknown and immeasurable, how to determine a composite particle is a fermion or a boson? A simple rule says: A composite particle is a fermion or a boson if containing odd or even number of elementary constituent fermions. Note: there is no need to measure OAMs and TAM of the FSPs as one identifies if a decaying composite particle is a fermion or a boson because the TAM of any composite particle is irrelevant to its FBS and composite particles do not spin. But as we know, elementary particles do spin and their spin values determine their FBS, like electrons and photons. Thus, for any decaying particle claimed to be elementary, the OAMs and the TAM of its FSPs must be measured in order to identify its FBS. Namely, J/Psi composite particles are found since the sharp humps were seen, the mass was measured, and its decay modes were observed regardless whether one could measure the OAMs and the TAM of the FSPs in each J/Psi decay event. It contains even number of elementary fermions in both the quark model without DS and the quark model with DS. So J/Psi particles are bosons no matter the J=1 label is right or not. On the contrary, the philosophy of massless photons, U(1) gauge invariance, the extension of U(1) gauge invariance to SU(2) and SU(3) gauge invariance require elementary Higgs particles to provide elementary force carriers (except photons and gluons) and elementary fermions such as electrons with masses while Higgs particles themselves do not need anything to get masses. Higgs couplings bring about many extra terms into quantum field theories. None of the numerical effect caused by these terms is theoretically calculated and verified experimentally. Even if one puts aside such problems and really believe that mass spectrum of elementary leptons/quarks are caused by phenomenological Higgs couplings with very heavy Higgs particles, he/she must measure the OAMs and TAM of the FSPs in order to prove that it was an elementary Higgs boson with spin 0 that showed up in the collision. It's irresponsible to claim a discovery of a spin 0 boson knowing that it's beyond experimenters' ability to measure the OAM of a FSP.

Example 2

ISPs in HE collisions have unknown and uncontrollable initial OAMs. Most head-on collisions weren't accurate MCtr to MCtr collisions. After collisions, initial KEs and consequently OAMs reduced. The lost OAMs reappeared in the forms of spins of FSPs, causing seemingly "**Spin Crises**". There is no crisis at all. TAM conserves in classical physics, quantum atomic physics, and quantum SA physics.

Example 3

In SM, muons and taus are treated as *elementary* fermions of *spin ½* without DS. This is called **MTH** (**M**uon-**T**au **H**ypothesis). Now one can easily recognize that empirical disability to measure the OAM of each FSP in decays of muons and taus make MTH empirically unprovable. Later one will see how the DS of muons and taus and MF as binding force explain every decay mode of muons and taus, masses of muons and taus, lifetime of muons longer than positronium with the same DS reason that makes charged pions living longer than a neutral pion.

Example 4

In SM, a hypothesis says that J=1/2 baryon formed by *sss* does not exist or hasn't been found while the found $\Omega^{\pm}$ are believed to belong to the well-known baryon decuplet. So J=3/2 label was set for $\Omega^{\pm}$. Now one can easily recognize that such TAM is empirically immeasurable and we must check all experimental evidences to decide whether the said hypothesis is correct. One can immediately notice the bizarre long lifetime of $\Omega^{\pm}$ in the decuplet, indicating that $\Omega^{\pm}$ do not have BC orbital excitation like other members in the decuplet. Experimenters could easily find so many hadrons far beyond the original meson octet, baryon octet, and baryon decuplet. This reminds us of the "backbone" of physics that has been seen since Newton, Einstein, and founders of quantum theory: structure and dynamics, not symmetries to predict the existence of composite particles. It's even worse to assume that elementary stable particles can be predicted by using a symmetry theory. Now we see why supersymmetry particles have never been found since the beginning of particle physics. SDL

III. DS-Explanations of Experimental Facts

HMC $u\bar{u}$ in any meson dooms to annihilate so that all mesons have short lifetimes without a single exception. DS and MF do not allow $d\bar{d}$ to form a particle or a mixture component of a particle because the DS electron and positron in $d\bar{d}$ MF-repel too severely. Then 0-pion $\pi^0 = u\bar{u}$ lives much shorter than $\pi^{\pm}$ for both the d-electron in π^- and the $\bar{d}$ -positron in π^+ slow down the $u\bar{u}$ -annihilations. SDL on page 84 and page 279.

BC uuu in every baryon (including nucleons) is a bound 3-body system with no possibility of SA annihilation inside. This is one of the reasons of BNC. Full explanation of BNC in all events (reactions and collisions) can be given only after FQA fact is explained in all events. SDL for how to explain FQA fact.

All non-nucleon baryons have short lifetimes without a single exception since all of them contain excited DSLs according to DS and SA de-excitations take place quickly without exception.

DS tells: every quark and every anti-quark contains one u and one $\bar{u}$ respectively. **QE (quark excitation)** means **UE (*u* excitation)**: Either u - excitation OR $\bar{u}$ **-**excitation. Because neither u nor $\bar{u}$ can appear alone, every **UE** is an excitation of BC, uuu antiBC $\bar{u}\bar{u}\bar{u}$, or HMC $u\bar{u}$. Due to strong interactions at SA short distances, a hadron in UE returns to ground state or lower level state very quickly. An s-quark is a d-quark with the DSLs in excitations but without any UE. A c, b, or t quark contains DSLs in higher excitations. Excited DSLs also return to ground or lower level states quickly. It's unclear at the present whether a UE involves in a b-quark or c-quark. This DS reality picture explains immediately all the hadrons that have short or very short lifetimes including non-nucleon baryons with their q-cores never annihilate. Please note, many hadrons have q-cores not in QE states. But once they contain quarks other than u-quarks and d-quarks, the DSLs inside are in excitation states. They also have short lifetimes for excited DSLs return to ground states quickly without exception. SA composite particles, other than Δ^{++} (uuu), its antiparticle ($\bar{u}\bar{u}\bar{u}$), and π^0 ($u\bar{u}$), contain both quarks and DSLs. Excited SA composite particles other than (uuu), ($\bar{u}\bar{u}\bar{u}$), and ($u\bar{u}$) contain excited DSLs, meaning **LE (leptonic excitation)**. Many excited composite particles in SA world are in both LE and QE.

< Examples of DS-Explanations of **Bizarre** Phenomena >

1. The 1st bizarre **uniqueness** among all hadrons is: ***Unbound*** protons are ***stable***. Among all the photos that have recorded cosmic shower events and collision events in accelerators all over the world since the beginning of SA physics no single one has showed an unbound proton decaying. All free hadrons (mesons and baryons) are not stable except free protons. This ironic fact has a simple DS-explanation: Both the BC uuu and the d-leptons inside any proton are in ground bound states with the TME and the ME of each constituent particle inside proton ***negative at all times*** so that any QT channel of decay is closed and no any de-excitation nor annihilation can happen to trigger a decay process.

DS also explains such fact: ***bound*** protons may change to neutrons. In β^+ -

decay nuclear *transmutation* occurs so that mother nucleus lost energy. The lost energy re-emerges in the form of PAP(s). DS proves that **one** e^+e^- **pair** and **one** $\nu_e\bar{\nu}_e$ **pair** are produced in β^+-decay so that e^+ and ν_e move away while a *u*-quark in proton captures e^- and $\bar{\nu}_e$ turning itself/proton into d-quark/neutron so that a β^+-decay takes place

$$energy + p \rightarrow n + e^+ + \nu_e$$

Now we see: ***DS forbids beta-plus decay if available energy is less than*** $2m_ec^2$.

In e^--capture bound atomic K-/L-electron falls down to SA ground state in the non-symmetric potential energy region created by u-quark or baryon core to bind d-electron. Such shell-electron's stepping down is part of nuclear transmutation. Here released energy takes the form of **one** $\nu_e\bar{\nu}_e$ **pair without** e^+e^- **pair** while three u-quarks and e^- join Wk forces to capture and bind $\bar{\nu}_e$ so that u-quark and proton become d-quark and neutron respectively, releasing ν_e:

$$p + (K/L\text{-}Shell)\, e^- \rightarrow n + \nu_e$$

It happens even if the available energy is less than $2m_ec^2$ for no e^+e^- pair is needed to nor is to be produced to complete e^--capture process.

2. The 2nd bizarre **uniqueness** among all hadrons is: Unbound neutrons decay 10^8 ***times slower*** than the slowest decaying hadrons other than neutrons. DS reveals ENM inside a neutron. Via ENM, the "stronger" d-electron slowly pushes away the "weaker" one. No any fast SA de-excitation is involved in neutron decay.

3. The 3rd bizarre phenomenon is the well-known **continuous** energy spectrum of beta particles. DS together with the **MBT** (**many body theorem**) and **MBP** (**many body postulate**) on pages 35, 36 explain it. MBT reveals time-varying ME of a constituent particle in time-varying resultant force field produced by other rapidly moving particles bound in the same composite SA particle. A constituent particle is bound if its ME, as the sum of its KE without rest energy and the potential energy is negative at all times. It's quasi-bound if its ME is positive sometimes that may trigger an escaping into free region with the same positive ME via QT. MBP further claims continuous ME of a constituent particle once the rapid motions of other constituent particles have significant time-varying impact. In addition, MBP adopts the discrete quantum levels for the TME of all the constituent particles in the same composite particle.

Remark: SA composite particles are MB-systems where each individual constituent particle is moving in the field of the resultant force produced by others in the system. At least, the two d-electrons in a neutron are in rapid motions and close to each other. This makes the resultant force field time-varying. Although quantum eqs in such time-dependent field have not been solved, the solution must be time-dependent (time-independent wave functions obviously do not satisfy the said eqs). We know that eigen-states with discrete constant energies are the solutions of quantum eqs in time-independent field. For the MB-system of a SA composite particle, the masses of its ground and excited states have fixed discrete constant empirical values. This supports the feature of TME described in the MBP.

In a hydrogen atom, the rapid motions of Qs/d-e (quarks/d-electron) inside the nucleus do not have any significant time-varying impact on the motion of the atomic electron for its MPR is far away from the Qs/d-e in local rapid motions.

4. The 4th bizarre phenomenon is the well-known fact that chemical elements on **stability island** live long enough to be called stable though their nuclei contain neutrons that may be more than protons bound in the same nucleus. In I. Introduction, we have shown its DS-explanation.

5. The 5th bizarre phenomenon is seen in the celebrated **anomalous MMs** of nucleons. It's impossible to explain MMs of nucleons accurately due to insolvability of **MB** problems. DS may offer half-quantitative explanations. Let us ignore mass contributions of d-leptons. Then proton's mass is the sum of the masses of the three u-quarks plus the TME of them. Stable proton requires negative TME. A composite particle's rest mass is the sum of its constituent particles' rest masses plus their TME. This is called **CME (composition mass eqn)** on page 32. Then u-quark's mass must be larger than 1/3 of proton's mass. This is called **HQMT (heavy quark mass theorem)**. Such heavy u-mass makes the empirical values of nucleons' MMs understandable. A good work along this direction was done by D.H. Perkins in his book "Introduction to High Energy Physics" (4th Edition, World Web, p. 130-132). He got $\mu_n = -(2/3)\mu_p$. The empirical ratio is -0.685. The relative error is less than 3%. For other baryons, his calculated MMs agree with observations with discrepancies at 10%-20% level. To obtain such good results, he has set $m_u = m_d = 336$ MeV and $m_s = 509$ MeV. He ignored DS electrons. It may be OK if OSOD leads to very small net MM of a d-electron. The OAM of any DS electron is theoretically unknown and empirically immeasurable. Unsolved MF-bound MB systems forbid revealing the theoretical value of an OAM. The empirical disability to measure the OAM of a SA constituent particle, a MSP, a FSP, or an ISP in HE collisions is crystal clearly seen in the celebrated reduced Planck constant

$$\hbar/2 = 1.93\cdot 10^{-11} cm\cdot m_e c = 0.986\cdot 10^{-11} cm\cdot Mev/c$$

For any above mentioned HE particle with near-c speed, if the deviation of its linear momentum line from the mass center is about $10^{-11} cm$ and the linear momentum is about 1 Mev/c, its OAM is about $\hbar/2$. Such small deviations have never been measured since the beginning of particle physics, and may remain immeasurable in the coming 100 years. Just by reminding us of the OAM issue in SA physics, we have explained "spin crises" before. The unknown value of a DS electron's OAM seems to imply another possibility: u-mass might be much heavier than Perkins's mass-setting. Then the empirical MMs of nucleons are mainly due to net MMs of d-electrons. However, the fact that beta-particles' MMs had preferential spin direction parallel to the EMF indicates that the magnitude of the OAM of the d-electron must be larger than the spin if the net

MM of the d-electron dominates the net MM of the BC. The author does not know for sure which may be the case.

6. The 6th bizarre phenomenon is the observed **preferential spin direction** of electrons and neutrinos emitted in β-decay, parallel to the EMF. That is: the spin MMs of most emitted electrons were bizarrely pointing at directions nearly opposite to the EMF. CED and experiment have proved that e.m. force turns a classical MM into the direction of EMF while a free or atomic electron in an EMF has two possible directions for its spin as seen in Stern-Gerlach experiment, none is preferential. How can beta particles fly out from Cobalt with spin MMs preferably pointing at directions nearly ***opposite*** to the direction of EMF? Besides, if electrons and neutrinos emitted in β-decay reversed both their spin and moving directions, handedness in the standard model and parity violation would remain true. But it never happened to the bizarre preferential direction. Here is a sample DS explanation:

By MM, if not specified, we mean MM associated with electric charge. The MM of a neutron is the vector sum of the net MM of its BC and the net MMs of its 2 d-electrons. SA MF-binding requires MM-correlation hence **SC (spin correlation)** so that the net MM of the BC that binds d-electrons and the spin MM of each d-electron are in opposite directions. That is, d-electron's spin is originally in the same direction of the net MM of the BC. A neutron contains two d-electrons moving in R1 at the net MM of the BC with their spin MMs parallel to each other but opposite the direction of the MM of the BC. One d-electron is closer to the BC. Due to OSOD, small net MMs of the two d-electrons only cause small effects on each other and the MM of a neutron is in the same direction of the net MM of its BC. Thus, a d-electron's spin is originally in the same direction of the MM of the neutron. Any nucleus of Cobalt $^{60}C_0$ in Wu's experiment was very heavy, more classical than quantum so that its net MM has a preferential direction nearly parallel to the EMF. The MT-C ratio in (6) indicates domination of St force in binding nucleons in nuclei with negligible role of MFs. Namely, nucleons are not required to have **SC (spin correlation)** to yield MF-attraction btw neighboring nucleons in order to be bound within a nucleus, though the ground state of the nucleus MB bound system with lowest possible negative TME may require SC and MM correlation of nucleons in the absence of an EMF, such as opposite directions of two neighboring MMs assumed in the celebrated shell model of nuclei. An EMF enforces a preferential direction of the MMs of the nuclei of Cobalt. Even if nucleons are not more classical than quantum, the preferential direction of the nuclei' MMs could enforce the same preferential direction of the MMs of the nucleons ***statistically***. Therefore, in the presence of an EMF and at super-low temperature, the preferential MM direction of Cobalt nuclei is also the preferential MM direction of the neutrons in the nuclei, which is in the same direction of the spins of the d-electrons. Thus, the d-electrons in Cobalt originally had a preferential spin direction parallel to the EMF. When they fly out from the neutrons, most of them do not have enough time to change their spin directions. To obey the SM handedness and parity violation, beta particles then have to have a preferential moving direction that is in the opposite direction of the EMF. The charge extension in DS explains the behavior of the antineutrinos emitted in beta decay. Neutrinos and beta particles (electrons) have opposite moving directions due to conversation of linear momentum.

We'll explain other bizarre phenomena soon, such as low masses of pions (under heavy u-mass setting needed to explain stable free protons and anomalous MMs of nucleons), masses of neutral/charged pions, different lifetimes of the neutral Kaon, and ironic FQA fact. We'll see abundant examples of how DS and MFs are able to explain, qualitatively and half-quantitatively nearly all facts of SA world, including mass spectrum, lifetimes, branch ratios, all reactions, hundreds of types of MSPs, and thousands of decay modes.

Every BC 3B-system is unsolvable even in CI. HMC $u\bar{u}$ in π^0 is a 2B-system. To explain the low mass of π^0 under a heavy u-mass setting, we adopt suppression of MF, realized in OSTP.

< OSTP >

OSTP imitates a hypothetic quantum ground state of HMC $u\bar{u}$ with MMs of $u\bar{u}$ parallel at all times, their θ-ZLs being R0 at each other and very close to 0R-MF angle at all times, and MF btw them attractive but greatly suppressed at all times. Two spin MMs tilted against orbit and change directions in precession to maintain MM-correlation and relative θ-ZLs near 0R-MF angle. In circular motion CI, two magnetons $u\bar{u}$ move along the same circle on the same orbit plane, revolving about their mass center with the same angular velocity. At every instant time, the center of circular orbit is the middle point of the segment joining the two magnetons. Such special motion is the simplest classical solution of any equal mass 2B-system. When a skater makes circular motion on the ice floor with body always stretched, not vertical, always forming the same angle with the floor in the way that the projection of body-line on the floor at every instant time passes through the center of the circle. Let 2 skaters represent $u\bar{u}$ in OSTP. Body-vectors from feet to head represent 2 MM-vectors of $u\bar{u}$. Their body-lines are not vertical but parallel at all times as they move alone the same circle on the ice floor opposite to each other, forming the same angle with the floor-plane at all times. The projection lines of the 2 parallel body-lines coincide on the floor-plane, rotating about the center of the circle and passing through that center at all times. From this picture we may see exactly what does it mean by 'OSTP'. As we use relativistic mechanics to rigorously calculate classical circular motion in CI for HMC $u\bar{u}$ 2B-system in ground state, such 2 MMs exert greatly suppressed attractive MF on each other. Calculation explains mass of π^0 under <u>heavy</u> u-mass setting, which is needed to simultaneously explain nucleons' masses, their MMs, and stability of free proton.

Note: Precession in OSTP is needed for spin MMs to change their directions in order to maintain orbit-spin tilting near 0R-MF angle ZL simultaneously. Orbit-spin tilting alone cannot fix two magnetons in the same near 0R-MF angle ZL all the times. In OSTP, two MMs with directions changing all the times remain parallel at all times so that spins of two magnetons remain antiparallel at all times to <u>**obey conservation of total angular momentum**</u>.

OSTP cannot happen to LMC e^+e^- 2B-bound system. As OSTP suppresses

MF btw e^+e^- greatly, Coulombian force is unable to bind e^+e^- within SA small space due to their small mass. It then becomes positronium e^+e^- of atomic size with MF at such atomic distances too weak to cause significant orbital motion for e^+e^- to make the ground state depart S-state. Without OSTP and with MF-binding domination and its large positive TME feature, muon-mass eqn is deduced to explain empirical value of heavy muon-mass, compared with much smaller total mass of all its leptonic constituents. In the following we will explain masses of pions and muons using DS with MFs.

< HMC $u\bar{u}$ and π^0-Mass Eqn >

Asymmetric MF field depends on r and θ. We may have discrete values of most probable angle θ while ground state may not correspond to 0R-MF angle exactly. Now consider **suppressed MF** nearby **0R-MF** angle for MF btw $u\bar{u}$ plus St Coulombian force btw them. We modify MF $\vec{f}_1$ in (5a) and r_{m1} in (6) with a suppressing factor k_{MF} when OSTP occurs and suppresses MF btw $u\bar{u}$. Write

(18) $$f_{MF}^{Sup} = 1.24k'e^2\frac{(3\cdot 10^{-11}cm)^2}{(2r)^4}\frac{1}{k_u}\frac{1}{k_u} \qquad (k_u \equiv m_u/m_e,\ k' \equiv k_S/k_{MF})$$

$$k_S = \begin{cases} 1 \ \ if \ \nexists \ St \ MF \\ \eta_C^{st} \ \ if \ \exists \ St \ MF \end{cases}, \qquad k_{MT-C}^{\sup} \equiv \frac{f_{MF}^{\sup}}{\eta_C^{st}e^2/4r^2} = (\frac{r_m^{\sup}}{r})^2$$

$$r_m^{\sup} = 1.67\sqrt{\frac{k'}{\eta_C^{st}}\cdot\frac{1}{k_u}}\cdot 10^{-11}cm$$

(18a) $$f = f_C^{st} + f_{MF}^{\sup} = \frac{\eta_C^{st}e^2}{4r^2} + \frac{\eta_C^{st}e^2}{4r^2}\frac{(r_m^{\sup})^2}{r^2} = m_u\gamma a = m_u\gamma u^2/r$$

$$\beta^2 = k\gamma^{-1}, \qquad k = \frac{\xi_u}{4}(\eta_C^{st} + 3513.98724k'\xi_u^2)$$

$$\xi_u \equiv \frac{r_{sa}}{k_u r}, \qquad r_{sa} \equiv \frac{e^2}{m_e c^2} = 2.8177471\cdot 10^{-13}cm, \qquad k' \equiv \frac{k_S}{k_{MF}}$$

$$\beta^2 = \frac{2}{1+\sqrt{1+4/k^2}}, \qquad \gamma = \frac{k+\sqrt{k^2+4}}{2}.$$

where the suppressing factor $k_{MF} \gg 1$ due to small deviation from 0R-MF angle in OSTP. If St MF never exists, $k_S = 1$ due to the existence of MF. Otherwise, $k_S = \eta_C^{st}$ is the coefficient of St-Coulombian force. Then

(18b) $$2(f_C^{st} + f_{MF}^{\sup})dr = dU\,, \qquad U = -\eta_C^{st} e^2/2r - \eta_C^{st} e^2 (r_m^{\sup})^2/6r^3$$

One can check the following PE and TME eqs:

(18c) $$U = -m_u c^2 (2k - 1171.32908\, k' \xi_u^3)$$

$$T_{ME} = 2m_u c^2(\gamma - 1) + U = m_u c^2(\sqrt{k^2+4} - k - 2 + 1171.32908 k' \xi_u^3)$$

At large enough r, ξ, k, and T_{ME} almost vanish. As r decreases, ξ and k increase while T_{ME} decreases to the minimum $T_{ME}^{\min} < 0$ and then increases. At r_{0TME}, T_{ME} is zero. At shorter distances, T_{ME} becomes positive.

T_{ME} in (18c) and CME lead to π^0 **-mass-eqn** with $u\bar{u}$ content only without DSLs or with small contribution from DSLs omitted:

(18d) $$m_{\pi^0} = 2m_u + T_{ME}/c^2 = m_u(\sqrt{k^2+4} - k + 1171.32908 k' \xi_u^3)$$

$$k = \frac{\xi_u}{4}(\eta_C^{st} + 3513.9872 k' \xi_u^2)$$

$$\xi_u \equiv \frac{\xi}{k_u} \equiv \frac{r_{sa}}{k_u r}, \qquad r_{sa} \equiv \frac{e^2}{m_e c^2} = 2.8177471 \cdot 10^{-13}\,cm, \qquad k' \equiv \frac{k_S}{k_{MF}}$$

$$k_u \equiv \frac{m_u}{m_e} = \frac{m_{\pi^0}}{m_e(\sqrt{k^2+4} - k + 1171.32908 k' \xi_u^3)}$$

$$m_u = \frac{m_{\pi^0}}{\sqrt{k^2+4} - k + 1171.32908 k' \xi_u^3}$$

$$m_{\pi^0} = m_u(\sqrt{k^2+4} - k + 1171.32908 k' \xi_u^3)$$

We see that once 3 parameters k', ξ_u (equivalently, product $k_u r$ or $m_u r$), and η_C^{st} are determined, k, γ, k_u, and m_u are also determined, while r_0, as radius of circular motion in ground state, can also be determined:

(18e) $$r_0 = r_{sa}/k_u \xi_u = e^2/m_u c^2 \xi_u$$

Following **sample settings** of k', ξ_u, and η_C^{st} with **results** (k, γ, k_u, m_u, r_0) for *ground* state are put into table below, called **HMC $u\bar{u}$ Ground State Table**. All parameter-settings are for ground state only. $r_{b,u\bar{u}}^{Dirac}$ is the DBR of $u\bar{u}$ 2B-system bound solely by St Coulombian force, with η_C^{st} and m_u given in the same row.

$r^{\sup}_{m,u\bar{u}}$ is the MD btw $u\bar{u}$ with suppressed MF & corresponding setting of k' and consequent values of k_u, m_u, r_0, OAM, IPE-size $r^{Dirac}_{b,u\bar{u}}$, and the MD $r^{\sup}_{m,u\bar{u}}$ with MFs suppressed greatly in the same row. The table below uses sample setting η^{st}_C =130. Unit for all distances/sizes is *cm*. (Same on p.79)

HMC $u\bar{u}$ Ground State Table

k'	ξ_u	k	k_u	$m_u c^2$	r_0	OAM ($\hbar/2$)	$r^{Dirac}_{b,u\bar{u}}$	$r^{\sup}_{m,u\bar{u}}$
.013	.176	5.74	626.9	320.4 MeV	2.55 10^{-15}	0.48	2.05 10^{-14}	2.66 10^{-16}
0.01	0.2	6.57	675.1	345.0 MeV	2.09 10^{-15}	0.49	1.90 10^{-14}	2.17 10^{-16}
10^{-6}	1.0	32.50	4217	2155 MeV	6.68 10^{-17}	0.47	3.04 10^{-15}	3.47 10^{-19}
10^{-8}	5	162.5	19184	9803 MeV	2.94 10^{-18}	0.47	6.69 10^{-16}	7.64 10^{-20}
10^{-10}	15	487.5	58736	30.01 GeV	3.20 10^{-19}	0.47	2.19 10^{-16}	1.49 10^{-22}

It shows that **QKoE (quark knockout energy)** $E(uKo) = 2m_u c^2 - m_{\pi^0} c^2 =$ 505.8, 555, 4174.6, 19470.8, 59893 MeV respectively. Even smallest QKoE above is larger than the total energy needed to produce three pions. But even the highest QKoE above is below or far below the highest energies of knocking particles seen in accelerators or cosmic rays. Then how to explain FQA fact? **AVI (absolute vacuum inelasticity)** with absolute priority for **PAPP (PAP production)** to consume impact energy and PQTR explain FQA fact if **QNC (quark non-confinement)** is the reality under such low QKoE:

Now that at least one bound $u\bar{u}$ pair is produced right before $u\bar{u}$ in pion or any u-quark in a hadron is knocked out and each one in new born bound $u\bar{u}$ pair is very sensitive to any empty **PW (potential well**, meaning region of low negative potential energy, not necessarily of well-shape) wrapping each naked one knocked out from $u\bar{u}$ pair in a pion or from any hadron. Naked ones and new born ones emerge at nearly the same place and time so that PQTR easily makes any hole left behind by a knocked out one stuffed AND no naked single unbound quark could appear among FSPs.

In cases of very heavy u-mass and very large QKoE, *quark confinement* can be true if QKoE is larger than **NLVUC (non-LV ultraviolet cutoff)**. DI collisions feature MSP productions, caused by PAPP and PQTR and showing **VI (vacuum inelasticity)** so that impact energies are consumed by PAPP instead of knocking. It may lead to the existence of NLVUC, which refers to an upper limit of all the energies transferred ***elastically*** such as seen in a process of knocking non-deeply bound quarks out & making them free. More details of NLVUC and DS explanation of FQA fact will be seen later. Besides, the V-shape of TME-*r* curve around r_c may create a series of bound states with binding sizes smaller than r_c. Excited states in inward direction may be another reason of FQA Fact: Impacts in collisions may push hadrons to inward excited states with less and less sizes of binding instead of knocking out quarks in outward direction. SDL for NLVUC.

OAM in the table is for reference only. OAM obtained in circular motion CI may be meaningless. For example, OAM vanishes in S-state while it's nonzero in any classical circular motion. We don't know if there is an example of PCQM that includes a better match in OAM.

Remark: If sample settings and outcomes in 1st two rows are good, u-mass is larger but not much larger than 1/3 of proton's mass. Pure St Coulombian force without any MF and Mag-forces could only bind heavy $u\bar{u}$ with the said u-mass within SA small space with $r_{b,u\bar{u}}^{Dirac}$ (DBR), about $2 \cdot 10^{-14} cm$, to be its size. Suppressed MF equals and is stronger than St Coulombian force at MD $r_{m,u\bar{u}}^{\sup}$ about $2 \cdot 10^{-16} cm$ and shorter distances respectively. However, its large influence reaches longer distances so that the binding size reduces from DBR $r_{b,u\bar{u}}^{Dirac}$ to r_0 about $2 \cdot 10^{-15} cm$, which is btw IPE-binding size $r_{b,u\bar{u}}^{Dirac}$ and MD $r_{m,u\bar{u}}^{\sup}$. If sample settings/outcomes in last two rows are better, we have much smaller binding sizes/heavier u-quark.

At the present, we do not know if LV space force plays a role in 3rd **GS (gamma sector)** $10^2 < \gamma < 10^3$. Once LV space force enters the stage of quark dynamics, Pion-mass eqn must be modified. Unsolvable MB-systems do not allow us to get OAM quantitatively. OSOD, at the present, is unable to provide with quantitative result. So mysterious u-mass and many consequences remain unknown. But DS provides physical pictures so close to reality that it has shown its power to explain huge number of SA facts and events.

If energy released in SA de-excitation or HE collision is large enough, excited HMC $u\bar{u}$ with larger OAM and higher j-label $J=1$ can be born and HMC for Rho ρ and other mesons with label $J=1$ may appear. CI of such quantum states might be **elliptical** motion. Such classical motion for equal-mass 2B-system may be unsolvable even in classical theory. Absence of classical and quantum solutions, however, does not cover physical picture of excited states of HMC $u\bar{u}$ and more complicated excited states of meson-systems with DS-leptons involved.

DRGGD (***De Rujula-Georgi-Glashow*** **D**escription) of Hadrons:

They consider excitation states. Quark model with a broken symmetry but without excitation states is equivalent to abandoning basic dynamics corner stones in all branches of modern classical and quantum physics. The broken symmetry was proposed when dense 'fog' covering dynamics in SA world didn't seem to be penetrable. We expect the original thinker to put aside his idea that's stood for so long because of unsuccessful potential models and other dynamical models as all of them had missed crucial effects of DS with MFs.

DS and MFs support DRGGD of hadrons. We've seen that pion mass is never accidentally low but is explained as ground states with DS and heavy u-mass that simultaneously explain nucleons' MMs as well as heavier mesons with label $J=1$ as excited states just as indicated in DRGGD.

< $\pi^{\pm}$-Mass Eqn >

We know that $m_{\pi^{\pm}} - m_{\pi^0} = (139.5702 - 134.9766) MeV/c^2 = 4.5936 MeV/c^2$. Such small difference indicates small contribution DS-leptons make to $\pi^{\pm}$ that contain them. Since that difference is larger than the sum of rest masses of DS-leptons in a charged pion, ***ME*** of DS-leptons must be ***positive*** but small. **PME (positive ME)** is due to MF and TME theorem. PME opens **QTC (QT channel)** for decay allowing bound DSLs to penetrate PW fleeing to free region through QT. DS predicts decay modes:

$$\pi^+ \to e^+ \nu_e \pi^0 / \pi^- \to e^- \bar{\nu}_e \pi^0$$

It is proved by experimenters with small branch ratio ($1.036 \cdot 10^{-6}$). In this example, again, HMC $u\bar{u}$ annihilation channels dominate QTC in the sense that $u\bar{u}$ annihilation most probably occurs before bound DSLs with positive MEs would escape via QT. DS further explains difference btw pions' masses as shown below with very small contribution from any **NFM (neutrino family member)** omitted:

HMC $u\bar{u}$ Ground State Table tells: **HMC size** ($<2.6 \cdot 10^{-15} cm$) is **much less** than the **MD** of MF btw ue^- or $\bar{u}e^+$ deduced from (6). We may expect much larger orbital sizes of DS-leptons than the orbital size of HMC $u\bar{u}$. Thus, to DS-electron/positron in $\pi^{\pm}$, HMC looks like a tiny particle with 0-net charge so that Coulombian forces on DSLs in $\pi^{\pm}$ are negligible at all places far away from HMC. The MF exerted on d-electron or $\bar{d}$ - e^+ dominates Coulombian force almost completely with **positive ME**. Here we use term ME instead of TME for tiny heavy HMC $u\bar{u}$ can be treated as a fictitious one particle that is at rest while TME of a charged pion equals the ME of d - e^- in π^- or $\bar{d}$ - e^+ in π^+, that moves around the said rest fictitious particle.

Antiparallel spins of $u\bar{u}$ in ground state HMC yield a net spin MM twice of u-quark's spin MM effective for any spin MM far away from $u\bar{u}$ 2B-system. It can MF-bind one DS-electron or one DS-positron to form π^- or π^+. To be bound by such net spin MM of HMC in ground state to form π^- also in ground state, d-electron must have its spin in the direction parallel to that of the net MM of HMC and stays in R1 at the net MM such that it together with u-quark in HMC can form d-quark in π^- though it is actually bound by entire HMC 2B-system $u\bar{u}$. Similarly, DS-positron stays in R1 at that net MM with its spin in opposite direction of HMC's net MM. If this spin correlation and MPR-location is correct, $d\bar{d}$ cannot exist as a particle or meson because e^+e^-, i.e., d - e^- and $\bar{d}$ - e^+ would have parallel spin MMs and they would severely repel each other if they were simultaneously bound by HMC. Later we will study spin-correlation of HMC $u\bar{u}$ in kaons.

Using u-mass in 1st row of HMC $u\bar{u}$ Ground State Table, we get net spin MM of HMC $u\bar{u}$ in ground state effective for far away spin MM such as that of a d-electron or that of a $\bar{d}$ -positron, by using quantum connection for MM with RC ignored:

(18f) $$\mu(\pi - HMC) = 2\mu_u = 2\mu_{\bar{u}} = \frac{2e\hbar}{3m_u c}$$

MF formula (2) gives an approximation for charged DS-lepton (e^- or e^+) with $\theta \approx 90^0$ bound in charged pion by net spin MM of pion's HMC:

(18g) $$f(e \;\; in \;\; \pi) = \frac{3\mu(\pi - HMC)\mu_e}{r^4} = \frac{m_e c^2 \xi^3}{k_u \alpha^2 r},$$

$$\xi \equiv \frac{r_{sa}}{r}, \qquad r_{sa} \equiv \frac{e^2}{m_e c^2} = 2.8177471 \cdot 10^{-13} cm$$

Here zenith axis (z-axis) is in the same direction of the net MM of pion's HMC that's at rest at the origin. This new coordinate system is for observation of charged DSLs, not for observation of $u\bar{u}$ in HMC. Orbit-spin tilting in OSTP tells us that orbit-plane of $u\bar{u}$ motion in CI now is not the xy-plane on which charged DSL moves about the net MM of $u\bar{u}$ in CI. A big problem is that precession in OSTP means: Spins and MM-vectors of $u\bar{u}$ must rotate about normal vector of $u\bar{u}$ orbital plane. Even classical motion of charged DSL about such net MM of $u\bar{u}$ in rotation may not be solvable. To penetrate dense fog and see something we fix z-axis on rotating net MM vector and use this partially co-moving coordinate system to describe charged pion approximately.

Circular motion CI leads to eqn of motion in relativistic mechanics:

(18h) $$f(e \;\; in \;\; \pi) = \frac{m_e c^2 \xi^3}{k_u \alpha^2 r} = m_e \gamma u^2 / r, \quad k\gamma^{-1} = \beta^2,$$

$$k \equiv \frac{f(e \;\; in \;\; \pi) r}{m_e c^2} = \frac{\xi^3}{k_u \alpha^2}, \quad \beta^2 = \frac{\sqrt{k^4 + 4k^2} - k^2}{2}$$

PE of $e^{\pm}$ in $\pi^{\pm}$ and composition mass of $\pi^{\pm}$ ($\pi^{\pm}$-mass eqn) is written

(18i) $$U = -\frac{\xi^3}{3k_u \alpha^2} m_e c^2 = -\frac{1}{3} k m_e c^2, \qquad m_{\pi^{\pm}} = m_{\pi^0} + m_e \gamma c^2 - k m_e c^2 / 3$$

With possible values of $k_u \equiv m_u / m_e$ shown in HMC $u\bar{u}$ ground state table we now put corresponding results of proper sample setting of parameter ξ of orbital size of charged DS-lepton in $\pi^{\pm}$ into $\pi^{\pm}$-Mass Table below:

< $\pi^{\pm}$ -Mass Table >

k_u	626.9	675.1	4216.8	19183.9	58736.34
ξ	0.764	0.783	1.4422	2.39	3.471
k	13.3583	13.3532	13.3587	13.3637	13.3698
β^2	0.994458	0.994454	0.9944583	0.99446238	0.98586298
γ	13.4328	13.42796	13.433169	13.43812	13.44421
r	3.688 10^{-13} cm	3.599 10^{-13} cm	1.954 10^{-13} cm	1.179 10^{-13} cm	0.8118 10^{-13} cm

Contribution from NFM is ignored. Orbital size r of charged DS-lepton in charged pion proves it to be a SA size. The γ -value is in 2nd GS so that we can indeed use SR's relativistic mechanics with LV omitted in CI.

Remark: From γ -values and orbital sizes r in the above table, OAM of d-electron is smaller than expected: It's less than 26% of electron's spin **AM** (angular momentum). This may be due to deviation from circular motion. Even in spherosymmetric Coulomb field, deviation of atomic electron's motion from S-state could get larger OAM such as $L=\sqrt{l(l+1)}\hbar$ with $l=1$. Now that ground state of d-electron in non-spherosymmetric MF field is not S-state, its classical motion in CI should not be circular. Once in more general closed classical motion, its OAM, as a constant of integration of the eqn of motion in SR with tiny LV ignored for classical planar motion on a plane where force field is central, can have a series of values larger than all OAMs of circular motions of radius noncircular closed classical motion's orbit covers.

Again, CI itself is unable to determine exact value of larger OAM. OSOD of d-electron with a small net MM is a must for DS to agree with observed small values of nucleons' MMs. Large net MM of d-electron, if any, would require large spin MM of u-quark which would demand small u-mass. Then heavy proton and its stability could not be explained. In π^-, d-electron is moving in nearly pure MF-field for net charge of HMC $u\bar{u}$ is zero. But d-electrons in nucleons are bound by both MF and Coulomb force. This should be leading to larger OAM. Another possible reason is that we get OAM of d-electron in π^- smaller than expected for precession of the net MM of HMC $u\bar{u}$ is ignored to simplify the problem as we try to know the motion of d-electron in π^-. OSTP and fast orbital motion of $u\bar{u}$ circling each other makes the net MM tilted forming an angle with the normal vector of $u\bar{u}$ orbital plane but rapidly rotating about It. Although motion of OSTP may make motion of d-electron in π^- bound by the net MM of $u\bar{u}$ unsolvable, we may expect larger OAM of d-electron due to such motion of the net MM of $u\bar{u}$. More radically, *OAM obtained in circular motion CI cannot be trusted.* Example of PCQM shows perfect match btw the average value of γ in quantum ground S-state and the γ -value obtained in circular motion CI using classical mechanics in SR. No match for OAM: OAM vanishes in S-states and is nonzero in classical circular motions. At the present, the author does not know if there is an example of PCQM that also includes perfect match btw OAMs.

< DS of **Muons** and **Taus** >

Among all leptons other than NFMs, only e^+ and e^- are elementary. DS of muons and taus is:

$$\mu^+ = e^+e^-e^+\nu_e\overline{\nu}_\mu,\ \mu^- = e^+e^-e^-\overline{\nu}_e\nu_\mu,\ \tau^+ = (e^+e^-)_2 e^+\nu_e\overline{\nu}_\tau,\ \tau^- = (e^+e^-)_2 e^-\overline{\nu}_e\nu_\tau$$

with heavy LMC e^+e^- bound by MF and born to have large ***positive*** TME with large γ-value. Here $(e^+e^-)_2$ means higher excitation of LMC e^+e^-. LMC in muon and tau can be denoted as $(e^+e^-)_1$ and $(e^+e^-)_2$ respectively. We may also use e^+e^- to denote the LMC of muon. The DS-explanation of all tau decay modes will be seen later after we study < 4TDSP >. Such DS explains all three confirmed decay modes of muons naturally:

1. In decay $\mu^+ \to e^+\nu_e\overline{\nu}_\mu$, e^+e^- that form heavy LMC annihilate into virtual γ and/or virtual Z-bosons for other leptons in the muon to absorb and gain KEs fleeing out.

2. In decay mode: $\mu^+ \to e^+\nu_e\overline{\nu}_\mu\gamma$, LMC e^+e^- annihilate into two real (not virtual) photons. One flies out and one hits the DS-positron (photoelectric encounter), either being absorbed or becoming a non-observable soft photon. Please note, excited atomic electrons, when stepping down to lower levels, emit photons that do not have a charged particle nearby to hit. They preserve atomic spectrum. In SA world, photons are produced in SA small spaces with other charged particles such as DS-electrons or DS positrons nearby. SA photoelectric encounters are inevitable and are part of MB-interactions so that all radiative decay modes have proved the emitted photons spectrum-less.

3. In decay mode: $\mu^+ \to e^+e^-e^+\nu_e\overline{\nu}_\mu$, e^+e^- that form heavy LMC with positive TME penetrate the deep PW through QT and flee to free region with the same positive TME. This causes the "dismembering". DS predicts impossibility for some unconfirmed decay modes. For example, DS predicts that $\mu^+ \to e^+ + \gamma$ can never happen, for μ^+ is not an excited state of e^+.

To produce heavy LMC in muon, needed energy may come from collision such as $e^+ + e^- \to \mu^+ + \mu^-$ or from $u\overline{u}$-annihilation such as seen in the pion-decay $\pi^+ \to \mu^+ + \nu_\mu$: $u\overline{u}$-annihilation in $\pi^+(u\overline{d} = u\overline{u}e^+\nu_e)$ produces two pairs (heavy LMC e^+e^- and $\nu_\mu\overline{\nu}_\mu$) with $\overline{\nu}_\mu$ and relics $e^+\nu_e$ (left over after $u\overline{u}$-annihilation) captured by LMC e^+e^- to form μ^+ and ν_μ flying away.

We have seen that the <u>1st application</u> of TME theorem leads to an **interpretation** of pions' **<u>masses</u>** with heavy u-mass, serving as example of mass of composite particle smaller than mass-sum of its constituent particles. Large *u*-mass explains MMs of nucleons. DS with OSOD proves small d-electrons' contributions to MMs of nucleons. The said 1st application is powerful enough to explain masses of pions and nucleons and nucleons' MMs simultaneously and half-quantitatively.

Now the 2nd application of TME theorem yields **sample interpretation** of *heavy* LMC e^+e^- & **muon-mass** in terms of DS $\mu^+/\mu^- = e^+e^-e^+\nu_e\bar{\nu}_\mu / e^-e^+e^-\bar{\nu}_e\nu_\mu$ and with MF-binding turned on ***without OSTP***, as an example of **RCMI (reversed composition mass inequality)**, where a composite particle is heavier than the mass-sum of all its constituent particles.

< **LMC** e^+e^- and **Muon-Mass Eqn** >

DS-leptons in muons and pions are in ground states, only making very small contributions to their masses. Then $m(LMC\ \ e^+e^-) = m_\mu$ is accurate enough in our half-quantitative treatment of the MB system. Heaviness of muons is due to heavy LMC e^+e^- . Heavy HMC $u\bar{u}$ provides enough energy released after annihilation. Such large positive TME can be a reality for composite muons that live long enough to leave long tracks. Settings, eqs, and results for LMC e^+e^- equal-mass 2B-system can be obtained from their counterparts in pion 2B-system dynamics with the following replacements

(19) $$u\bar{u} \to e^+e^-,\ \ m_u \to m_e,\ \ k_S,\ k_u,\ k_{MF} \to 1,\ \ \eta_C^{st} \to \eta_C^{em} = 1,$$

$$k' = k_S / k_{MF} \to 1,\ \ f_C^{st} \to f_C^{em},\ \ m_{\pi^0} \to m_\mu \approx m(LMC\ \ e^+e^-).$$

We obtain TME of LMC e^+e^- 2B-system without OSTP and muon-mass eqn:

(20a) $$T_{ME} = m_e c^2(\sqrt{k^2+4} - k - 2 + 1171.32908\xi^3),$$

$$k = \frac{\xi}{4}(1 + 3513.9872\xi^2),\ \ \xi \equiv \frac{r_{sa}}{r},\ \ r_{sa} \equiv \frac{e^2}{m_e c^2}$$

$$m_\mu \approx m(LMC\ \ e^+e^-) = m_e(\sqrt{k^2+4} - k + 1171.32908\xi^3)$$

The muon-mass eqn above is unsolvable non-linear eqn of 1 independent variable r. Solution exists but is unknown algebraically even if we know mass of LMC e^+e^- exactly. Sample setting and numerical solution are

(20b) $$m(LMC\ \ e^+e^-) = 105.12\,MeV/c^2,\ \ \xi = 0.56,\ \ k = 154.418,$$

$$\gamma = 154.425,\ \ r = r_{sa}/\xi = 5.03\cdot 10^{-13}\,cm$$

From (6) one can see: MD btw e^+e^- is about $3\cdot 10^{-11}cm$, much longer than the binding size of LMC e^+e^- shown above. IPE-binding size of e^+e^- 2B-system with **MF turned off** equals the size of a **positronium** about $10^{-8}cm$. At such distance, much longer than the MD btw e^+e^-, MF is almost zero. It seems impossible for such weak MF to cause a bound state of e^+e^- 2B-system with a size btw the IPE-size and the MD. On the contrary, much heavier u-quark's IPE-binding size is much smaller. Besides, DBR is shorter than SBR due to small factor $\sqrt{1-\alpha_s^2}$ in St force case. At distances btw the MD with suppressed MF and the IPE-binding distance of $u\bar{u}$ 2B-system, MF still has large magnitude though suppressed. It may cause a bound state with a binding size btw the IPE-binding distance and the very short MD btw $u\bar{u}$ with OSTP.

Note: (20a) can be used to calculate positronium's mass as ground state mass of 2B-system e^+e^- as well as mass of LMC e^+e^- in tau, which is in higher excitation than LMC e^+e^- in muon.

< **Positronium** e^+e^- and **Positronium-Mass Eqn** >

Although at atomic distances MF btw e^+e^- is not exactly zero, we still have very accurate result after we ignore MF to get the binding size: 2B-system e^+e^- bound by Coulombian force with MF ignored is of binding size $2r_b$ proved by rigorous solution of equal-mass 2B-system's Schrödinger eqn. Then we set radius r of classical circular motion in classical imitation to be equal to Bohr radius ($r = r_b$). From (20a) we immediately obtain positronium mass eqn:

(20c)
$$\xi \equiv \frac{r_{sa}}{r} = \frac{r_{sa}}{r_b} = 5.325\cdot 10^{-5} \quad (r_{sa} \equiv \frac{e^2}{m_e c^2})$$

$$k = \frac{\xi}{4}(1+3513.9872\xi^2) \sim 10^{-5},$$

$$m(Positronium \quad e^+e^-) = m_e(\sqrt{k^2+4} - k + 1171.32908\xi^3) = 2m_e$$

Accurate empirical measurement of the energies of 2 photons, produced as a positronium decays into them, proves above accurate mass value. Each photon's energy equals 0.511 MeV if viewed in the CMF (center-of-mass frame).

< **LMC** e^+e^- in **Higher** Excitation and **Tau-Mass Eqn** >

At atomic distances, e^+e^- in a positronium are bound by Coulombian force. MF only causes hyperfine effects. Such IPE-binding yields larger binding sizes for higher excitations. In small SA spaces, MF plays crucial roles, even dominates Coulomb force. MF-binding features smaller binding size for higher excitation, which can be seen clearly in the proof of TME theorem, together with amazing positive TME at binding sizes less than 0TMED. LMC in muon and tau can be denoted as $(e^+e^-)_1$ and $(e^+e^-)_2$ respectively. From (26a) we make the following sample r-setting and get mass eqs for tau's LMC e^+e^- and for tau.

$$\text{(20d)} \quad \xi \equiv \frac{r_{sa}}{r} = 1.42, \ (r_{sa} \equiv \frac{e^2}{m_e c^2}, \ r = \frac{r_{sa}}{1.42} = 1.984 \cdot 10^{-13} cm)$$

$$k = \frac{\xi}{4}(1 + 3513.9872\xi^2) = 2515.744,$$

$$m_\tau \approx m(Tau's \ \ LMC \ \ e^+e^-) = m_e(\sqrt{k^2+4} - k + 1171.32908\xi^3)$$

$$= 1713.82 \ \ MeV / c^2$$

Note: Masses of positronium, muon, and tau are explained via classical circular motion imitation once dynamical effect of the resultant force of Coulombian force and MF is calculated and DS of muons and taus puts light on their physical pictures. Classical imitation brings about perfect match btw QM and CI only if quantum wave eqn is solved to provide with the binding size for us to determine r-setting or r-inputting in CI. Since quantum eqs in MF field are not solved and may be unsolvable, there is no guarantee for sample r-settings in muon-mass eqn and tau-mass eqn to be alright.

Remark: We can get accurate agreement with empirical values of masses of pions and muons by more carefully inputting values of some parameters. But that does not mean we can get accurate quantitative results. By its very nature, MB-systems in SA world are not solvable and 2B-systems with MFs are not solved and may be unsolvable too. Accurate and quantitative theoretical results of many important things in SA world will remain unknown and uncertain in the future. Only thing we and future generations can do is to get better ideas and pictures of reality and get rid of wrong ideas and wrong pictures.

In decay mode $\pi^+ \to \mu^+ + \nu_\mu$, HMC $u\bar{u}$ in π^+ annihilates to produce two PAPs: heavy LMC e^+e^- and $\nu_\mu\bar{\nu}_\mu$. Immediately, LMC e^+e^- captures relics $e^+\nu_e$ and $\bar{\nu}_\mu$ to form $\mu^+ = e^+e^-e^+\nu_e\bar{\nu}_\mu$. Even if we ignore NFMs to simplify problems, can we explain how LMC e^+e^- captures and binds DS-positron during microscopically long lifetime of muon? A sample ZL and SC of e^+e^- e^+ 3B-system is like this:

Before annihilation, $u\bar{u}$ in π^+ had antiparallel spins and parallel spin MMs to join MFs to bind the DS-positron with its spin MM antiparallel to net spin MM of $u\bar{u}$. OSTP of $u\bar{u}$ greatly suppresses MF and St MF btw $u\bar{u}$. Once the HMC $u\bar{u}$ has annihilated and LMC e^+e^- is born, e^+, originally in π^+, is captured by LMC, relocating via QT in R2 at e^+ in LMC. Both 2 positrons have parallel spins and spin MMs, which are antiparallel to that of the electron in LMC. Then Electron in LMC MF-binds 2 positrons which are in R1 at electron while 2 positrons MF-bind each other which are in R2 at each other. Roughly, Attractive MF in R1 causes circular motion while attractive MF in R2 causes oscillation along MM-line. Furious Coulombian repulsion btw 2 positrons within SA short distances is suppressed by attractive MF at all distances shorter than MD and they never annihilate during oscillation, making muon live long time. Such 3B-system (e^+ plus e^+e^- in Composite μ^+ with NFMs omitted) cannot be reduced to 2B-system or 1B-system as an approximation. But for 3B-system $u\bar{u}$ plus e^+ in π^+ with v_e omitted, we do have 2B-approximation for HMC $u\bar{u}$ with perturbation from light e^+ omitted and 1B-approximation for e^+ in π^+ where heavy HMC $u\bar{u}$ is treated as a single tiny particle with net spin MM and at rest and its physical existence can be replaced by effective MF field produced by its net spin MM. How to calculate TME of above said e^+e^- e^+ 3B-system is unknown. But we've just calculated TME of e^+e^- 2B-system produced in SA space before e^+ steps in.

Note: Before $u\bar{u}$ -annihilation, e^+ made small contribution to TME and mass of π^+ due to its small ME which is reserved as e^+ relocates into LMC e^+e^- via QT. QT preserves ME. Thus, once LMC e^+e^- captures e^+ to form 3B-system e^+e^- e^+, its mass and TME are close to what we have calculated for 2B-system LMC e^+e^-.

As e^+e^- circling each other, local electric currents, associated with opposite orbital velocities of e^+e^- , are always in parallel directions causing attractive magnetic forces if retardation is omitted. We have ignored such mag-forces so far. It has the same magnitude as Coulomb force due to near-c speed of large γ -value. MF-domination at short distances permits such neglect.

In <u>**positronium**</u> e^+e^- case, IC of e^+e^- <u>**validates**</u> solution of Dirac eqn for 2B-system in Coulomb field of interaction with mag-forces (large at short distances) ignored at atomic distances. Such solution matches empirical facts and data so well since wave function as part of such solution reveals very small probabilities for distance btw e^+e^- to be short enough like SA short distances.

Since neutrinos and antineutrinos only exert Wk forces on other particles, their existence inside SA composite particles, motions, spins, and Wk MMs have little influence on d-electrons and u-quarks. To figure out motions and spin directions of u-quarks and d-electrons, we can ignore DS-neutrinos and DS-antineutrinos.

Bohr's model was soon proved by Schrödinger and Dirac via theoretical and quantitative results, in good agreement with experimental data. However one can neither prove nor disprove DS by solving quantum eqs quantitatively due to the **MMD** (many-body and mag-force difficulty). Qualitatively/half-quantitatively speaking, there're no phenomena DS with MFs has failed to explain. Some facts do not depend on details about MB systems and denamics of MFs. Then DS can explain them even exactly.

DS is powerful enough to explain hundreds of middle stage particles and thousands of decay modes, showing only four types of DS processes, called **4TDSP**.

< 4TDSP >

1. DSE (**DS-E**xcitation) and **DSDE** (**DS-De-E**xcitation)

DS-leptons, sometimes uuu-**BC**, $\bar{u}\bar{u}\bar{u}$-**AntiBC**, $u\bar{u}$-HMC, or e^+e^--LMC get excited in HE collisions. DSE and DSDE always happen quickly **<u>without a single exception</u>**.
<u>Examples</u>: Target nucleon gets excited turning itself into a short-lived baryon then SA DS-de-excitation quickly happens. Neutrons slowly decay for d-quarks and neutrons themselves are not in DS-excitation.
DSLs as DS-contents add leptonic excitations to entire collection of SA excitations. MFs are θ-depending and non-spherosymmetric. Complicated excitations of MF-bound MB-systems with unknown details nicely and qualitatively explain empirical large collection of hadrons.

2. PAPP (**PAP P**roduction) and **PAPA** (**PAP A**nnihilation)

PAPP consumes energy released in c*ollision*, SA *de-excitation*, or *PAPA*. SA PAPA such as annihilations of HMC $u\bar{u}$ and LMC e^+e^-, either in ground or excited states, may produce real photons, virtual carriers of forces absorbed by nearby DSLs and hadrons to gain KEs and flee away, or PAP(s), leading to radiative decay modes, **KC** (**k**inetic **c**hannel) of decay, and **PPC** (**p**article **p**roduction **c**hannel) of decay. (See decay channels later)
<u>Examples</u>: e^+e^- collision produces PAPs needed to form $\mu^+\mu^-$ or $\tau^+\tau^-$. Mesons, muons, and taus show up as middle-stage particles in events of HE collisions due to the same type of DS-processes: PAPP with HMC $u\bar{u}$ and/or heavy LMC e^+e^- produced, w/o non-heavy PAPs e^+e^-, $\nu_e\bar{\nu}_e$, $\nu_\mu\bar{\nu}_\mu$, and so forth. All non-baryonic middle-stage particles (mesons, muons, and taus) are made out of produced elementary PAPs via the 3rd type of DS-processes described soon below.
De-excitation of excited d-leptons in composite s-quark in baryon Σ^+ produces bound PAP $u\bar{u}$ (HMC) to form π^0 with s-quark returning to d-quark and Σ^+ (*uus*) to proton (*uud*). This is DS-way decay mode $\Sigma^+ \to p\pi^0$ is going on. It's so natural and simple.

Annihilation of HMC $u\bar{u}$ in π^+ may produce 2 elementary PAPs: heavy LMC e^+e^- and $\nu_\mu\bar{\nu}_\mu$, followed by other DS-processes described before and below to cause decay mode $\pi^+ \to \mu^+ + \nu_\mu$.

Annihilation of excited LMC e^+e^- in tau may produce one or more HMC $u\bar{u}$, leading to production of meson(s) in hadronic decay modes .

3. PQTR (**P**article **Q**uantum **T**unnel **R**elocating) with SA particle(s) nearby in the same SA small space to capture relocating SA elementary particle(s).
PQTR creates MSPs and FSPs out from ISPs and from produced elementary PAPs so that the set of MSPs and the set of FSPs are usually not the original ones plus the produced elementary PAP(s).
<u>Examples</u>: The d-leptons emitted from a neutron relocate into a proton nearby

bound in the same nucleus, changing p/n into each other and causing stability seen on stability island. Many stable elements contain more neutrons than protons for neutrons do not decay at the same time.

HMC $u\bar{u}$ in π^+ annihilate leaving relics $e^+\nu_e$ and producing LMC e^+e^- plus another PAP $\nu_\mu\bar{\nu}_\mu$ then $e^+\nu_e$ and $\bar{\nu}_\mu$ relocate in LMC e^+e^- via QT to form muon μ^+ $(=e^+e^-e^+\nu_e\bar{\nu}_\mu)$, with ν_μ flying away so that decay mode $\pi^+ \to \mu^+ + \nu_\mu$ is said to occur.

4. SA **PR** (**P**hoton **R**adiation) Followed by SA Photoelectric-Encounters.

SA PR is an important channel for DSDE and PAPA to release energy. Spectrum-less SA PR and atomic PR with spectrum bring us to a unified picture for **QS** (**q**uantum **s**cattering) of photon by electric charge.
Examples: $J/\psi(c\bar{c}/J=1) \to \eta_c(c\bar{c}/J=0)+\gamma$. Small difference in mass btw η_c and J/ψ is not enough to produce HMC $u\bar{u}$ or π^0. As long as η_c appears as a MSP in J/ψ decay, no other meson can appear together with η_c. Too many other examples of PR after DSDE and PAPA seen among observed thousands of decay modes. Physicists call decay modes with PR ***radiative*** decay modes.

Note: Some DSLs among FSPs originally were constituent particles of initial SA composite particles while some of them were produced in middle stages as members of produced PAPs. Some were originally bound, even with negative MEs. To change to free FSPs with positive MEs (or with larger positive MEs than initial positive values, proper energies must be received. Some needed energies come from SA QS with photons in SA PR. Photons in PR are not virtual photons. Real photons in SA PR are born within small SA spaces loaded with u-quarks and DS electrons. QS is inevitable within crowded SA spaces. But needed energies may also be gained by absorbing virtual force carriers that are produced in PAPA as described below:

AVC (**a**nnihilation **v**irtual **c**hannel) may be another important channel for SA PAPA to release energy and to transfer energy to nearby constituent particles. AVC is seen when SA PAP annihilates into **virtual** γ and/or **virtual** Z-boson to be absorbed by SA elementary particle(s) nearby to gain KE and flee away, where virtual Z-boson is the only one NFM within the same small SA space can absorb. See examples of AVC later.

Energy released in SA de-excitation may be enough to produce 1, 2, or 3 bound $u\bar{u}$ HMCs plus some other PAP(s). PAPP channel in **ERM** (**e**nergy **r**elease **m**echanism) may suppress PR channel leading to small branch ratios of radiative decay modes. ERM claims: *energy released in any classical transition or quantum transition from higher to lower level or in any particle-antiparticle (not valence pair) annihilation may take form of one or many **particle-antiparticle** pairs, as long as permitted by energy conservation.*

Classical transition occurs when free incident particles lose or gain KEs as they pass by without colliding. In SA world, PAPPs may suppress PR-channel due to VI, which means that PAPP has priority to consume impact energy over knocking and production of real photon. ERM seen in atomic de-excitations seems to have PR-channel only. PAPs $u\bar{u}$ and e^+e^- can't be born in atomic de-excitations due to energy conservation. But unified ERM does predict radiation of low-energy NFMs in de-excitations of atoms. Such NFMs elude detection. May be, two hundred years later, experiment will be able to tell if unified ERM covers both atomic and SA world. If modern tech allows experimenters to confirm a single photon emission or absence of photon emission from a single hydrogen hit by a single incident photon with no chance to get excited from thermo-collision with other particles than a single incident photon, atomic emission of low energy NFM can be indirectly proved or disproved in 100 years.

Energies released in all SA PAPAs and SA *de-excitations*, which happed in decays, are not large enough to produce radiations of very heavy real gluons and real Z-bosons. Experimenters've proved their absence among FSPs of *decays*. If a PAP is formed by a free particle and its antiparticle in a head-on collision, such PAPAs can produce virtual gluons as seen in 2-Jet and 3-Jet events. Photons never decay. If real gluons & real Z-bosons were confirmed to decay, it would be too bizarre and we might need to rethink essence of strong and weak forces. A simple fact that no single stable free gluon has ever been found in all super-HE and HE collisions recorded in cosmic showers and accelerators in past 100 years may remind us of an unexpected fact: VI may cause the existence of a **NLVUC** that forbids any transferred energy in a single quantum process higher than NLVUC in any collision or PAP annihilation. If all the collision events recorded in past 100 years are indeed showing a NLVUC that is lower than the energy needed to produce a real gluon, then real gluons will never show up.

SA PR may be followed immediately by QS btw radiated real photon(s) and nearby elementary constituent particle(s).

Examples of AVC:

In decay $\pi^+(u\bar{d} = u\bar{u}e^+\nu_e) \to e^+\nu_e$ (0.0123 %), $u\bar{u}$ annihilate into virtual force carrier(s) to be absorbed by $e^+\nu_e$ to gain energy and flee away. Due to VI, this KC is greatly suppressed by the following PPC:

$$\pi^+(u\bar{d} = u\bar{u}e^+\nu_e) \to e^+e^-\nu_\mu\bar{\nu}_\mu e^+\nu_e \to \mu^+(e^+e^-e^+\nu_e\bar{\nu}_\mu) + \nu_\mu \quad (99.98770\ \%)$$

In decay mode $\mu^+(e^+e^-e^+\nu_e\bar{\nu}_\mu) \to e^+\nu_e\bar{\nu}_\mu$ (~100%), LMC e^+e^- annihilate into virtual force carrier(s) to be absorbed by above DS-leptons nearby. This KC of decay could dominate for PPC (LMC $e^+e^- \to$ HMC $u\bar{u}$) is forbidden in muon decay by energy conservation.

We'll soon see examples of how to use 4TDSP to explain SA events (hundreds of sorts of MSPs and thousands of decay modes).

Remark: Within SA composite particles, MB-interactions in crowded tiny space may lead to production of many virtual force carriers in a single DS PAPA or a single DS de-excitation. Energy-momentum distribution of FSPs obtained via large number of decay events of the **same mode** may indicate such production of multiple virtual force carriers. Please note, SA PR events may look like AVC events if radiated photons lose too much of energies eluding detection after SA photoelectric encounters.

< DS-Explanation of Tau Decay Modes >

Samples of how DS with 4TDSP easily explains tau ($\tau^- = (e^+e^-)_2 e^- \bar{\nu}_e \nu_\tau$) decay modes are shown below:

In decay mode $\tau^- \to \mu^- \bar{\nu}_\mu \nu_\tau$ (17.39%), de-excitation of tau's LMC $(e^+e^-)_2$ turns it into muon's LMC $(e^+e^-)_1$, releasing energy in the form of $\nu_\mu \bar{\nu}_\mu$ and with ν_τ flying away. For some reason ($(e^+e^-)_1$ is unable to bind ν_τ OR ν_τ absorbs virtual Z-boson released together with $\nu_\mu \bar{\nu}_\mu$ gaining enough energy to flee to free region). At the same time, ν_μ relocates into $(e^+e^-)_1$ so that they together with $e^- \bar{\nu}_e$ in tau form μ^- and let $\bar{\nu}_\mu$ go.

In decay mode $\tau^- \to e^- \bar{\nu}_e \nu_\tau$ (17.82%), $(e^+e^-)_2$ annihilates into virtual force carriers absorbed by relic DSLs in τ^- and sending them free.

In decay mode $\tau^- \to \pi^- \pi^0 \nu_\tau$ (25.51%), $(e^+e^-)_2$ annihilates into 2 $u\bar{u}$ -pairs. One captures DSLs $e^- \bar{\nu}_e$ in tau, forming π^- while another $u\bar{u}$ pair becomes π^0. At the same time, ν_τ, losing attraction from LMC $(e^+e^-)_2$, goes away.

In decay mode $\tau^- \to \pi^- \pi^+ \pi^- \nu_\tau$ (9.31%), $(e^+e^-)_2$ annihilates into 3 $u\bar{u}$ -pairs, one e^+e^- pair, and one $\nu_e \bar{\nu}_e$ pair. Produced DSLs $e^- \bar{\nu}_e$ and $e^+ \nu_e$ relocate in 2 $u\bar{u}$ -pairs turning them into $\pi^- \pi^+$ respectively while relic DSLs $e^- \bar{\nu}_e$ left after $(e^+e^-)_2$ -annihilation relocate in 3rd $u\bar{u}$ -pair turning it into another π^-. Once again, ν_τ, losing attraction from LMC $(e^+e^-)_2$, goes away.

In decay mode $\tau^- \to \pi^- \nu_\tau$ (10.91%), $(e^+e^-)_2$ annihilates into a $u\bar{u}$ -pair that captures $e^- \bar{\nu}_e$ to form π^- and let ν_τ go.

< **NFMs** (Neutrino Family Members) >

Among electrons, muons, and taus, only electrons are elementary, according to DS. The author does not know if among NFMs $\{\nu_l\}$ only ν_e is elementary. Such NFM-issue, including neutrino oscillations, is open for future study.

< Three **Non-De-Excitation** DS-Channels of Decay: **KC**, **PPC**, **QTC** >

Pions $\pi^0 \pi^\pm$ are in ground states and must decay via non-de-excitation channels. Somehow one may treat LMC e^+e^- in muon as in deep SA MF-excitation state while positronium e^+e^- as in ground state with MF playing small hyperfine effect at atomic distances. LMC e^+e^- in muon is produced in deep SA space with e^+e^- so close to each other that annihilation channel dominates. Please remember, at distances shorter than 0TMED, the higher the DSE, the larger the positive TME and the smaller the binding size.

Three types of non-de-excitation DS-cannels of decay are

1. KC (**k**inetic **c**hannel) of PAPA: PAP may annihilate into ***real*** carrier(s) of force (so far only real photons are seen) and/or timelike ***virtual*** force carrier(s) for some ***bound*** particle(s) to absorb, gain KE, change ME to positive, and become free FSP(s).

Remark: Energy conservation forbids free elementary particle to absorb or emit any timelike ***real*** force carrier and only to gain or lose KE. SR and QFT without LV proved long time ago: A free electron cannot absorb nor emit massless real photon to gain or lose KE for conservation of 4-momentum forbids such event. Theories with LV and without LV reach the same conclusion in this issue: A free elementary particle can't absorb nor emit a real force carrier, as experiment shows. Elastic scattering, however, allows an elementary particle other than a force carrier to absorb or emit a ***virtual*** force carrier and change its 4-momentum when another elementary particle is present to emit or absorb the said virtual force carrier. Transferred 4-momentum then must be **spacelike**. Each force carrier is either massless in massless QFT or massive in massive QFT. The transferred p^{μ}, serving as a label for an internal line in Feynman diagram is forbidden to be carried by any individual force carrier. It's just an effect of virtual force carrier exchange. QFT explicitly shows: p^{μ} of each individual virtual force carrier appear in 3-dim propagator. The 4-dim version of propagator contains a mathematical parameter irrelevant to the energy carried by any individual force carrier. Use of such mathematical parameter is due to the application of integration of complex function. But it's not a physical quantity though it has energy dimension. This view of QFT immediately explains why we see examples of neutral currents with transferred energy much smaller than Z-boson's rest energy.

Photoelectric effect proves that a bound electron can absorb a real photon to change its ME from negative to positive and become free. Atomic physics and QM proves: an electron bound in atom can absorb a real photon to jump to a higher level while an excited electron bound in atom can emit a real photon to return to ground state or a lower-level state.

2. PPC (**p**air **p**roduction **c**hannel) of PAPA: PAP may annihilate and turn itself into another PAP or more than one new PAPs. Note: PPC of SA de-excitation is an example of PAPP. But PPC of PAPA is a non-de-excitation channel and is another example of PAPP.

3. QTC: In many cases, MFs create bound or quasi-bound states for DSLs with positive ME at all times or sometimes so that they can penetrate deep PW through QT fleeing into free regions with the same positive ME. QTC may be dominating or be dominated by other channels of decay.

< Sharp **Contrast** btw **Muon**'s Decay and **Charged Pion**'s Decay >

PPC is ***shut*** in muon-decay so that it cannot produce mesons, due to energy conservation. The energy released in muon-LMC e^+e^- annihilation is not enough to produce even the lightest $u\bar{u}$ 2B bound system. It's open in tau-decay for tau-LMC (e^+e^-)$_2$ is much heavier.
PPC is ***open*** and ***dominating*** in $\pi^\pm$ -decay. (SDL)
KC ***dominates*** in $\mu^\pm$ -decay for **PPC** is ***shot*** in $\mu^\pm$ -decay. (SDL)
KC is ***suppressed*** in $\pi^\pm$ -decay for **PPC**'s ***open*** and ***dominating*** in $\pi^\pm$ -decay.
QTC is ***suppressed*** in $\mu^\pm$ -decay: It has little chance ($\sim 3.4\cdot 10^{-5}$) for LMC e^+e^- bound and with positive TME to penetrate deep PW and become free with released $e^+\nu_e\bar{\nu}_\mu$ becoming free too and finishing the muon-decay. This rare decay mode of muon is a DS-process that belongs to one of the 4TDSP: PQTR.
QTC for DS-leptons is open but ***suppressed*** in $\pi^\pm$ -decay: $\pi^+ \to e^+\nu_e\pi^0$ and $\pi^- \to e^-\bar{\nu}_e\pi^0$ $(1.036\cdot 10^{-6})$.
QTC for HMC $u\bar{u}$ is ***shut*** in any meson decay for $u\bar{u}$ are bound with ME of each one negative enough at all times. Since $u\bar{u}$ in any pion tilt their spins to turn off MFs nearly completely, IPE binding dominates leading to negative TME of $u\bar{u}$ in any pion. Heavy u-mass required by stable BC and proton as well as by small MMs of nucleons leads to very large energies needed to produce such inward exited HMC $u\bar{u}$ with positive TME. Due to VI and/or NLVUC, such particles, just like very heavy real gluons, are forbidden to produce. Thus, no free quarks can appear among FSPs via QTC.
QTC for DSLs in $\pi^\pm$ is narrowly open, suppressed by PPC of $u\bar{u}$ -annihilation.
QTC in ***muon*** decays, leading to free e^+e^- among other FSPs is ***open*** but ***narrow***, suppressed by LMC e^+e^- annihilation channel.
PPC of $u\bar{u}$ -annihilation in ***pion*** decay is ***allowed*** by energy conservation. Once PPC is open, it immediately dominates **KC** of $u\bar{u}$ -annihilation due to **SAVI** with **PAPP** having priority to consume energies released in SA de-excitations, collisions, and annihilations as effects of virtual force carrier exchanges. Inside π^+ , HMC $u\bar{u}$ has very small chance (~0.013%) to annihilate into virtual photon and/or Z-boson absorbed by e^+ and/or ν_e . Charged pion's decay mode via KC $\pi^+ \to e^+ + \nu_e$ is rare.

Now let's use abbreviations **PiD** (pion decay), PiDKC, PiDPPC, PiDQTC, **MuD** (muon decay), MuDKC, MuDPPC, MuDQTC to write π^+/μ^+ decay modes (channels) in terms of DS:

PiD

$$\pi^+(u\bar{u}e^+\nu_e) \to \text{the following}$$

PiD**KC**, rare 0.0123% :

$$\pi^+(u\bar{u}e^+\nu_e) \to e^+\nu_e,\ (u\bar{u} \to \text{ virtual } \gamma/Z \text{ absorbed by } e^+\nu_e),$$

PiD**PPC**, dominating 99.99%:

$\pi^+(u\bar{u}e^+\nu_e) \rightarrow \mu^+(e^+e^-e^+\nu_e\bar{\nu}_\mu)\nu_\mu$, ($u\bar{u} \rightarrow$ virtual $\gamma/Z \rightarrow e^+e^-\nu_\mu\bar{\nu}_\mu$),

Pi**DQTC** (for DSLs), rare $1.036 \cdot 10^{-6}$:

$\pi^+(u\bar{u}e^+\nu_e) \rightarrow u\bar{u}e^+\nu_e$, (Bound $u\bar{u}e^+\nu_e$ →Bound $u\bar{u}$ (π^0) + Free $e^+\nu_e$

MuD

$\mu^+(e^+e^-e^+\nu_e\bar{\nu}_\mu) \rightarrow$ the following

Mu**DKC**, Dominating ~100%:

$\mu^+(e^+e^-e^+\nu_e\bar{\nu}_\mu) \rightarrow e^+\nu_e\bar{\nu}_\mu$, ($e^+e^- \rightarrow$ Virtual γ/Z Absorbed by $e^+\nu_e\bar{\nu}_\mu$),

Mu**DPPC,** Forbidden 0%:

$\mu^+(e^+e^-e^+\nu_e\bar{\nu}_\mu) \rightarrow \pi^0(u\bar{u})e^+\nu_e\bar{\nu}_\mu$, (LMC $e^+e^- \rightarrow$ Virtual $\gamma/Z \rightarrow u\bar{u}$),

Mu**DQTC**, Rare $\sim 3.4 \cdot 10^{-3}$ %:

$\mu^+(e^+e^-e^+\nu_e\bar{\nu}_\mu) \rightarrow e^+e^-e^+\nu_e\bar{\nu}_\mu$, (Bound $e^+e^-e^+\nu_e\bar{\nu}_\mu$ →Free $e^+e^-e^+\nu_e\bar{\nu}_\mu$)

< **NLVUC** (Non-LV Ultraviolet Cutoff) >

If an elementary particle, free or bound, gets struck in a collision and changes (usually increases) its energy, we call such process an **ES** (**elastic scattering**). Please note, in many cases an incident particle may reduce its energy in an amount larger or much larger than the energy gained by the struck elementary particle while many particles are produced in the collision. Thus, an ES event may be a sub-event coexisting with many particle-production sub-events in a DI collision. The new definition of ES includes the classical ES without particle production as a special case. As the energies of incident particles get higher and higher to the largest energy ever recorded in prime cosmic rays much higher than what the largest man-made accelerators can reach, do the energies transferred in ES sub-events have an upper limit? If, yes, we call the upper limit or the upper bound NLVUC. To prove or disprove the existence of NLVUC and get its magnitude empirically, no new experiment is needed. Billions of photos that have recorded collision events in cosmic showers and accelerators all over the world since the beginning of the 20th century are enough. We expect that physicists can verify it as soon as possible. No tracks of individual quarks can be seen. By checking the tracks of electrons before and after the ES and measuring the transferred energies, NLVUC will be verified and its value nearly equals the ***largest energy of non-incident electrons flying out from collision centers***. Those collision events should be reviewed by using all the photos taken since the beginning of cosmic shower recordings and accelerator collision recordings with impact energies forming a nice and natural spectrum from 0.1Gev, 1Gev, 10Gev, 100Gev, 1000Gev, …, 10^9Gev and beyond.

VI, as the reason of the existence of NLVUC, is seen every day in DI Scattering/

collision. Energy-consuming in PAPP breaks down collision process into many sub-processes of lower energy scales that make effective couplings in LH-model perturbative or, at least, not terribly non-perturbative. If cosmic shower events and HE collisions in accelerators have proved NLVUC for 4-momentum transfers to ***real*** elementary particles unchanged before and after the transfers, the NLVUC must also be a reality for ***virtual*** force carriers responsible for any t-channel 2nd-order Feynman diagram, whether or not it is part of higher-order Feynman diagram. If NLVUC exists, it forbids energy transfer larger than it in each ***single*** quantum energy transfer. Every quantum energy transfer satisfies uncertainty principle. Every classical scattering is formed by super-large number of quantum scatterings, allowing much longer period of time for small amount of quantum energy transferred. Super-large number of quantum energy transfers leads to classical and gradual energy transfers. If NLVUC is about 1 GeV, repeated and gradual energy transfers taking place in accelerators can easily produce HE particles with KEs beyond or far beyond NLVUC. If universe does not have super-accelerators, where did super-HE protons in prime cosmic rays come from? See < Origin of Super-HE Protons in Prime Cosmic Rays > on page 7 for details.

< DS-Explanation of **FQA** Fact >

DSE of FQA fact is seen on pages 42-43 and on pages 79, 261 after the HMC $u\bar{u}$ Ground State Table.

< **Slowly** and **Fast** Decaying Mesons >

SDMs (Slowly **D**ecaying **M**esons) with lifetime ranging from $10^{-8}s$ to $10^{-13}s$ and **FDMs** (**F**ast **D**ecaying **M**esons) with lifetime ranging from $10^{-17}s$ to $10^{-24}s$ can be explained via DS;

Without exception, FDMs are either mesons with J=1 label showing orbital excitation of their HMC $u\bar{u}$ always returning to lower level very quickly like baryons of J=1 label (see discussion of long lived Ω later). FDMs with J=1 Label are $\rho^{\pm},\rho^{0},\omega,\phi,J/\psi,Y,K^{*\pm},K^{*0}$, and all D*-mesons. In view of DS, 'vector mesons' are FDMs for de-excitation of orbital excitation of HMC $u\bar{u}$ goes very quickly.

Note: Asterisk D-mesons and asterisk B-mesons are not determined for J-label while asterisk B-mesons' lifetimes are not yet measured. Please see **PDG** on Web. PDG refers to **Particle Data Group**.

FDMs with J=0 label $\pi^{0},\eta,\eta',\eta_{c}$. In view of DS, they are all DS-symmetric, lacking significant e^{-}-$\bar{u}$ or e^{+}-u electric repulsion to slow down $u\bar{u}$-annihilation. Such repulsions are seen clearly in *DS-asymmetric* mesons such as $\pi^{\pm}$.

Some FDMs with J=1 label are also DS-asymmetric similar to $\pi^{\pm}$, such as $\rho^{\pm}$. Pion and rho with the same non-zero charge even have the same quark content as well as the same DS-content. DS-asymmetric mesons decay fast if having J=1 label and slowly if having J=0 label. This is because excited HMC of

J=1 label quickly return to ground state to finish decay and to produce DS-asymmetric mesons of J=0 label. Further decay of DS-asymmetric mesons of J=0 label as middle stage particles is not counted as part of their decay process. The e^- -$\bar{u}$ or e^+-u repulsion that exists in all DS-asymmetric mesons is much weaker in the time period of de-excitation of HMC since excitation of HMC from J=0 to J=1 state and de-excitation of HMC from J=1 to J=0 state is sort of spin-flipping forth and backwards or some unknown de-excitation. It happens and finishes very quickly. The said repulsions have no way to slow down motion of de-excitation but they do slow down motion of annihilating $u\bar{u}$.

As an outstanding example, **d-electron** in DS-asymmetric π^- **slows down** $u\bar{u}$ **-annihilation** as shown below:

Due to OSTP of $u\bar{u}$ in a pion, the HMC $u\bar{u}$ can be treated as a tiny particle with a net MM. The d-electron in π^- mainly stays in R1 at the MM. The MPR of d-electron in π^- is almost centrosymmetric (not spherosymmetric) in places far away from $u\bar{u}$ while not centrosymmetric in places outside HMC $u\bar{u}$ but close to it. At places outside HMC $u\bar{u}$ but close to it, *electron cloud* is **denser** on u -side producing net Coulombian *attractive* force on u and *repulsive* force on $\bar{u}$ to resist and slow down their annihilation. Asymmetric distribution of electron cloud outside and near $u\bar{u}$ is due to a simple fact that u Coulomb-attracts *electron cloud* while $\bar{u}$ Coulomb-repels it and the norms of 2 forces differ significantly at places on or near $u\bar{u}$ -line and not too far from $u\bar{u}$.

< **Lifetimes** of **Muon** and **Positronium** 2s-Ps/*o*-Ps/*p*-Ps >

Longer lifetime of muon than positronium e^+e^- is also due to repulsion from DS-asymmetric content e^+ or e^- in muon. It slows down annihilation of LMC e^+e^- in muon, just like e^+ or e^- in charged pion slowing down HMC $u\bar{u}$ -annihilation. Pion-life is shorter than muon simply because HMC has much smaller size than LMC, the St force btw $u\bar{u}$ is much stronger than the e.m. force btw e^+e^-, and the Coulombian repulsion from e^+ or e^- has less effect to slow down annihilation of heavy $u\bar{u}$.

In sharp contrast, life-inequalities for Ps 2B-system e^+e^- in 2s-Ps/*o*-Ps/*p*-Ps states are

$$1.1\mu s > 1.42 \cdot 10^{-7} s > 1.244 \cdot 10^{-10} s$$

They are partially due to distance inequalities:

d(2s-Ps)>d(*o*-Ps)>d(*p*-Ps)

Ignition time of e^+e^--annihilation is sensitive to distance. Large distance in 2s state yields microscopically long life of 2s-Ps (1.1μs). Strangely, distance btw e^+e^- in LMC is much shorter than what's btw e^+e^- in *p*-Ps & *o*-Ps while muon-decay is even slower ($2.19 \cdot 10^{-6} s > 1.42 \cdot 10^{-7} s$). It happens solely because DS-asymmetric content (e^- in μ^- and e^+ in μ^+) slows down annihilation of LMC e^+e^-, just like DS-electron/DS-positron in π^-/π^+ slows down $u\bar{u}$ -annihilation of HMC $u\bar{u}$ making charged pions live longer than 0-pion.

From (6) we know that at distances about $10^{-8}cm$, MF btw e^{-}/e^{+} is about 10^{-5} times weaker than Coulomb force btw them, playing small role in spectrum issue and the issue of geometric shape of MPR. Ground state of positronium is close to S-state. At SA distances, MF and its non-spherosymmetric and θ-depending feature play important roles in binding and annihilation. In R1, MF btw e^{-} and e^{+} is attractive and repulsive for *o*-Ps (anti-parallel MMs with parallel spins) and *p*-Ps (parallel MMs with anti-parallel spins) respectively. In R2, MF btw e^{-} and e^{+} is repulsive and attractive for *o*-Ps and *p*-Ps respectively. $e^{+}e^{-}$-annihilation in *o*-Ps goes through R1 while it in *p*-Ps goes through R2. Shorter life and faster annihilation of *p*-Ps than *o*-Ps are due to larger attractive MF in R2.

For positronium, MF brings about measurable effect such as difference btw its two states: *p*-Ps and *o*-Ps. For hydrogen atom, MF seems to have caused negligible effects or hyperfine effects only. The reason is obvious: The magnitude of proton's MM is much less than that of the electron and positron in a positronium.

< Bizarre Long Life of Ω >

Quantum excited states of MF-bound particles are unknown & complicated. For instance, we don't know the details of non-radial (orbital) excitations. We adopt scenario of sample **spin flip**. Spin flip of $u\bar{u}$ 2B-system leads to **non-radial** excitation of HMC core $u\bar{u}$ with opposite spin directions of u and $\bar{u}$ and with $\bar{u}$ in V-vane at u, or equivalently, u in V-vane at $\bar{u}$. Such $u\bar{u}$ spin-flip takes place as a pion meson turns itself into a rho meson after receiving certain amount of energy and gets excited. It also occurs as $\eta/\eta'/\eta_c/\eta_b$ are turned into $\omega/\phi/J/\psi/\Upsilon$ mesons respectively after receiving certain energies and get excited. Spin flip may be quantized, somehow θ-depending now that MF field is θ-depending. We do not know quantitative details.

Spin flip of d-electron turns d-quark into s-quark by making it relocating from R1 to R2 at the u-quark or quark core that binds it. Common feature of **spin-flip UE** is <u>extremely fast de-excitation</u>, leading to such short lifetimes of vector mesons of $J=1$ label and baryons of $J=3/2$ label, ranging from 10^{-20} to 10^{-24} sec. The **only exception** is Ω's lifetime ($8.21\cdot10^{-11}$ sec). It may be an evidence that Ω has no *u*-core excitation while delta and Σ^{*} baryons with 3/2 J-label do have *u*-core excitation. In addition, evidence of J=3/2 label for Ω doesn't exist for composite particles don't spin and TAM of their constituent particles is not measurable, as carefully shown before in the topics < Spin/TAM **Misconception** and Empirical **Disability** to Measure OAM/TAM >. Symmetry, thought to be powerful enough to predict and explain existence of composite particles. An extension of such idea is to propose some symmetries, like *supersymmetry*, that has been used to predict existence of new particles never discovered. Among all the photos, taken since the beginning of particle physics, no single one was found to support the existence of such hypothetic particles. Symmetry in group theory may not be true mathematical principle of natural philosophy regarding composite particles and elementary particles. Dynamics and structures that have played back bone role in physics since Newton, Einstein, and founders of QM and QFT w/o **RC (radiative correction)** cannot be replaced with symmetries. Quark-model is most important milestone on the way to reveal SA secretes because of its quark content scenario so right everywhere in SA world for all hadrons (even DS cannot change quark content of any hadron though DS further considers deeper structure of d, s, c, b, and t quarks), not because of hypothetic relation btw the existence of hadrons and the group-symmetries. The well-known truth is: experimenters had easily discovered so many new hadrons far beyond the original meson octet, baryon octet, and baryon decuplet since long time ago.

Atomic de-excitations happen to atomic electrons, not to SA electrons (DS-electrons). By *SA de-excitations* we mean de-excitations of excited nuclei, excited DS-leptons in s, c, b, t quarks, excited q-cores (HMC/BC/Anti-BC) in meson/baryon/anti-baryon, possibly in t, b, c quarks also, and excited LMC in tau. Higher J-labels are for hadrons with non-radially excited q-cores. LMC in tau is radially excited. Non-radial excitations would make e^+e^- in LMC depart R1 at each other so that MF attraction may greatly be reduced, unable to bind light e^+e^- within SA small spaces. If experiment shows non-existence of non-radially excited state of muon, it also may imply that non-radially excited state of LMC e^+e^- cannot live long enough to serve as a LMC to stick some other leptons to form muon or tau. Non-radial excitation in R2 would force e^+e^- to oscillate and end up with annihilation immediately so that it could not be treated as particle. Some states of hadrons may be due to radially excited q-cores.

K and L shell electrons are atomic electrons. **Electron-capture** is not atomic nor SA de-excitation: Shell-electrons step down to low potential regions around u-quark cores of protons. It can be called **ASA** (**A**tomic-**S**ub-**A**tomic) de-excitation.

< Bizarre **Different Lifetimes** of K_L^0 and K_S^0 >

K_L^0 and K_S^0 have the same quark content $d\bar{s}$ hence the same DS content, but K_S^0 decays 571 times faster than K_L^0, even if the FSPs form **EXACTLY** the same set. Here is a sample DS-explanation of this bizarre phenomenon.

Decay mode $K_S^0 \to \pi^+\pi^-$ can be written in terms of quark model:

$$K_S^0(d\bar{s}) \to \pi^+(u\bar{d}) + \pi^-(d\bar{u})$$

In terms of DS, it's written

$$K_S^0(d\bar{s} = ue^-\bar{\nu}_e\bar{u}e^+\nu_e \uparrow) \to \pi^+(u\bar{d} = u\bar{u}e^+\nu_e) + \pi^-(d\bar{u} = ue^-\bar{\nu}_e\bar{u})$$

Let's see the details of this decay process. Excited anti-d-leptons $e^+\nu_e\uparrow$ in anti-s-quark step down to ground state, turning $\bar{s}$ into $\bar{d}$ and K_S^0 into $d\bar{d}$ system and releasing energy in the forms of a bound $u\bar{u}$ pair. The $d\bar{d}$ system may not exist as a particle if OSTP of HMC $u\bar{u}$ is real and causes severe MF-repulsion btw e^- in d and e^+ in $\bar{d}$. But before $d\bar{d}$ could be dismembered, $e^+\nu_e$ or $e^-\bar{\nu}_e$ in $d\bar{d}$ immediately **QT-relocate** in the produced bound $u\bar{u}$, turning $u\bar{u}$ into $\pi^+(u\bar{d} = u\bar{u}e^+\nu_e)$ or $\pi^-(d\bar{u} = ue^-\bar{\nu}_e\bar{u})$ and $d\bar{d}$ into $\pi^-(d\bar{u} = ue^-\bar{\nu}_e\bar{u})$ or $\pi^+(u\bar{d} = u\bar{u}e^+\nu_e)$. Fast **QT-relocating** makes $d\bar{d}$ middle stage system having no time to annihilate nor to leave recognizable track experimenters can measure or see. PAPP of $u\bar{u}$ bound pair belongs to strong interaction and goes faster than e.m. interaction process in different decay mode of neutral K-meson such as the one written below.

$$K_L^0 \to \pi^+ e^- \bar{\nu}_e \text{ or } K_L^0 \to \pi^- e^+ \nu_e$$

DS-explanation of the above slower decay mode of the same neutral K-meson, for instance, is written:

$$K_L^0(d\bar{s} = ue^-\bar{\nu}_e\bar{u}e^+\nu_e\uparrow) \to \pi^+(u\bar{d} = u\bar{u}e^+\nu_e) + e^-\bar{\nu}_e$$

Let's see the details. Excited DSLs $e^+\nu_e\uparrow$ in anti-s-quark step down to ground state, turning anti-s-quark into anti-d-quark and K_L^0 into $d\bar{d}$ 2B-system and releasing energy in the form of virtual force carriers γ and/or Z-boson. DS-leptons $e^-\bar{\nu}_e$ in $d\bar{d}$ absorb them, gain KE, and fly away, turning $d\bar{d}$ into $\pi^+(u\bar{u}e^+\nu_e)$. Much slower decaying ($\sim 5.1\cdot 10^{-8}s$) is due to much slower EW (Electro-Weak) interaction process.

Note: PDG in Particle Summary claims a slow decay mode $K_L^0 \to \pi^+\pi^-\pi^0$, where the MSP π^0 may not be there. It is well known that π^0 lives too short to show any tracks that are recognizable with measurable lengths. The author thinks that this much slower mode is actually showing much slower non-strong interaction process without π^0 MSP but with photons and/or leptons among final state particles, which were assumed by original experimenters/reporters to be produced in a π^0-decay. De-excitation of s-quark releases energies for production of $u\bar{u}$ pair and pushing $\pi^+\pi^-$ to fly away from each other as well as for production of photons and/or some leptonic PAPs, which are assumed to come from π^0-decay by original experimenters/reporters. We can expect a larger **TKE (Total Kinetic Energy)** for $\pi^+\pi^-$ system in CMF and in decay mode $K_S^0 \to \pi^+\pi^-$ than TKE for $\pi^+\pi^-$ in CMF and in decay mode $K_L^0 \to \pi^+\pi^- +$ photons and/or leptonic PAP(s), which are ***assumed*** in PDG's summary to come from π^0-decay. Such ***assumption*** does not match the fact of slow decaying now that it belongs to fast strong interaction process to produce π^0, just like shown in the fast decay mode $K_S^0 \to \pi^0\pi^0$. Similarly, PDG decay mode $K_L^0 \to 3\pi^0$ may be just $K_L^0 \to$photons and/or leptonic PAP(s), which're assumed by PDG to come from decay of three neutral pions. In such decay mode, de-excitation of s-quark that takes place in decay mode $K_S^0 \to \pi^+\pi^-$ may halt, allowing HMC $u\bar{u}$ in K^0 to annihilate before excited d-leptons in s-quark could return to ground state. As we said, anti-d-leptons in anti-s-quark stay in R2, allowing d-electron in d-quark to slow down $u\bar{u}$ -annihilation and prevent fast e^+e^- annihilation and fast $u\bar{u}$ -annihilation taking place in π^0 -decay and other DS-symmetric mesons' decays if they do contain PAPs like $d\bar{d}$ and $s\bar{s}$. Once HMC $u\bar{u}$ in K^0 annihilate, neutral pion production may or may not happen. Due to DS, ERM, & PAPP, K^0 -decay via $u\bar{u}$ -annihilation channel may produce the same final state particles w/o π^0 in the middle. Lifetime of π^0 is about $8.4 \cdot 10^{-17} s$, always leaving a track too short to observe. Descriptions of events w/o π^0 in many cases heavily depend on theoretical assumptions. If DS, ERM, PAPP, and PQTR are ignored, is there any way to understand lifetime difference of K^0 with the same quark content? Or, is there any explanation of faster decay mode $K_S^0 \to \pi^+\pi^-\pi^0$ with smaller branch ratio than $K_L^0 \to \pi^+\pi^-\pi^0$ of larger branch ratio under the condition that initial particles have the same quark-content and the final state particles form two equal sets in the two decay modes? DS provides a lot of brand new ways to explain huge number of empirical events. MB-systems with many kinds of forces acting on each u-quark and/or anti-u-quark and on each DS-electron and/or DS-positron cause difficulty to know every detail of SA events. We all have fundamental difficulty to figure out theoretically and empirically what can happen as a MB-system bound by more than one kind of forces and in excitation steps down to ground state. MB-system bound by the same sort of forces is not solvable. Quantum bound MB-system with more than 1 kind of forces binding the same constituent particle makes difficulty severer.

< Miscellany and Reactions >

Non-existence of $uu / \bar{u}\bar{u}$ 2B-systems and existence of $uuu / \bar{u}\bar{u}\bar{u}$ 3B-systems and of 'valence'-quark meson systems suggest **magnetism-like** dynamics: 2 St **currents** in **opposite** directions **repel** while St magnetic-like force is comparable with St force when both 2 interacting St charges move at **near-c** speeds with large γ -value due to binding at small SA distances, similar to magnetic force that is comparable with electric force only in near-c speed cases. This fact needs more than color-particle index coupling factor to explain (**open**). Note, powerful magnetic forces exerted on accelerated charged particles in all accelerators are due to extremely large number of slow electrons inside wires. But for 2 charged particles, magnetic or St magnetic-like forces btw them are comparable with electric or St forces btw them if and only if both of them move at near-c speeds.

Triplet uuu in each baryon is able to form BC never affected by any decay because **no** two u-quarks in the triplet move in **opposite** directions while in CMF **two** permanently/temporarily bound particles with same mass must move in **opposite** directions as they revolve about their mass center.

A d-quark ($d = ue^-\bar{\nu}_e$) is a SPS (sun-planet-satellite) system where the 2 d-leptons form a PS system.

Without DS, u-quark in Δ^{++} is believed to be able to make an **eerie escape** from Δ^{++} for new 'companion' $\bar{d}$ to form π^+ by just emitting a virtual gluon that is assumed to be able to decay into such real $\bar{d}$ plus a real d. The latter occupies the space escaped u-quark left, fixes *uuu*-core damaged by a *u*'s leaving, and changes Δ^{++} into a proton. This terribly wrong physical picture is immediately disproved by an ironic fact that no u-quark in a free proton or neutron is ever seen to be able to make such **eerie escape** for new 'companion' $\bar{d}$ to form π^+ by just emitting a virtual gluon that could decay into such $\bar{d}$ plus a d while d is able to occupy the space the escaped u-quark left behind and fix BC *uuu* damaged by u's hypothetic leaving under free-proton condition without being knocked so that the postulated process above could turn the free proton into a neutron and release π^+. Without DS, the explanation runs into fetal conflict with experimental facts.

In sharp contrast, quark model with DS offers a natural explanation for the above decay mode. In decay events without collisions, no u-quark is able to escape from a baryon where it was bound with all-time negative ME and zero-chance to escape in any reaction, decay, or event without larger than QKoE **KkE (knocking energy)** received. In view of DS, decay mode $\Delta^{++} \to p\pi^+$ is due to simple DS-processes below ($uuu\uparrow$ means excited BC)

$$\Delta^{++}(uuu\uparrow) \to uuu + PAPs(u\bar{u}\,/\,e^+e^-\,/\,\nu_e\bar{\nu}_e) \to p(uud = uuue^-\bar{\nu}) + \pi^+(u\bar{d} = u\bar{u}e^+\nu_e)$$

The 1st step is de-excitation releasing 3 PAPs. The 2nd step is PQTR. So simple, natural, and coherent without any conflict with any experimental fact. Such DS-reality picture is fully consistent with quark's bound state and the fact of stable free proton with all its *u*-quarks stable.

As baryons switch btw ground and excited states (LE or UE), *uuu* core-structure remains the same in the sense that no one has ever been detached or replaced. **BNC (Baryon Number Conservation)** is a book-keeping label of bound states of quark-cores and ERM. PAP production in ERM preserves baryon number & all other quantities of which each particle and its antiparticle have opposite numerical values. For example, every time when a meson's found, a $u\bar{u}$ PAP is produced. Since ERM makes PAPP and PAPP conserves baryon number, no one in the 4TDSP could ever change baryon number. The fact that one and only one nucleon appears among FSPs of every baryon-decay serves as an ironic piece of evidences of bound state of the constituents of BC. Even if quarks are knocked out millions of times every minute, DS with 4TDSP explains FQA fact and BNC simultaneously: PAPP and PQTR that guarantee instant re-stuff of every hole left behind by any knocked out u-quark with one u-quark born in PAPP just explain BNC. Instant capture of one anti-u-quark by any knocked out u-quark to form HMC for a meson to be born and to appear as a MSP simultaneously explains FQA fact. That anti-u-quark is produced in the PAPP and its companion u-quark just fills in the said hole.

Bound neutrons in nuclei are **not stable** either just like free neutrons since repulsion btw each 2 identical d-leptons and ENM is an *inner mechanism* whether neutron is free or not. Beta-decay with one $e^-\bar{\nu}_e$ pair emitted into outside free region is the **only channel** for quasi-bound d-leptons to go under repulsion in **free neutron**. DSL-emitting pathway in beta-decay is **suppressed** by **QT-relocating** in **elements** not above optimum line. Larger Z elements on stability island require N>Z for stability because such larger N is able to offset electric repulsion btw protons (well-known explanation). **N>Z** on stability island may **maintain stability** under the condition that every neutron decays into a proton and $e^-\bar{\nu}_e$ pair because beta-decays of many neutrons do not occur at the same or nearly same time so that no equal number of protons are needed to take released and relocating d-leptons.

Energy conservation of *isolated* composite system means constant TME (sum of KEs of constituent elementary particles and PE of entire MB-system), from initial to middle-stage to final state, before and after PAPP. For $u\bar{u}$ pair produced in pair production, their *TME* is *negative* so that relatively small amount of impact energy or energy released in SA de-excitation can afford to produce one or even more $u\bar{u}$ pairs to form HMC(s) as bound system(s), even if available energy consumed in a pair production is much less than the total rest energy of one u and one $\bar{u}$.

At least many lepton/anti-lepton pairs among produced PAP-pairs are bound systems. Such leptons are born to be able to relocate through QT to nearby places with equal or close PE values. As they relocate in nearby nucleons, if any, they turn them into middle-stage baryons. As they relocate in nearby produced $u\bar{u}$ pairs, various mesons are born.

ERM is seen in every decay and HE collisions. ERM is also seen in $\nu\bar{\nu}$ annihilation: e^+e^-, $q\bar{q}$, or other type of $\nu\bar{\nu}$ pair are found as products of $\nu\bar{\nu}$ annihilation though original $\nu\bar{\nu}$ particles only respond to and exert Wk forces. In SA world, ERM applies to three types of processes:

1. De-excitations of excited nuclei and excited composite SA particles.
2. Inelastic collision with PAPP.
3. PAP-annihilation.

DS of quarks/leptons and 4TDSP have power to explain all reactions. **More examples** are seen as follows: In a celebrated collision event or reaction

$$K^-(\bar{u}s) + p(uud) \rightarrow \Omega^-(sss) + K^+(u\bar{s}) + K^0(d\bar{s}),$$

DS-leptons $e^-\bar{\nu}_e$ in the s-quark in K^-, attracted by the BC in p as collision makes them so close to one another, relocate there (a place with same or lower PE) through QT, captured by a u-quark and turning it into an s-quark and at the same time the u-quark core in proton may be pushed to higher quantum level with J=3/2 (see remark below) while $e^-\bar{\nu}_e$ PS-system in p gets excited turning the d-quark into an s-quark so that p changes to X*i* Ξ^{*0}(uss) as a MSP more spacious than neutron ready for its naked u-quark to take one $e^-\bar{\nu}_e$ pair to form a third s-quark turning Ξ^{*0} into Ω^-. The process goes like this: Ξ^{*0} (lifetime

$7.2 \cdot 10^{-23}$ sec) makes no visible trajectory and has no time to decay since the KE lost in *K-p* inelastic collision re-emerges immediately right before the Ξ^{*0} was born, taking the form of five PAPs: $2e^+e^-2\nu_e\bar{\nu}_e\ u\bar{u}$, immediately followed by leptonic re-pairing $e^+e^-\nu_e\bar{\nu}_e \rightarrow e^-\bar{\nu}_e e^+\nu_e$ and capture or PQTR process: One $e^-\bar{\nu}_e$ pair is captured by the naked u-quark in Ξ^{*0} and u/Ξ^{*0} become s/Ω^- . At the same time, one $e^+\nu_e$ pair is captured by $\bar{u}$ in the produced $u\bar{u}$ pair and $\bar{u}$ changes to $\bar{s}$ while another $e^-\bar{\nu}_e$ pair is captured by u in the produced $u\bar{u}$ pair and it changes to d and K^0 $(d\bar{s})$ is born. At the same time, $\bar{u}$ in the naked $u\bar{u}$ core, left by K^- after its original s-quark lost its d-leptons, captures another $e^+\nu_e$ pair and turns itself into $\bar{s}$ so that $K^+(u\bar{s})$ is born. Now the original K^- and proton disappeared, and the produced Ω^- , K^+ , and K^0 start to move away making tracks of visible lengths before they end up with decays. Ξ^{*0} 's track is shorter than 22 fm, invisible and undetectable.

Remark: Since Ω^- is the only one in the baryon-decuplet that has a lifetime much (about 10^{13} times) longer than others, the BC in Ω^- may not be in excitation. The DS-explanation of reaction $K^- + p \rightarrow \Omega^- + K^+ + K^0$ is surely independent of whether Ω^- should have label $J = 3/2$.

Kaon(s) and leptons produced in decay of **D-meson** suggest that c-quark not only contains all the constituents in s-quark with higher LE or with 1st or 2nd UE but also contains $e^+\nu_e$ pair so that c-quark/u-quark have the same electric charge.

The following decay mode suggests that **b**-quark is either LE of **s**-quark, higher than ***s*** or UE while J/Psi is produced solely in accordance with ERM:

$$\Xi_b^- \rightarrow \Xi^- + J/\psi \ \ (b \rightarrow s + c\bar{c}\ )$$

Charm and bottom related details are open for **LKoE** (**lepton knockout energy**) issue is open. More details about the excitations in s, c, b, and t-quarks are discussed below.

Once d-electron's spin flips over with its spin vector anti-parallel to the u-quark's spin, it becomes excited, jumping from R1 (H-Vane) to R2 (V-Vane). Then the d-quark becomes an s-quark. An s-quark is just a d-quark in the 1st **LE** (**l**eptonic **e**xcitation).

A c-quark contains all constituents of *d*-quark plus an $e^+\nu_e$ pair, called c-leptons, which are exactly the same as $\bar{d}$ -leptons or anti-d-leptons. The d-leptons $e^-\bar{\nu}_e$ inside a c-quark are in the 2nd LE, or, in the 1st **LUE** (**l**eptonic-**u** **e**xcitation) if the excitation energy is lower, or, higher than **LKoE**. Due to heavy u-quark mass, LE puts insignificant impact on u-quark, but not vice versa. Any u-quark always stays together with a $\bar{u}$ in a HMC or with other two u-quarks in a BC. **UE** always means excitation of HMC or BC. We may let UE include anti-BC excitation. Due to influence of UE on light DS-leptons, UE may have large impact on SA leptons' states causing complexity of excitation states of hadrons.

Inside a c-quark, d-electron and c-positron may stay in H-vane and V-vane respectively with their spins parallel to u-quark's. In a c-quark, the d-electron is in radial excitation staying and bound in R1 while c-positron stays in R2 temporarily (attractive coulomb force from d-electron does not allow c-positron staying in R2 for long).

A **b**-quark, having the same constituents as d, s-quarks, is the 3rd LE of d-quark if *below* LKoE *otherwise* it's the 1st or the 2nd LUE depending on whether c-quark is the 2nd LE or the 1st LUE.

A ***t***-Quark is in higher excitation state than ***c***-quark with its constituents the same as what form a c-quark.

Lifetimes of nuclei up to millions through billions of years are **quantitative** results of **unknown** solutions of MB-systems with time-varying, non-central force fields associated with electric, Wk, and magnetic/magnetic-like forces in tiny space binding without a rest source particle and with charge currents time-dependent.

Simple **extensions** to **Wk** charge/magneton/magnetic-like force immediately show that Wk magnetic-like force btw **Wk-magnetons** (spinning Wk charges) is surely **attractive** in some directions and repulsive in some different directions no matter they have the same or opposite Wk charges. Its magnitude is much larger than Wk force at every distance much shorter than MD. Thus, three Wk charges, two're the same, can stay together for very short period of time.

Lepton number conservation is a book-keeping label of DS, ERM, and PAPP. PAPP preserves lepton number and every such quantity of which every particle and its antiparticle have opposite numerical values. Taus, muons, and electrons have exactly the same lepton number simply because the DS of muons and taus proves such sameness.

When energy released in DS-de-excitation and/or $u\bar{u}$ -annihilation in a meson decay is large enough, it may take the form of **more** PAPs such as seen in $K^{\pm} \to \pi^{\pm}\pi^{\pm}\pi^{\mp}$.

In decays $K^{\pm} \to \pi^{\pm} + \pi^{0}$, the energies released in transition of excited lepton in $\bar{s}/s$ to ground state in $\bar{d}/d$ ($\bar{s}/s \to \bar{d}/d$ +Energy) take the form of **one** $u\bar{u}$ pair to form π^{0} (or **two** $u\bar{u}$ pairs of lower kinetic energies **plus** Some leptonic PAPs to form π^{0} plus $d\bar{d}$ pair in the mixture of π^{0}, if $d\bar{d}$ can survive for long enough to justify the SM mixture of π^{0}). De-excitation of DSLs in $K^{\pm}$ leads to transition $\bar{s}/s \to \bar{d}/d$ +Energy and turns $K^{\pm}$ into $\pi^{\pm}$. The energy released in de-excitation is transferred into the produced π^{0} in decay-mode $K^{\pm} \to \pi^{\pm} + \pi^{0}$ or into the produced $\pi^{+}\pi^{-}$ in decay-mode $K^{\pm} \to \pi^{\pm}\pi^{+}\pi^{-}$.

Free ν_e *rarely* causes *inverse* process of e^{-}-capture due to **repulsion**: the u-quark and the d-electron bound in a nucleon carry the same weak charge as ν_e and join Wk forces to repel any intruding ν_e, dominating Wk attractive force exerted on ν_e by the bound $\bar{\nu}_e$ and preventing ν_e from annihilating the bound $\bar{\nu}_e$ so that ν_e can **penetrate earth** with little chance to interact even on their antiparticles bound in nucleons. Such free ν_e keeps distances larger than Wk

MDs, having almost zero Wk-MF with those bound particles inside nucleons. No spin-correlation could turn almost vanishing Wk MFs far beyond Wk MDs to large enough ones so that ν_e could be attracted via Wk MF and move close enough to bound SA $\bar{\nu}_e$ to trigger $\nu_e\bar{\nu}_e$-annihilation. Even if a ν_e hits a nucleon so close to its DS content $\bar{\nu}_e$ and $\nu_e\bar{\nu}_e$-annihilation has happened, it may produce a new $\nu_e\bar{\nu}_e$ pair with much larger branch ratios than to produce an e^+e^- pair. Consequently, ν_e in the produced new $\nu_e\bar{\nu}_e$ pair flies away while $\bar{\nu}_e$ in the new pair fills the hole left by the annihilated $\bar{\nu}_e$.

Inverse reactions of e^--capture as ***rare events*** may have occurred in *solar*-neutrino experiment. Reactions

$$\nu_e + {}^{37}Cl \to {}^{37}Ar + e^- \text{ and } \nu_e + {}^{71}Ga \to {}^{71}Ge + e^-$$

Reveal the *inverse* process of e^--capture: $\nu_e + n \to p + e^-$. Here e^- is free while it is bound in e^--capture. In view of DS, e-neutrino hits the less tightly bound d-$\bar{\nu}_e$ in a neutron and they annihilate, producing a virtual photon absorbed by the less tightly bound d-electron in the neutron sending it free and the neutron becomes a proton as it loses its quasi-bound d-lepton pair $e^-\bar{\nu}_e$. Neutrinos ν_e can easily penetrate the earth but they may be unable to penetrate the sun without observable losses. The sun and the sun's core are much larger and denser than the earth.

Note: SM without DS adopts W-scenario that an event of charged current is postulated: a hypothetic intermediate boson W^-, exchanged btw a d-quark in n and solar ν_e, turns that ν_e into e^- and d-quark into u-quark. But the concept of charged current with a charged intermediate W-boson exchanged could not explain why one of the 2 d-quarks in a neutron dooms to decay into a u-quark by emitting a charged intermediate W-boson while d-quarks in free protons have never been found to emit such charged intermediate W-bosons and decay. Before quark model, the W-scenario seemed to be able to explain why a neutron could change to a proton while emit e^- and $\bar{\nu}_e$. But stable elements in stability island do not support W-scenario. After quark model was proposed, stable free protons and then stable d-quarks in free protons have disproved W-scenario once for all and one can see it clearly that W-scenario and quark model cannot co-exist. Only quark model with DS and without W-scenario can explain all experimental facts without a single exception.

In reaction $p + \bar{\nu}_e \to n + e^+$, a **free** $\bar{\nu}_e$ with enough initial energy hits a proton losing enough KE and part of the lost energy reappears in the form of an e^+e^- pair. Then $\bar{\nu}_e$ and e^- are captured by a u-quark in the proton and turn it into a d-quark while e^+ moves away so that a proton capturing a free $\bar{\nu}_e$ as a relatively rare event is said to occur. Another part of the lost energy re-appears in the form of the recoil of the nucleon.

In reaction *free* $e^- + p \to n + \nu_e$, a **free electron** with enough energy hits a proton and loses certain KE, part of which re-appears in the form of a $\nu_e\overline{\nu}_e$ pair. Then a u-quark in the proton captures $\overline{\nu}_e$ and the intruding e^-, turning itself into a d-quark while ν_e moves away: a proton p capturing a free e^- is said to occur.

In reaction $\gamma + p \to \Sigma^+ + K^0$, an energetic **photon** hits a proton inelastically, loses certain amount of KE, gets absorbed by the d-electron in the proton, pushing it to an excited state, turning d-quark and proton into s-quark and Σ^+ respectively. While part of the lost KE takes form of 3 PAPs to re-emerge: $u\overline{u}e^+e^-\nu_e\overline{\nu}_e$. Then PQTR follows: u captures $e^-\overline{\nu}_e$ changing to d-quark while $\overline{u}$ captures $e^+\nu_e$ changing to $\overline{s}$ so that $K^0(d\overline{s})$ is born and a *photoproduction* is said to occur. Here $e^+\nu_e$ pair captured by $\overline{u}$ has proper energy to turn $\overline{u}$ into $\overline{s}$, not $\overline{d}$. If the produced pair $e^+\nu_e$ does not have proper energy to turn $\overline{u}$ into $\overline{s}$ after being captured by $\overline{u}$, then they simply turn $\overline{u}$ into $\overline{d}$. Consequently, some particles other than K^0 appear among FSPs. Note: Some details of MB-PQTR process are permanently unknown due to MBD.

A free photon without interaction never make itself disappearing and turns itself into a PAP such as e^+e^-, no matter how energetic it is. However, in reaction $\gamma \to e^+e^-$ w/o elusive low-energy photon, after a γ**-photon** with its energy larger than $2m_ec^2$ hit a nucleus, lost all or almost all of its energy, it simply disappeared or became a low-energy photon that eluded detection. Part of the lost energy took the form of one e^+e^- pair to re-emerge and a photoproduction of e^+e^- PAP was said to happen.

In reaction $\gamma + p \to n + \pi^+$, an energetic γ**-photon** hit a proton, lost nearly all or all of its energy, became a low-energy photon that eluded detection or disappeared. Part of the lost energy took the form of one e^+e^- pair, one $u\overline{u}$ pair (definitely bound with TME less than $2m_uc^2$), and one $\nu_e\overline{\nu}_e$ pair to re-emerge. The produced $\overline{u}$ in bound $u\overline{u}$ pair captured e^+ & ν_e via QT, turned itself into $\overline{d}$ while the produced $u\overline{u}$ captured the relocating e^+ and ν_e and became π^+.
At the same time, a u-quark in proton captured $e^-\overline{\nu}_e$ and turned itself into a d-quark and the proton into a neutron. A photoproduction of π^+ is said to occur w/o elusive low-energy photon.

In reaction $\gamma + n \to p + \pi^-$ w/o elusive low energy photon, energetic γ-**photon** hit a neutron, lost nearly all or all of its energy and became a low-energy photon that eluded detection or disappeared. Part of the lost energy took the form of one bound $u\overline{u}$ pair to re-emerge with quasi-bound d-leptons in n relocating in the produced $u\overline{u}$ pair so that the neutron changed to a proton and π^- was born. Thus, a photoproduction of π^- was done.

On the Web, the author of "*The Sigma Baryon*" says: "According to the Particle Data Book, the branching ratio for the decays of the Sigma-plus is

51.57% for the $p\pi^0$ pathway and 48.31% for the $n\pi^+$ pathway. This near equivalence is really surprising me–the neutron pathway looks a lot harder."

In view of DS we see no surprise: When excited d-electron in the s-quark in a Sigma-plus returned to ground state turning the s-quark into a d-quark and the Sigma-plus into a *middle-stage* particle *proton*, energy was released in the form of PAPs as long as the law of energy conservation permits:

In $p\pi^0$ pathway, $u\bar{u}d\bar{d}$ ($2u\bar{u}$, $e^+e^-\nu_e\bar{\nu}_e$) were produced to form a zero-pion mixture, or one $u\bar{u}$ pair was produced to form a zero-pion without mixture, while the *middle-stage* particle *proton* was also a FSP.

In $n\pi^+$ pathway, $u\bar{u}e^+e^-\nu_e\bar{\nu}_e$ PAPs were produced and re-paired such that a u-quark in the MSP proton captures $e^-\bar{\nu}_e$ pair turning itself/proton into a d-quark/neutron respectively. At the same time the produced $\bar{u}$ captured an $e^+\nu_e$ pair turning itself into a $\bar{d}$ staying with the produced u to form π^+.

None of the above two pathways has a reason to dominate the other. The $n\pi^+$ pathway is so natural and easy in view of the quark model with DS.

< **CC** (**C**urrent **C**onservation) >

We have seen that DS undoes transition btw elementary particles so that each elementary particle can only disappear by annihilating one of its antiparticles and could be created only together with one of its antiparticles. Any 4-current, if written in terms of elementary particle, is automatically conserved.

When one d-quark in a neutron changes to a u-quark and $e^-\bar{\nu}_e$ are emitted so that the neutron becomes a proton, the current of d changing to u is not written in terms of elementary particles since d-quark is no longer an elementary particle in view of DS. If a d-quark were elementary just like a u-quark and transition btw elementary particles were permitted, the d-quark would have to emit or absorb something to be able to transform itself to a u-quark. However, an ironic fact is: the d-quark in any free proton has never been found to be able to transform itself to a u-quark by just emitting something. Therefore, quark model w/o DS cannot co-exist with W-scenario. Once $e^-\bar{\nu}_e$ are identified as constituent particles of composite d-quark, 4-current is guaranteed to be conserved as long as written in terms of an elementary particle in DS-model. LV, massive force carriers, and DS give coherent picture of reality in physical world. QFT based on such foundation is renormalizable, without IR catastrophe and with all quantities canceled in renormalization being finite.

More……

References

PDG (Particle Data Group), World Web.
D.H. Perkins, "Introduction to High Energy Physics" (4th Edition, World Web, p.130-132)

Abbreviation

AM	angular momentum
AVC	annihilation virtual channel
AVI	absolute vacuum inelasticity
BC	baryon core
BI	backside interpretation
BNC	baryon number conservation
btw	between
CBR	cosmic background radiation
CC	current conservation
CD	characteristic distance
CED	classical electrodynamics
CHSP(s)	cosmic high-speed particle(s)
CI	classical imitation
CME	composition mass equation
CMF	center of mass frame
CMI	composition mass inequality
CRs	classical replacements
CT	cosmological term
DBR	Dirac-Bohr radius
DE fact	different energy fact
DI	deep inelastic
DM	dark matter
d-Q	d-quark
DS	deeper structure
DSDE	deeper structure de-excitation
DSE	deeper structure excitation
DSE	deeper structure explanation
DSL(s)	DS lepton(s)
d-e	d-electron
EF	Einstein-Friedman
EDF	Einstein-Dirac-Feynman
EFD	Einstein-Friedman-Dirac
e.m.	electromagnetic
EMF	external magnetic field
ENM	eagle nest mechanism
EqM	equal-mass
eqn	equation
eqs	equations
ERM	energy release mechanism
ES	elastic scattering
ET	energy theorem
FBS	fermion/boson statistics
FCIs	flavor and color indices
FDMs	fast decaying mesons
5BF	five basic facts (of nucleons)

FIMC	frame-invariance of *mass* and *charges*
FQA	free quark absence
FSP(s)	final state particle(s)
4TDSP	four type of DS processes
GB	gluon ball
GF, *G*-field	gravitational field
GI	gauge invariance
GR	general relativity
GS	gamma-sector (γ -sector)
GS	gauge-self
HE	high-energy
HM	Higgs mechanism
HMC	hadronic meson core
HPs	Higgs particles
HQM	heavy quark mass
HQMT	heavy quark mass theorem
H-Vane	horizontal vane
IC	initial condition
IDP	inverse Dirac problem
IES	inward excited state
IPE	inverse potential energy
IPEBF	inverse potential energy binding feature
IP(s)	initial particle(s)
IR	infrared
IRC	infrared cutoff
IRCF	infrared convergent factor
ISP	initial state particle
KC	kinetic channel
KE	kinetic energy
KkE	knocking energy
LE	leptonic excitation
LH-model	large strong coupling and heavy gluon model
LHS	left hand side
LIDE	Lorentz invariant Dirac eqn
LKoE	lepton knockout energy
LM	Lorentz manifold
LMC	leptonic meson core
LV	Lorentz violation
LUE	lepton-u excitation
LY	Lee-Yang
mag	magnetic
mag force	magnetic force
mag-force	magnetic force
mag field	magnetic field
mag-field	magnetic field
MB	many-body
MBD	many-body difficulty
MBP	many-body postulate

MBT	many-body theorem
MCC	mass/charge conservation
MCtr	mass center
MD	magneton distance (distance where MF=Coulombian force)
ME	mechanic energy
MF(s)	magneton force(s)
MFBF	magneton force binding feature
MFP	most fundamental principle
MM	magnetic moment
MMD	many-body/mag-force difficulties
MPR	most probable region
MSP(s)	middle-stage particle(s)
MSS	minimum spacetime sizes
MT	magneton
MT-Force(s)	magneton force(s)
MT-C	magneton-Coulomb
MTH	muon-tau hypothesis
NCB	near-critical bound
NFM(s)	neutrino-family member(s)
NG	non-gravitational
NLVUC	non-LV ultraviolet cutoff
NSC	necessary and sufficient condition
NSL	nature of the speed of light
OAM	orbital angular momentum
OSOD	opposite spin-orbital directions
OSTP	orbital-spin-tilting-precession
PAP	particle-antiparticle pair
PAPA	particle-antiparticle pair annihilation
PAPP	particle-antiparticle pair production
PCQM	perfect classical-quantum match
PDG	Particle Date Group
PE	potential energy
PNMI	proton-neutron mass inequality
PPC	particle production channel
PQTR	particle quantum tunnel relocating
PR	photon-radiation
PS	planet-satellite
PSE	planet-satellite excitation
PW	potential well
QC	quark confinement
Q-core	quark core
QCD	quantum chromodynamics
QD	quark dynamics
QFT	quantum field theory
QE	quark excitation
QED	quantum electrodynamics
QKoE	quark knockout energy
QM	quantum mechanics

QNC	quark non-confinement
QS	quantum scattering
QT	quantum tunnel
QTC	quantum tunnel channel (of decay)
RC(s)	radiative correction(s)
RCMI	reverse composition mass inequality
RD	radial domination
RFF	resultant force field
RHS	right hand side
RQC	regular quark core
RS-D	red shift-distance
RS-B	red shift-brightness
SA	subatomic
SAVI	sub-atomic vacuum inelasticity
SBd-e	single bound d-electron (in proton)
SBR	Schrödinger-Bohr radius
SC	spin correlation
SD	size-density
Sd-e	'stronger' d-electron (bound tighter/closer to Q-core in neutron)
SDL	see details later
SDMs	slowly decaying mesons
SIC	standard initial condition
SM	standard model
SPS	sun-planet-satellite
SQC	small quark core
SR	special relativity
SS	spacetime singularities
St	strong
STB	spacetime bag
STD	speed-test disability
TAM	total angular momentum
3B	3-body
2B	2-body
TE	total energy (including rest energy)
TKG	Tycho-Kepler G (constant G since Tycho-Kepler)
TME	total mechanic energy (not including rest energy)
TMET	TME theorem
UC	ultraviolet cutoff
UE	u-excitation
UE	unusual explanation
u-Q	u-quark
VI	vacuum inelasticity
VSS	velocity shell-space
1B	1-body
1BEq	1-body equivalence
1BEx	1-body extraction
Wd-e	'weaker' d-electron (bound weaklier/farther away from Q-core in

	neutron with its ME positive sometimes that may make Wd-e escape via QT and trigger beta-decay)
Wk	weak
w/o	with or without
YF	Yukawa factor
0TMED	zero TME distance (where TME=0)
ZL(s)	zone location(s) (location(s) associated with the θ-coordinate)

Corrections

In the following, the line spaces must not be counted as one counts any number of rows.

1. On page i, 2nd row: Change

6.625x10.25

to

8.5x11

2. On page i, 13th row: Change

unmoved

to

not removed

3. On page i, between the 1st and 2nd paragraphs: Insert the following new paragraph

Many words, terminologies, and phrases are replaced with their abbreviations. The Abbreviation (page 292 through 296) is put after the Appendix. It contains all the abbreviations used in this book, together with the corresponding words, terminologies, or phrases they represent.

4. On page xiii, last row: Change

In the text,

to

In the text (not including the Appendix),

5. On page 33, 28th and 29th row: Change

$r=$

to

$r =$

6. On page 34, 18th row: Change

<u>Negative</u>

to

<u>Very Negative</u>

7. On page 34, 21st row: Change

OSTP:

to

OSTP.

8. On page 40, 13th row: Change

$e^+e^-u\bar{u}\nu_e\bar{\nu}_e\nu_\mu\bar{\nu}_\mu\nu_\tau\bar{\nu}_\tau$

to

$e^+e^-u\bar{u}\nu_e\bar{\nu}_e$

9. On page 76, 11th row: Change

MFs Off Completely

to

Radial MFs Off

10. On page 128, 24th row: Change

of ***U*** for any ***V***

to

of ***V*** for any ***V***

11. On page 136, after the last row: Add the following

Remark: Many things in chapter 1 through chapter 6 can be seen in a paper (Rudolf Schmid and Qicun Sun, "Relativity without the First Postulate", Proceedings of Institute of Mathematics of NAS of Ukraine, 2002, Vol. 43, Part 2, 577-588). In this book, the author has changed the 2nd postulate too, replaced it with the c-postulate, and revealed the nature of the speed of light. Although the new transformations and the new eqs of the laws of physics co-variant under the new transformations seen in this book look exactly the same as what are in the above mentioned paper, some notations have changed. Most importantly, the constant c in the new transformations and in the new eqs of the laws of physics is no longer the speed of light anymore. It's just the limiting speed all bodies and particles can approach but cannot reach. One can find the renewed interpretations, the statement about the nature of the speed of light, and all relevant empirical evidences in the Highlights and other parts of the book.

12. On page 140, 8th row: Change

$$m'_{\pi}+k^2$$

to

$$m'^2_{\pi}+k^2$$

13. On page 142, 14th row: Change

all forces.

to

all forces. We omit the details. Obviously, LV leads to no divergent pole.

14. On page 156, 42nd row: Change

annihilation (not $e^{\pm}$ annihilation) of electrons into the core

to

the relocating of atomic electrons into protons

15. On page 157, 6th-7th row: Change

by nucleons by chance and destroyed (annihilated), but there is no way to force them stay closer to the nucleons in stable states.

to

by protons by chance.

16. On page157, 8th row: Change

bound electrons are annihilated

to

atomic electrons relocate into protons

17. On page 157, 9th row: Change

grand

to

ground

18. On page 160, 30th row: Change

$\underline{h_{\mu\nu}\text{ are all}}$

to

$h_{\mu\nu}$ are all

19. On page 161, 16th-20th row: Delete the following entire paragraph

Small quantities……as one wants….

20. On page 162, 12th row: Change

$t_{\min} \approx \hbar/2Nmc$

to

$t_{\min} \approx \hbar/2Nmc^2$

21. On page 162, the last row: Change

E^{-N}

to

$$E^{-n}$$

22. On page 178, after the 4th row, add the following:

Remark: The flavor indices in eqn (45) were introduced before the DS model was proposed. Because only $u\bar{u}$ are elementary, the flavor indices become redundant for the strong forces btw elementary quarks. Again, we see a historical record of wrong concept not removed from this book.

23. On page 179, 3rd row: Change

$$p \rightarrow$$

to

$$\boldsymbol{p} \rightarrow$$

24. On page 183, 11th row: Change

the QSI

to

many empirical

Note: QSI is the only abbreviation not seen in the Abbreviation.

25. On page 275, 4th row: Change

cannels

to

channels

26. On page 284, 21st row: Change

$$\bar{\nu}$$

to

$\overline{V}_e$

Note: The original document of this book was created more than ten years ago with MS word 1996 and 2000. It has been modified and edited too many times with MS word 2003, 2010, and MS equation editor which is part of the soft wares. Last year the author downloaded WPS Office and the advertisement guaranteed 100% compatibility with MS word. A WPS file of this book was created and a PDF file was also made. Lulu printer does not accept WPS files while PDF files can not be changed in any proof-reading. Each time I opened the WPS file of this book with MS word 2010 and 2016, too many deformations appeared. It's too difficult to complete a proof-read with such severely deformed MS word document. Lulu publishing allows me to use the above list of changes for reader's reference. The author apologize for any inconvenience it may cause.

www.ingramcontent.com/pod-product-compliance
Lightning Source LLC
Chambersburg PA
CBHW081106170526
45165CB00008B/2349
* 9 7 8 1 4 8 3 4 8 2 1 2 5 *